Lecture Notes in Computer Science 16041

Founding Editors

Gerhard Goos
Juris Hartmanis

Editorial Board Members

Elisa Bertino, *Purdue University, West Lafayette, IN, USA*
Wen Gao, *Peking University, Beijing, China*
Bernhard Steffen, *TU Dortmund University, Dortmund, Germany*
Moti Yung, *Columbia University, New York, NY, USA*

The series Lecture Notes in Computer Science (LNCS), including its subseries Lecture Notes in Artificial Intelligence (LNAI) and Lecture Notes in Bioinformatics (LNBI), has established itself as a medium for the publication of new developments in computer science and information technology research, teaching, and education.

LNCS enjoys close cooperation with the computer science R & D community, the series counts many renowned academics among its volume editors and paper authors, and collaborates with prestigious societies. Its mission is to serve this international community by providing an invaluable service, mainly focused on the publication of conference and workshop proceedings and postproceedings. LNCS commenced publication in 1973.

Rajeev Gupta · Deepak P. · Jun Shen ·
Srinath Srinivasa · Vinu Ellampallil Venugopal ·
Uttam Kumar

Editors

Big Data and Artificial Intelligence

13th International Conference, BDA 2025
Bangalore, India, July 17–20, 2025
Proceedings

 Springer

Editors
Rajeev Gupta
Microsoft India
Hyderabad, Telangana, India

Jun Shen
University of Wollongong
Wollongong, NSW, Australia

Vinu Ellampallil Venugopal
International Institute of Information
Technology Bangalore
Bengaluru, Karnataka, India

Deepak P.
Queen's University Belfast
Belfast, UK

Srinath Srinivasa
International Institute of Information
Technology Bangalore
Bengaluru, Karnataka, India

Uttam Kumar
International Institute of Information
Technology Bangalore
Bengaluru, Karnataka, India

ISSN 0302-9743   ISSN 1611-3349 (electronic)
Lecture Notes in Computer Science
ISBN 978-3-032-15133-9   ISBN 978-3-032-15134-6 (eBook)
https://doi.org/10.1007/978-3-032-15134-6

This Springer imprint is published by the registered company Springer Nature Switzerland AG
The registered company address is: Gewerbestrasse 11, 6330 Cham, Switzerland

If disposing of this product, please recycle the paper.

Preface

With applications of big data and AI going from strength to strength, there is an ever-increasing need to facilitate deep conversations in this area. Big data and AI are increasingly diverse, not just in the kinds of data they process and the scales at which such processing takes place, but also in terms of the application domains, which range from commercial applications to public administration. We are also at cross-roads of various kinds, with tensions emerging between the interest in general purpose technologies and the necessity to address very specific needs within niche domains. It is against this backdrop that BDA 2025 was held in what is inarguably the technology capital of India, Bangalore.

The 13th International Conference on Big Data and Artificial Intelligence (BDA 2025) was held in physical mode from July 17–20, 2025 within the campus of the International Institute of Information Technology Bangalore (IIITB). Located in Electronics City, one of the earliest technology centres, the IIITB campus hosted a number of participants from across academia, industry and government for a vibrant technical programme. The conference featured 28 peer-reviewed research papers, 6 invited/keynote talks, 3 tutorials and 3 workshops. The conference covered a wide variety of topics across the various programme parts, which notably included topics relating to graph data processing, deep learning architectures, ethical issues, as well as applications of AI within non-commercial settings. Many of the sessions had very rich Q/A segments, enabling substantive interaction between the presenters and the audience.

The BDA 2025 conference committee received 104 valid research paper submissions for the contributed programme. The papers were subject to a two-tier single-blind reviewing process where reviewers rigorously evaluated the research papers for scientific merit and rigour while being mindful of aspects such as contemporary relevance and societal significance. Many of the paper submissions were from colleges from across India, and the committee recognised that many of the authors may be taking their first steps towards entering the research community through their BDA 2025 submissions. Accordingly, the PC chairs made it a point that constructive and detailed feedback be provided to authors even when acceptance could not be recommended, in order to encourage them to develop their research further for a stronger submission in appropriate avenues. The PC chairs made it a point to attend to each paper, and in many cases, intense discussions followed the review phase. Based on such careful evaluation, 15 papers were chosen as regular-track research papers, with a further 13 being chosen as short research papers. This yielded a 27% acceptance rate across both paper categories.

The conference featured four keynotes by eminent scholars and professionals in the field of Big Data and AI. The speakers were Yogesh Simhan (Indian Institute of Science, Bangalore), Biplav Srivastava (University of South Carolina), Kalika Bali (Microsoft Research India, Bangalore) and Anvar Sadath (Kerala Infrastructure and Technology for Education). The topics covered across the keynotes ranged from distributed systems, multiculturalism and sustainability in AI, and the successes and challenges in using

AI to facilitate primary education. In addition to the keynotes, the conference featured invited talks by Prasad Deshpande (DataBricks, Bangalore), Manas Gaur (University of Maryland Baltimore County) and Karthik Ramachandra (Microsoft).

BDA 2025 also included a rich conference programme comprising three workshops which were all held during the first day of the conference. These were: (1) 2nd Indo-French Workshop on Scalable Technologies for Real-time Event Analysis and Management (STREAM), (2) Beyond Generation - Advanced AI Technologies for Stimuli Design, Conducting Interviews and Generating Synthetic Data for Research, and (3) Workshop on Novel Algorithms, Models, Libraries and Tools for Achieving Sustainability and Energy Efficiency in Computing Paradigms. The conference programme also included three tutorials, viz. Big Graph Mining and LLMs, Extracting Scientific Knowledge from Research Articles, and Geo-spatial Data Analysis.

The PC chairs would like to thank all the PC members for their hard work in meticulously going through research papers and providing thoughtful comments and feedback, and in particular, providing constructive comments to authors of papers which were not recommended for acceptance. The authors who enthusiastically submitted their research for consideration by the BDA 2025 conference form the backbone of this conference, and they enable this conference to reflect the cross-section of high-end technological research taking place across the length and breadth of India and beyond. The Steering Committee and the Organizing Committee deserve boundless praise for the support they provided all through the planning and organisation of the technical program for the conference. We thank all the keynote/invited speakers, workshop and tutorial speakers for their unwavering support for the conference, as well as their thoughtful insights. We thank all our sponsors whose support enabled us to provide the attendees a rich experience. The biggest applause is reserved for the various student volunteers who were always available to ensure that the conference went on smoothly and that any technical glitches and co-ordination issues were spotted early and resolved promptly. Along with that, we express our thanks to everybody associated with International Institute of Information Technology Bangalore for providing us the wonderful venue and facilities. We hope the readers of this volume will find the contents interesting, insightful and inspiring.

July 2025

Rajeev Gupta

Deepak P.

Jun Shen

Srinath Srinivasa

Vinu Ellampallil Venugopal

Uttam Kumar

Organization

Patron

Rajeev Gupta	IIIT Bangalore, India

Honorary Chairs

Mahadeva Prasanna	IIIT Dharwad, India
Jayant R. Haritsa	IISc Bangalore, India
Manohar Swaminathan	Microsoft Bangalore, India

General Chair

Srinath Srinivasa	IIIT Bangalore, India

Program Committee Chairs

Rajeev Gupta	Microsoft Bangalore, India
Deepak P.	Queen's University Belfast, UK
Jun Shen	University of Wollongong, Australia

Steering Committee

P. Krishna Reddy	IIIT Hyderabad, India
Srinath Srinivasa	IIIT Bangalore, India
Mukesh Mohania	IIIT Delhi, India
Rajkumar Buyya	University of Melbourne, Australia
Sharma Chakravarthy	University of Texas at Arlington, USA
Philippe Fournier-Viger	Shenzhen University, China
Sanjay Kumar Madria	Missouri University of Science and Technology, USA
Masaru Kitsuregawa	University of Tokyo, Japan
Raj K. Bhatnagar	University of Cincinnati, USA
Vasudha Bhatnagar	University of Delhi, India

Ladjel Bellatreche	ISAE-ENSMA, France
P. Radha Krishna	NIT Warangal, India
H. V. Jagadish	University of Michigan, USA
Ramesh K. Agrawal	Jawaharlal Nehru University, India
Divyakant Agrawal	University of California at Santa Barbara, USA
Arun Agarwal	University of Hyderabad, India
Jaideep Srivastava	University of Minnesota, USA
R. Uday Kiran	University of Aizu, Japan
Sanjay Chaudhary	Ahmedabad University, India
Jerry Chun-Wei Lin	Western Norway University of Applied Sciences, Norway
S. K. Gupta	IIIT Delhi, India
Satish N. Sreerama	University of Hyderabad, India
Krithi Ramamritham	IIT Bombay, India
Rajeev Gupta	Microsoft Research, USA
Praveen Rao	University of Missouri, USA

Workshop Chairs

Raghava Mutharaju	IIIT Delhi, India
Animesh Chaturvedi	IIIT Dharwad, India
Navneet Goyal	BITS Pilani, India

Industry Track Chairs

Vijay S. Agneeswaran	Microsoft Bangalore, India
Atul Kumar	IBM Bangalore, India
Srinivas Karthik V.	Microsoft Azure SQL R&D, India

Tutorial Chairs

Sowmya S. Sundaram	Stanford School of Medicine, USA
Mauro Dalle Lucca Tosi	Luxembourg Institute of Science and Technology, Luxembourg
Sudarsun Santhiappan	IIT Madras, India

Publication Chair

Uttam Kumar IIIT Bangalore, India

Local Arrangements Chair

Sushree S. Behara IIIT Bangalore, India

Publicity Chair

Raj Sharma Walmart Global Tech Bangalore, India
Sanju Tiwari Sharda University, India

Registration Chairs

Tulika Saha IIIT Bangalore, India
Apurva Kulkarni IIIT Bangalore, India

Organizing Committee

Vinu E. Venugopal IIIT Bangalore, India
Uttam Kumar IIIT Bangalore, India

Web Chairs

Shivani Singh IIIT Bangalore, India
Neelam Gupta IIIT Bangalore, India

Sponsors

International Institute of Information Technology Bangalore, India
Microsoft Research
Data for Public Good

Contents

ML Frameworks and System-Level Intelligence

Deep Learning Architectures and Model Design

Deep Learning Architecture and Adaptations

Domain-Specific AI Models

Fine-Tuning and Generative Modeling Techniques

Language Understanding and Interactive AI

Long-Context Non-factoid Question Answering in Indic Languages

Ritwik Mishra[1(✉)] [iD], Rajiv Ratn Shah[1] [iD], and Ponnurangam Kumaraguru[2] [iD]

[1] Indraprastha Institute of Information Technology, Delhi, New Delhi, India
{ritwikm,rajivratn}@iiitd.ac.in
[2] International Institute of Information Technology, Hyderabad, Hyderabad, India
pk.guru@iiit.ac.in

Abstract. Question Answering (QA) tasks, which involve extracting answers from a given context, are relatively straightforward for modern Large Language Models (LLMs) when the context is short. However, long contexts pose challenges due to the quadratic complexity of the self-attention mechanism. This challenge is compounded in Indic languages, which are often low-resource. This study explores context-shortening techniques, including Open Information Extraction (OIE), coreference resolution, Answer Paragraph Selection (APS), and their combinations, to improve QA performance. Compared to the baseline of unshortened (long) contexts, our experiments on four Indic languages (Hindi, Tamil, Telugu, and Urdu) demonstrate that context-shortening techniques yield an average improvement of 4% in semantic scores and 47% in token-level scores when evaluated on three popular LLMs without fine-tuning. Furthermore, with fine-tuning, we achieve an average increase of 2% in both semantic and token-level scores. Additionally, context-shortening reduces computational overhead. Explainability techniques like LIME and SHAP reveal that when the APS model confidently identifies the paragraph containing the answer, nearly all tokens within the selected text receive high relevance scores. However, the study also highlights the limitations of LLM-based QA systems in addressing non-factoid questions, particularly those requiring reasoning or debate. Moreover, verbalizing OIE-generated triples does not enhance system performance. These findings emphasize the potential of context-shortening techniques to improve the efficiency and effectiveness of LLM-based QA systems, especially for low-resource languages. The source code and resources are available at https://github.com/ritwikmishra/IndicGenQA.

Keywords: Long-context · Non-factoid · Question-Answering · LLM · Indic-languages · Low-resource

1 Introduction

Question Answering (QA) represents one of the foundational tasks in Natural Language Processing (NLP) [22]. The objective of QA is to automatically generate or extract an appropriate answer to a given question. Factoid questions,

R. Gupta et al. (Eds.): BDA 2025, LNCS 16041, pp. 3–18, 2026.
https://doi.org/10.1007/978-3-032-15134-6_1

which can be answered with short spans of text, are a well-studied area in QA. For example, the question "*Where is the Taj Mahal located?*" is an example of a factoid question. Recent advancements in QA systems have demonstrated that factoid questions can often be addressed with near-human performance [18]. This work emphasizes on non-factoid questions which require detailed and descriptive responses. For instance, a non-factoid question might be, "*How was the construction of the Taj Mahal perceived by the citizens of Agra in the 17th century?*" Despite the importance of such questions, non-factoid QA has received relatively less attention since the early stages of QA research [51]. Non-factoid questions are prevalent on the web [7].

Transformer-based models are widely used in building QA systems [41]. Nevertheless, a notable limitation of Transformers is their quadratic memory complexity in the self-attention mechanism, which constrains their application in domains requiring the handling of longer sequences [53]. While multilingual models such as mT5 and mBART exist, they inherit the same limitations as their English counterparts, particularly in managing long sequences effectively [57]. Even modern Large Language Models (LLMs) are shown to be struggling with long-context [32]. In this study, we investigate the use of Answer Paragraph Selection (APS) models, Open Information Extraction (OIE), and coreference resolution to reduce the context length associated with non-factoid questions. Our objective is to address the research question: *Which is the best technique to shorten the context for the given question?* Additionally, we aim to provide an explanation of the functioning of the most effective technique. OIE refers to the task of extracting n-ary tuples from a given text. For instance, consider the sentence: *Helen, John Wick's wife, gifted him a beagle puppy named Daisy.* The extracted triples from this sentence might include: (i) {*Helen*, wife-of, *John Wick*}, and (ii) {*Helen, gifted, him*}. Notably, triple extraction is not a deterministic process, and different sets of triples may be derived from the same text. Coreference resolution, on the other hand, involves identifying text spans that refer to the same real-world entity. For example, in the aforementioned sentence, the terms *him* and *John Wick* both correspond to the same entity.

Similar to many subfields in computational linguistics, the majority of research on Question Answering (QA) has been conducted in English [49]. However, transferring knowledge from a high-resource language to a low-resource language has shown to be challenging for long-sequences [46]. Morever, non-factoid QA systems for English were shown to be achieving near-human performance whereas only a few resources exists for non-factoid QA in South Asian languages [18]. For our experiments, we focus on Indic languages such as Hindi, Urdu, Tamil, and Telugu, as publicly available open information extraction (OIE) tools exist for these languages [26,38]. The critical importance of non-factoid question answering (QA) systems and the limited research on Indic non-factoid QA have inspired us to develop models tailored for non-factoid QA in Indic languages. The primary contributions of this study are as follows:

1. We demonstrate that the performance of LLMs in non-factoid QA can be significantly improved by incorporating an Answer Paragraph Selection (APS) model to reduce the context size effectively.
2. We release the finetuned checkpoints for multiple LLMs trained on non-factoid question-answer pairs in Hindi, Urdu, Tamil, and Telugu.
3. We propose a novel Semantic Text Similarity score for Multilingual Texts (STS-MuTe), specifically envisioned to evaluate the quality of generated answers.
4. Our analysis reveals that when the APS model assigns a high score to a question-paragraph pair, nearly all tokens contribute to this prediction. Conversely, when a low score is predicted, only a few tokens are influential in the APS model's decision.

2   Literature Survey

We observed that contemporary language models, such as Mamba, which are not based on the Transformer architecture, exhibit inferior performance compared to Transformer-based LLMs on tasks requiring long-context understanding [28]. Similarly, in the task of open-domain non-factoid question answering, referred to as long-form question answering, it was seen that retrieval-based approaches outperform direct generative answering methods in this context [58]. Triples and knowledge bases have been extensively utilized in the literature to enhance question-answering pipelines [9,31]. The authors of K-BERT [33] incorporated knowledge derived from OpenIE triples to address various tasks, including question answering. Studies such as [6] and [47] have demonstrated that retrieving knowledge base triples, verbalizing them, and incorporating them as context significantly improves the performance of LLMs. It has been observed that incorporating coreference resolution as a preprocessing step prior to query-specific short context retrieval significantly enhances the performance of LLMs, even for QA tasks involving novel-length contexts [34]. Additionally, [14] proposed a pipeline leveraging Open Information Extraction (OIE) triples and coreference resolution to address QA tasks. Their approach further employed post-hoc explainable artificial intelligence (XAI) techniques, including LIME [45] and SHAP [35], to interpret the generated answers. However, this pipeline was not evaluated on any extensive dataset.

Non-factoid QA Systems: Answer ranking has been observed to be an important step to answer non-factoid queries [51]. Existing studies also reveal that semantic techniques, such as BM25, perform inadequately when applied to non-factoid questions [15,60]. Cohen et al. (2018) [16] employed various APS techniques specifically to address non-factoid questions. Glass et al. (2022) [21] demonstrated that the process of retrieving, re-ranking passages, and subsequently generating answers from them leads to improved performance. The approach of extracting a short context from a longer passage not only decomposes the long-context question answering problem into distinct steps but also enhances

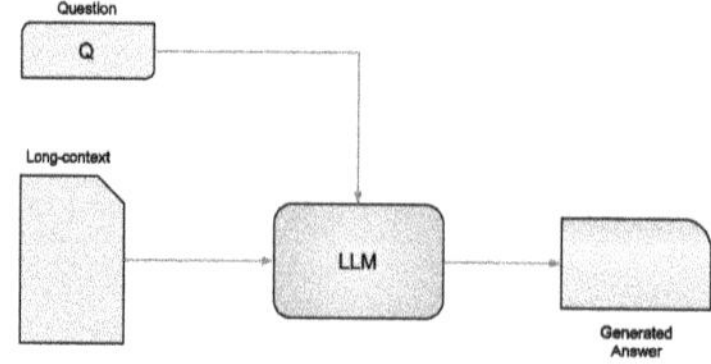

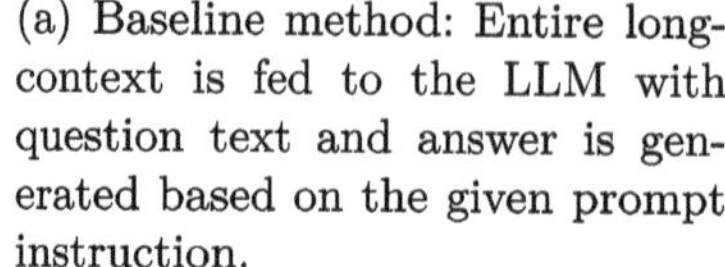

(a) Baseline method: Entire long-context is fed to the LLM with question text and answer is generated based on the given prompt instruction.

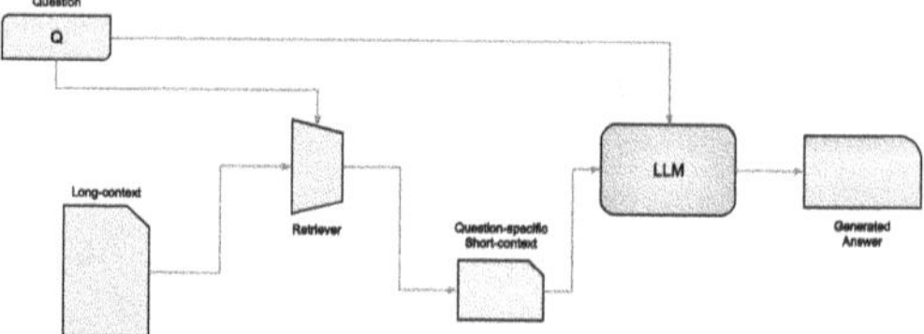

(b) Retrieval Method: A short-context is derived by the retriever using the long-context and the question text. This obtained short-context is subsequently provided as input to the LLM.

Fig. 1. Methods for long-context question answering using Large Language Models (LLMs).

interpretability in the results [12]. Singh et al. (2024) [48] introduced a QA benchmark for Indic languages by consolidating all publicly available QuAD datasets in Indic languages and augmenting them with synthetic data. Nonetheless, the representation of non-factoid questions within multilingual QuAD datasets remains an area that requires further exploration.

3   Dataset

In this study, we utilized the MuNfQuAD [39], the largest publicly available multilingual non-factoid question answering dataset. The question-answer pairs and their corresponding context were extracted from BBC news articles. The dataset is publicly available for research purposes with due permission from BBC. In our investigation of the potential of OIE and coreference resolution for long-context non-factoid question answering, we were constrained to languages for which such tools are available. The coreference resolution model of Transmucores [37] has been trained on 31 South Asian languages. However, there is a notable scarcity of multilingual OIE tools. To the best of our knowledge, IndIE [38] and Gen2OIE [26] are the only two multilingual OIE tools that can be used with four Indic languages: Hindi, Tamil, Telugu, and Urdu. Consequently, this study evaluates the proposed methodology primarily on these four languages. After excluding the APS model's training data from MuNfQuAD, the resulting dataset comprised over 40K question-answer pairs across four languages.

4   Methodology

A conventional method for answering a question from the provided context involves presenting the question-context pair to a LLM. The conventional approach to utilizing LLMs for question answering, contingent upon the given context, is visually depicted in Fig. 1a.

In this work, we explore the use of a retriever within the QA pipeline to shorten the long-context for answering non-factoid questions. As illustrated in Fig. 1b, this retrieval mechanism plays a central role in the overall methodology. To support this approach, Sect. 2 discusses the relevance of coreference resolution and OIE in enhancing QA systems. Building on these insights, we implement the following four distinct retrieval strategies:

A1. **Simple APS model:** We pass the paragraphs in long-context and question to the finetuned APS model of [39]. Top-5 paragraphs were selected as the short-context.

A2. **Ranking verbalized triples (OIE+APS):** We extracted triples from each paragraphs of the given long-context. A triple is consisted of a head, relation, and a tail. During the verbalization phase, short sentences were generated by concatenating the extracted head, relation, and tail. Owing to the free word order characteristic of Indian languages [52], the resulting sentences retained their semantic coherence. We take the verbalized triples and concatenate it with the question to pass them to the APS model. Using the APS model output we rank and filter top-10 verbalized triples and use them as the short context.

A3. **Coreference resolution and APS ranking (coref+APS):** We run the multilingual coreference resolution from [37] on the long-context to get the coreference links between paragraphs. After selecting the paragraph which gives highest score by APS model for the given question, we filtered all the paragraphs that contains coreference links to this paragraph. Top-5 paragraphs are selected based on their APS scores.

A4. **Ranking coreference chains of verbalized triples (OIE+ coref+APS):** Using the predicted coreference chains and verbalized triples, we constructed clusters of verbalized triples according to the coreference chains. Figure 2 shows the visual representation of the four approaches used in this work.

4.1 Implementation

In this work, we use the finetuned APS model from [39] to predict a score that indicates whether the given text can answer the given question. Based on these scores, paragraphs are ranked, and the top five passages ($k = 5$) are selected. We have chosen checkpoint with the base encoder of XLM-R [17]. Morever, the GPU memory footprint and inference time on XLM-R based APS model is significantly less than other multilingual encoders.

We compare our approaches (A1-A4) with the baseline (B) shown in Fig. 1a where no retriever is used in the QA pipeline. We used 2 billion and 7 billion parameter model of instruction finetuned Gemma [54]. We also used 8 billion parameter model of instruction finetuned Llama 3.1 [19]. We opted for these LLMs since their effectiveness is documented in prior studies on question answering in low-resource languages [2, 43, 56]. For each approach (A1-A4) and baseline

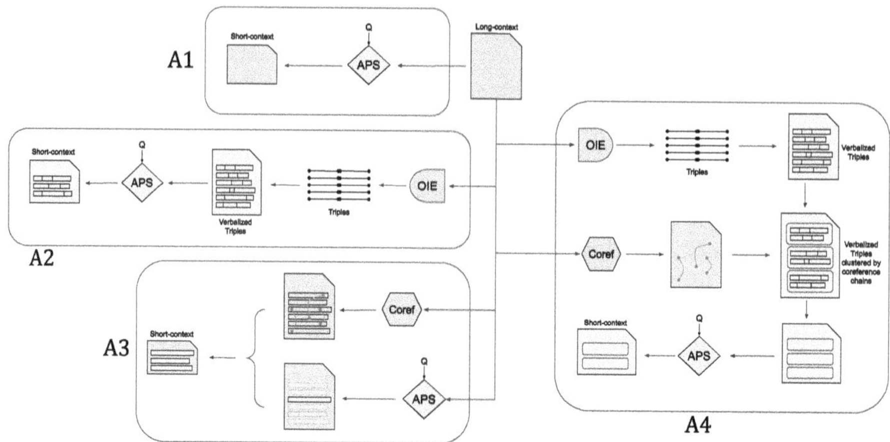

Fig. 2. Illustration of the four approaches used to construct the retriever. The verbalization of triples refers to the process of transforming a triple (composed of a head, relation, and tail) into a short sentence by combining its constituent components.

(B), we finetuned the aforementiond LLMs on a training set of more than 29K QA pairs. We used parameter efficient finetuning [36] with LoRA [23] rank and alpha of 32 to develop the *QA pipeline containing finetuned LLMs* ($QA_{finetuned}$). The prompt used for finetuning LLMs is shown in Fig. 3a. We ran zero-shot inferences while running the models on the test set. Four bit quantization[1] was used to run inferences with *QA pipeline containing base LLMs* (QA_{base}). All the experiments were performed on two NVIDIA A100 cards with each card having 40 GB of GPU RAM. In order to ensure reproducibility, we used low temperature value (0.001) for local LLMs and API calls.

4.2   Evaluations

In order to compare the LLM generated answers with silver answers of MuN-fQuAD, we used ROUGE score [30] to quantify the similarity between the two answers. However, [50] has shown that ROUGE metric cannot be trusted to evaluate long sequences. Furthermore, while LLM-generated answers may perform poorly on lexical metrics, they often convey the same semantic information as the ground truth answers [25]. Hence we propose a semantic text similarity score for multilingual texts (STS-MuTe). For a given text pair (t_1, t_2), STS-MuTe is computed as the arithmetic mean[2] of the cosine similarity (cos) between various multilingual embeddings and the BERTScore [61], which itself is derived from token-level pairwise cosine similarity and yields a single floating-point value representing semantic similarity. We used the following models to calculate the

---

[1] https://pypi.org/project/bitsandbytes/.
[2] We also experimented with taking harmonic mean instead arithmetic mean. But no significant change in results was observed.

<table>
<tr>
<td>

Answer the question based on the given context.
##Question
{question}
##Context
{context}
##Answer
{answer}

</td>
<td>

Given the following question, you are given a ground-truth answer and two options. Choose the option that is closest to the ground truth. You are only allowed to choose one option. Print either "option1" or "option2". Print nothing else.
##Question
{question}
##Ground_Truth: {ground_truth}
##Option1: {option1}
##Option2: {option2}

</td>
</tr>
<tr>
<td>

(a) Prompt used to finetune LLMs in the QA pipeline. During the inference stage, {answer} was omitted from the prompt.

</td>
<td>

(b) Prompt used to query ChatGPT model to perform qualitative analysis.

</td>
</tr>
</table>

Fig. 3. Different prompts used in this work for LLMs.

cosine similarity: (a) USE [13], (b) LaBSE [20], and (c) LASER [4]. Therefore, we used token-level and semantic-level measures together for quantitative evaluation of LLM-generated responses. Qualitative evaluation was conducted using the LLM-as-a-judge framework, detailed in the next section.

5 Results

Our test set consisted of more than 6k questions. However, due to resource constraints, we had to limit the inference on a subset of 1100 questions from the test set. Average context length in these 1100 examples was 1070 tokens. We ensured that the four languages in consideration are all represented in this subset of the test set. Table 1 illustrates the performance achieved by our QA pipeline with different approaches. We ensured that the test samples on which QA_{base} is evaluated are the same test samples on which $QA_{finetuned}$ is evaluated. As compared to the baseline (B), it was observed that A1 outperformed other approaches on nearly all evaluation metrics. The instances where performance of B was shown to be slightly better than A1 (gemma-7b-it) even in those instances A1 was seen to be giving competitive values. It was observed that IndIE method performed better than Gen2OIE in A2 whereas Gen2OIE performed better in A4. We attribute this to the fact that IndIE has shown to generate more fine-grained triples than Gen2OIE [38]. Our results indicate that APS model is able to extract better short-context from verbalization of fine-grained triples. Whereas coarse triples of Gen2OIE are more likely to contain intact coreferring mentions, and hence the resulting clusters of verbalized triples are shown to be giving better results. After comparing the performance gain of A1 over B across three LLMs, we observed that the average percentage improvement of STS-MuTe and ROUGE scores was 4% and 47%, respectively for QA_{base} whereas the improvement was 2% on both the metrics for $QA_{finetuned}$.

In order to perform a qualitative analysis on the answers generated by QA pipelines with B and A1, we decided to use LLM-as-a-judge. Since human annotators are time intensive and costly, LLM-as-a-judge has been used in prior

Table 1. Performance of QA pipeline with different LLMs on a subset of test set. Since A2 and A4 depends on an OIE tool, we used red color to highlight the where IndIE tool performed better than Gen2OIE and green color to highlight where Gen2OIE performed better than IndIE. We used F1 score of ROUGE metric. Numbers written inside parenthesis represents the performance of $QA_{finetuned}$ whereas numbers written outside represents the performance of QA_{base}. It can be seen that A1 has consistently performed better than others.

	Gemma 2 Billion Instructional (gemma-2b-it)								
	Semantic-level					Token-level			
	BERTScore	USE	LaBSE	LASER	STS-MuTe	R1	R2	R3	RL
B	0.63 (0.75)	0.38 (0.80)	0.43 (0.58)	0.58 (0.83)	0.50 (0.74)	0.10 (0.44)	0.04 (0.31)	0.02 (0.29)	0.08 (0.34)
A1	**0.66 (0.78)**	**0.39 (0.81)**	**0.46 (0.70)**	**0.56 (0.85)**	**0.52 (0.79)**	**0.12 (0.51)**	**0.08 (0.44)**	**0.07 (0.42)**	**0.11 (0.39)**
A2	0.62 (0.61)	0.31 (0.53)	0.38 (0.51)	0.50 (0.74)	0.46 (0.60)	0.07 (0.14)	0.02 (0.04)	0.01 (0.02)	0.05 (0.11)
A3	0.57 (0.64)	0.18 (0.61)	0.32 (0.50)	0.49 (0.77)	0.39 (0.63)	0.03 (0.21)	0.01 (0.09)	0.00 (0.07)	0.03 (0.14)
A4	0.65 (0.75)	0.37 (0.77)	0.45 (0.67)	0.55 (0.83)	0.50 (0.75)	0.11 (0.43)	0.07 (0.35)	0.06 (0.33)	0.10 (0.33)

	Gemma 7 Billion Instructional (gemma-7b-it)								
	Semantic-level					Token-level			
	BERTScore	USE	LaBSE	LASER	STS-MuTe	R1	R2	R3	RL
B	0.66 (**0.81**)	**0.52 (0.84)**	0.53 (0.72)	**0.70 (0.87)**	0.60 (**0.81**)	0.16 (**0.55**)	0.07 (**0.47**)	0.05 (**0.46**)	0.11 (**0.49**)
A1	**0.69** (0.79)	**0.52** (0.81)	**0.56 (0.74)**	0.66 (0.86)	**0.61** (0.80)	**0.19** (0.53)	**0.13** (0.46)	**0.11** (0.44)	**0.16** (0.42)
A2	0.65 (0.63)	0.44 (0.58)	0.46 (0.56)	0.59 (0.77)	0.53 (0.63)	0.10 (0.17)	0.03 (0.06)	0.01 (0.03)	0.07 (0.12)
A3	0.63 (0.70)	0.43 (0.69)	0.45 (0.62)	0.63 (0.81)	0.54 (0.71)	0.09 (0.33)	0.03 (0.21)	0.01 (0.17)	0.07 (0.24)
A4	0.68 (0.76)	0.49 (0.77)	0.53 (0.70)	0.63 (0.83)	0.58 (0.77)	0.16 (0.45)	0.11 (0.36)	0.09 (0.34)	0.14 (0.35)

	Llama 3.1 8 Billion Instructional (llama3.1-8b-it)								
	Semantic-level					Token-level			
	BERTScore	USE	LaBSE	LASER	STS-MuTe	R1	R2	R3	RL
B	0.69 (**0.78**)	0.61 (0.77)	0.57 (0.72)	0.69 (**0.82**)	0.64 (0.77)	0.28 (0.46)	0.18 (0.39)	0.15 (0.38)	0.22 (**0.41**)
A1	**0.70 (0.78)**	**0.68 (0.78)**	**0.62 (0.73)**	**0.76 (0.82)**	**0.69 (0.78)**	**0.34 (0.49)**	**0.26 (0.43)**	**0.24 (0.41)**	**0.27 (0.41)**
A2	0.61 (0.65)	0.56 (0.63)	0.49 (0.57)	0.71 (0.74)	0.59 (0.65)	0.17 (0.22)	0.06 (0.08)	0.02 (0.04)	0.11 (0.15)
A3	0.61 (0.71)	0.51 (0.70)	0.49 (0.63)	0.68 (0.78)	0.57 (0.70)	0.16 (0.32)	0.06 (0.20)	0.03 (0.16)	0.11 (0.24)
A4	0.69 (0.75)	0.65 (0.75)	0.59 (0.69)	0.74 (0.80)	0.67 (0.75)	0.30 (0.43)	0.21 (0.35)	0.19 (0.33)	0.23 (0.36)

works [42]. Our preliminary experiments on a subset of test set showed that base `gpt-4o-mini` from [24] is able to outperform all $QA_{finetuned}$ on all the quantitative evaluation metrics. We used the same prompt shown in Fig. 3a. Therefore, we decided to use ChatGPT LLM as a judge. We used `gpt-4o` to evaluate the answers generated from $QA_{finetuned}$ pipeline with all three LLMs. The prompt used for this task is shown in Fig. 3b. The percentage of questions where answers generated by A1 are preferred over the answers generated by B are as follows: 73% for gemma-2b-it, 93% for gemma-7b-it, and 51% for llama3.1-8b-it. Thus illustrating the better qualitative performance of A1 over B.

As an ablation of A1 approach, instead of fine-tuned APS model from [39] we used LangChain[3] vectorstore retriever with Huggingface embeddings [59] from the `paraphrase-multilingual-MiniLM-L12-v2` model [44]. We also used BM25 to select answer paragraphs for a given query from the corresponding context. It was observed that A1 approach with fine-tuned APS model ($A1_{APS}$) outperformed the LangChain variant and BM25 variant of A1 on all our evaluation metrics. Notably, 75% of the top-k paragraphs selected by $A1_{APS}$ differed from

---

[3] https://github.com/langchain-ai/langchain.

those retrieved by BM25, suggesting that, unlike BM25, the APS model captures information beyond bag-of-words features and probabilistic ranking.

The A1 approach not only outperforms method B in terms of evaluation metrics but also demonstrates greater computational efficiency. When using base LLMs, A1 consumed approximately 50% less GPU memory and achieved a 30% reduction in inference time on average. Similarly, with fine-tuned LLMs, A1 required around 30% less GPU memory and was 18% faster. The QA pipeline with A4 using GenOIE tool has shown to be performing competitive to the best performing approach A1. However, it adds a compute overhead to the QA pipeline. The coref component takes 5GB on GPU and needs 0.14 s per article whereas Gen2OIE component takes 6.5GB on GPU and needs 15 s per article.

The resource intensive nature of B led to Out of Memory (OOM) errors on many test examples with the GPU cards we had at our disposal. Therefore, we compared B and A1 on the test examples common to both of them among the 1100 test examples. We observed that A1 outperformed B in majority of the test examples most of the time. Among all test examples where A1 successfully generated an answer but B resulted in OOM error, the average number of tokens in the (long) context was 1238, 1189, and 1069 for QA pipelines with gemma-2b-it, gemma-7b-it, and llama3.1-8b-it, respectively.

To assess the performance of B relative to $A1_{APS}$, a subset of non-factoid QA pairs from the NaturalQA dataset [27] was automatically translated into Hindi, Tamil, Telugu, Urdu, Marathi, and Bengali using the NLLB 1.3B model [55]. The selection focused on questions for which no short answer was available. The resulting dataset featured an average context length exceeding 8000 tokens. All associated resources will be made publicly accessible. Empirical results demonstrate that (with the exception of QA_{base} with Llama 3.1) $A1_{APS}$ consistently outperforms B across evaluation metrics, within the ChatGPT-as-a-judge framework, and in terms of memory efficiency across all LLMs.

6   Discussion

To better understand the APS model's role in context shortening, we analyzed its predictions using post-hoc explainability methods. Specifically, we employed LIME [45] and SHAP [35] using the Ferret library [5]. A sample rationale for a Hindi text, along with its English translation using the NLLB 1.3B model [55], is shown in Fig. 4. The APS model assigns high scores when it confidently identifies that a paragraph, combined with a question, is likely to answer the query. High-scoring predictions were associated with a greater number of tokens receiving high relevance values (in red), while low-scoring predictions showed fewer such tokens. This observation was empirically validated.

Rationale generation for APS predictions took 4–15 s depending on input length, so analysis was performed on a subset of over 8.5k question-paragraph pairs. As shown in Fig. 5, results confirmed that only a small number of tokens are needed for the APS model to recognize irrelevant paragraphs, supporting its effectiveness as a retriever in QA systems. Using question category annotations

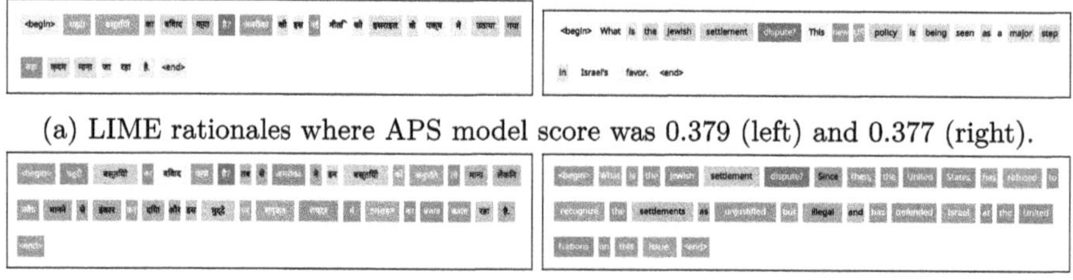

(a) LIME rationales where APS model score was 0.379 (left) and 0.377 (right).

(b) LIME rationales where APS model score was 0.574 (left) and 0.56 (right).

Fig. 4. Rationales generated by LIME for Hindi (left) text and its English translation (right). Tokens with high relevance values (red) indicate their significant contribution to the predicted logit. Masking these tokens is expected to reduce the model's confidence. Notably, as the APS model score increases, the rationales become more dispersed (shown in red). Similar patterns were observed with SHAP rationales. (Color figure online)

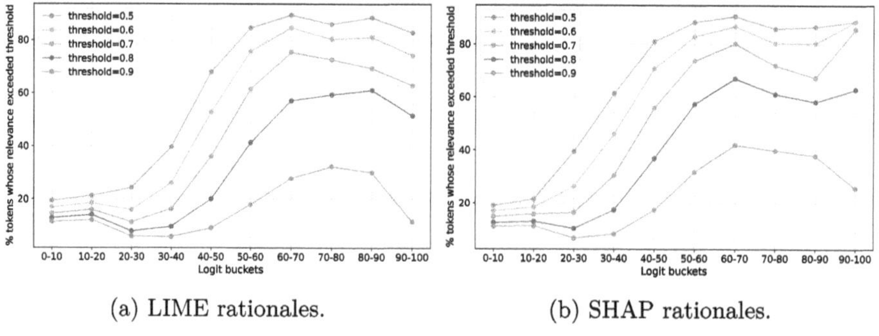

(a) LIME rationales.   (b) SHAP rationales.

Fig. 5. Consider an example to interpret the figure above: an APS model predicts a logit of 0.35 for a text containing 10 tokens, where 7 tokens exhibit a relevance value exceeding a specified threshold (say 0.5). In this case, the logit bucket would fall within the range of 30–40, and the percentage of tokens with relevance values surpassing the threshold (0.5) would be 70%. The figure illustrates that higher logit values are positively correlated with a broader distribution of relevance values across varying thresholds.

from MuNfQuAD [39], we analyzed the top 100 test examples ranked by the average of semantic-level (STS-MuTe) and token-level (ROUGE) scores. Table 2 demonstrates that the highest-quality answers are more frequently associated with factoid and evidence-based questions, as opposed to questions categorized under debate, reason, and experience, aligning with prior findings from [11].

Table 2. QA pipelines across various question categories. The format employed in the cells is as follows: "*Total* vs *Best_100 for* QA_{base} *(Best_100 for* $QA_{finetuned}$*)*" where *Total* and *Best_100* denote the percentage representation of a category among all generated answers and the best-100 answers, respectively. We selected best-100 answers by sorting the generated answers based on their average scores and selecting the top-100. If the *Best_100* percentage surpasses its corresponding *Total* percentage then it is highlighted in green ; otherwise, it is highlighted in red . It is evident that Factoid and Evidence-Based questions are more frequently observed in *Best_100* as compared to questions belonging to the Debate, Reason, and Experience categories.

QA pipeline		Factoid	Evidence-Based	Debate	Reason	Experience
gemma-2b-it	B	30 vs 28 (32)	26 vs 25 (32)	16 vs 21 (13)	12 vs 15 (10)	3.6 vs 3 (2)
	A1	31 vs 32 (40)	25 vs 28 (28)	20 vs 15 (18)	11 vs 11 (4)	3.9 vs 3 (2)
gemma-7b-it	B	30 vs 42 (40)	25 vs 30 (29)	19 vs 11 (17)	11 vs 8 (9)	4 vs 2 (2)
	A1	31 vs 25 (34)	25 vs 31 (32)	20 vs 22 (19)	11 vs 4 (4)	3.9 vs 6 (2)
llama3.1-8b-it	B	31 vs 41 (40)	25 vs 29 (22)	20 vs 18 (24)	11 vs 5 (6)	3.9 vs 1 (3)
	A1	31 vs 46 (37)	25 vs 24 (26)	20 vs 11 (19)	11 vs 5 (7)	3.9 vs 4 (4)

7 Conclusion

Question Answering (QA), one of the earliest tasks in NLP, has seen the development of numerous approaches to address the task. With the emergence of Large Language Models (LLMs) and their superior performance across various NLP tasks, the focus of QA has shifted toward overcoming the limitations of LLMs, particularly the restricted context window, as well as addressing the relatively underexplored area of non-factoid question answering in low-resource languages. This study investigates the use of a retriever mechanism to shorten the context associated with a question. We explore several techniques, including Answer Paragraph Selection (APS), Open Information Extraction (OIE), and coreference resolution (coref), to implement a question-specific retriever. As a baseline, the entire context is provided to the LLM without any context reduction to answer the posed question. To evaluate the quality of the generated answers, we propose a Semantic Text Similarity score for Multilingual Text (STS-MuTe). Our findings indicate that the APS-based approach (A1) outperforms the baseline and other methods on both semantic and token-level metrics. Moreover, the A1 approach is found to be less resource-intensive. Additionally, experiments with post-hoc explainability methods show that the APS model aligns with human intuition when generating a score for a given question-paragraph pair.

8 Limitations

The Gen2OIE method was evaluated only on Hindi and Telugu, with zero-shot experiments conducted on Urdu and Tamil due to their linguistic proximity to Hindi and Telugu, respectively [8]. As a result, the performance of Gen2OIE may not generalize consistently across all languages. Additionally, the ROUGE

Python package[4] used in this study is tailored for English and lacks native support for non-Roman scripts. To address this, all texts were transcribed into Roman script following the method in [10], which may influence the results, particularly in terms of stopword handling.

To assess potential data leakage in evaluation due to LLMs being trained on publicly available data [1], we filtered out test instances that were accessible prior to each model checkpoint's release, based on the last commit date in their HuggingFace repositories. This yielded 78, 197, and 18 QA pairs for gemma-2b-it, gemma-7b-it, and llama3.1-8b-it, respectively. Compared to results in Table 1, the QA_{base} performance with approach B dropped by 13%, 2.5%, and 22% for these models, respectively, suggesting a moderate decline in QA performance when evaluated on data unseen during pretraining.

We recognize that the context lengths used in this study are significantly shorter than the token limits supported by many modern LLMs, such as 8K for Gemma and 131K for Llama 3.1, as specified by the `max_position_embeddings` attribute in their `config.json` files on Huggingface. However, due to the lack of a multilingual non-factoid QuAD dataset with sufficiently long contexts, we consider MuNfQuAD the most appropriate choice for our experiments. Although context length could be artificially extended by concatenating contexts from other questions, prior work has shown that non-factoid answers are typically localized within a small section of the document [60], making such augmentation unnecessary. While our retrieval-based pipeline can theoretically handle arbitrarily long contexts, its performance on long multilingual gold contexts remains to be evaluated. Moreover, our experiments are limited to three LLMs, and further validation across a broader range of models is needed to confirm the generalizability of our findings.

9   Future Work

A promising direction for future work involves evaluating the proposed pipeline on multilingual datasets featuring longer contexts than those found in MuNfQuAD. Extracting text from school textbooks has been demonstrated as an effective approach for curating high-quality datasets in low-resource languages [3]. Furthermore, the APS model can serve as a reward model to align LLMs for generating more informative answers, facilitated by differential performance preference tuning algorithms [29, 40].

Acknowledgments. Ritwik Mishra acknowledges with gratitude the partial financial support provided by the University Grants Commission (UGC) of India through the UGC Senior Research Fellowship (SRF) program. He also expresses sincere thanks to IHub-Data at IIIT Hyderabad for the financial assistance extended during the course of this research. Rajiv Ratn Shah extends his appreciation for the partial funding received from the Infosys Center for AI, the Center of Design and New Media, and the Center of Excellence in Healthcare at IIIT Delhi.

---

[4] https://pypi.org/project/rouge-score/.

Disclosure of Interests. The authors have no competing interests to declare that are relevant to the content of this article.

References

1. Ahuja, K., et al.: Mega: multilingual evaluation of generative AI. In: Proceedings of EMLP 2023, pp. 4232–4267 (2023)
2. Al Nazi, Z., Hossain, M.R., Al Mamun, F.: Evaluation of open and closed-source LLMs for low-resource language with zero-shot, few-shot, and chain-of-thought prompting. Nat. Lang. Process. J. 100124 (2025)
3. Anand, A., et al.: Multilingual mathematical reasoning: advancing open-source LLMs in Hindi and English. arXiv preprint arXiv:2412.18415 (2024)
4. Artetxe, M., Schwenk, H.: Massively multilingual sentence embeddings for zero-shot cross-lingual transfer and beyond. Trans. ACL **7**, 597–610 (2019)
5. Attanasio, G., Pastor, E., Di Bonaventura, C., Nozza, D.: ferret: a framework for benchmarking explainers on transformers. In: Proceedings of EACL: System Demonstrations. Association for Computational Linguistics (2023)
6. Baek, J., Aji, A.F., Saffari, A.: Knowledge-augmented language model prompting for zero-shot knowledge graph question answering. In: Proceedings of the 1st Workshop on Natural Language Reasoning and Structured Explanations (NLRSE), pp. 78–106 (2023)
7. Bajaj, P., et al.: MS MARCO: a human generated machine reading comprehension dataset. arXiv preprint arXiv:1611.09268 (2016)
8. Beaufils, V., Tomin, J.: Stochastic approach to worldwide language classification: the signals and the noise towards long-range exploration (2021)
9. Berant, J., Chou, A., Frostig, R., Liang, P.: Semantic parsing on freebase from question-answer pairs. In: Proceedings of EMNLP, pp. 1533–1544 (2013)
10. Bhat, I.A., Mujadia, V., Tammewar, A., Bhat, R.A., Shrivastava, M.: IIIT-H system submission for fire2014 shared task on transliterated search. In: Proceedings of the Forum for Information Retrieval Evaluation, FIRE 2014, pp. 48–53. ACM, New York (2015). https://doi.org/10.1145/2824864.2824872
11. Bolotova, V., Blinov, V., Scholer, F., Croft, W.B., Sanderson, M.: A non-factoid question-answering taxonomy. In: Proceedings of the 45th ACM SIGIR Conference on Research and Development in Information Retrieval, pp. 1196–1207 (2022)
12. Caciularu, A., Dagan, I., Goldberger, J., Cohan, A.: Long context question answering via supervised contrastive learning. In: Proceedings of NAACL: Human Language Technologies, pp. 2872–2879 (2022)
13. Cer, D., et al.: Universal sentence encoder for english. In: Proceedings of EMNLP: System Demonstrations, pp. 169–174 (2018)
14. Chai, A., et al.: Eager: explainable question answering using knowledge graphs. In: Proceedings of the 6th Joint Workshop on Graph Data Management Experiences & Systems (GRADES) and Network Data Analytics (NDA), pp. 1–5 (2023)
15. Cohen, D., Croft, W.B.: End to end long short term memory networks for non-factoid question answering. In: Proceedings of the 2016 ACM International Conference on the Theory of Information Retrieval, pp. 143–146 (2016)
16. Cohen, D., Yang, L., Croft, W.B.: Wikipassageqa: a benchmark collection for research on non-factoid answer passage retrieval. In: The 41st ACM SIGIR Conference on Research & Development in Information Retrieval, pp. 1165–1168 (2018)

17. Conneau, A., et al.: Unsupervised cross-lingual representation learning at scale. In: Jurafsky, D., Chai, J., Schluter, N., Tetreault, J. (eds.) Proceedings of the 58th ACL Conference, pp. 8440–8451. Association for Computational Linguistics, Online (2020). https://doi.org/10.18653/v1/2020.acl-main.747
18. Cortes, E.G., Woloszyn, V., Barone, D., Möller, S., Vieira, R.: A systematic review of question answering systems for non-factoid questions. J. Intell. Inf. Syst. 1–28 (2022)
19. Dubey, A., et al.: The llama 3 herd of models. arXiv preprint arXiv:2407.21783 (2024)
20. Feng, F., Yang, Y., Cer, D., Arivazhagan, N., Wang, W.: Language-agnostic BERT sentence embedding. In: Proceedings of the 60th ACL Conference (Volume 1: Long Papers), pp. 878–891 (2022)
21. Glass, M., Rossiello, G., Chowdhury, M.F.M., Naik, A., Cai, P., Gliozzo, A.: Re2G: retrieve, rerank, generate. In: Carpuat, M., de Marneffe, M.C., Meza Ruiz, I.V. (eds.) Proceedings of NAACL: Human Language Technologies, pp. 2701–2715. Association for Computational Linguistics, Seattle, United States (2022). https://doi.org/10.18653/v1/2022.naacl-main.194
22. Green, B.F., Wolf, A.K., Chomsky, C., Laughery, K.: Baseball: an automatic question-answerer. In: Papers Presented at the May 9-11, 1961, Western Joint IRE-AIEE-ACM Computer Conference, pp. 219–224. IRE-AIEE-ACM 1961 (Western). Association for Computing Machinery, New York (1961). https://doi.org/10.1145/1460690.1460714
23. Hu, E.J., et al.: Lora: low-rank adaptation of large language models. arXiv preprint arXiv:2106.09685 (2021)
24. Hurst, A., et al.: GPT-4o system card. arXiv preprint arXiv:2410.21276 (2024)
25. Kamalloo, E., Dziri, N., Clarke, C., Rafiei, D.: Evaluating open-domain question answering in the era of large language models. In: Rogers, A., Boyd-Graber, J., Okazaki, N. (eds.) Proceedings of the 61st ACL Conference (Volume 1: Long Papers), pp. 5591–5606. Association for Computational Linguistics, Toronto, Canada (2023). https://doi.org/10.18653/v1/2023.acl-long.307
26. Kolluru, K., Mohammed, M., Mittal, S., Chakrabarti, S., et al.: Alignment-augmented consistent translation for multilingual open information extraction. In: 60th ACL Conference (Volume 1: Long Papers), pp. 2502–2517 (2022)
27. Kwiatkowski, T., et al.: Natural questions: a benchmark for question answering research. Trans. Assoc. Comput. Linguist. **7**, 452–466 (2019)
28. Li, T., Zhang, G., Do, Q.D., Yue, X., Chen, W.: Long-context LLMs struggle with long in-context learning. arXiv preprint arXiv:2404.02060 (2024)
29. Li, X., Yong, Z.X., Bach, S.: Preference tuning for toxicity mitigation generalizes across languages. In: Findings of the EMNLP 2024, pp. 13422–13440 (2024)
30. Lin, C.Y.: ROUGE: a package for automatic evaluation of summaries. In: Text Summarization Branches Out, pp. 74–81. Association for Computational Linguistics, Barcelona, Spain (2004). https://aclanthology.org/W04-1013
31. Litkowski, K.: Question-answering using semantic relation triples. In: TREC (1999)
32. Liu, N.F., Lin, K., Hewitt, J., Paranjape, A., Bevilacqua, M., Petroni, F., Liang, P.: Lost in the middle: how language models use long contexts. Trans. ACL **12**, 157–173 (2024)
33. Liu, W., et al.: K-bert: enabling language representation with knowledge graph. In: Proceedings of the AAAI Conference on Artificial Intelligence, vol. 34, pp. 2901–2908 (2020)
34. Liu, Y., et al.: Bridging context gaps: leveraging coreference resolution for long contextual understanding. arXiv preprint arXiv:2410.01671 (2024)

35. Lundberg, S.M., Lee, S.I.: A unified approach to interpreting model predictions. In: Advances in Neural Information Processing Systems, vol. 30 (2017)
36. Mangrulkar, S., Gugger, S., Debut, L., Belkada, Y., Paul, S., Bossan, B.: Peft: state-of-the-art parameter-efficient fine-tuning methods (2022). https://github.com/huggingface/peft
37. Mishra, R., Desur, P., Shah, R.R., Kumaraguru, P.: Multilingual coreference resolution in low-resource South Asian languages. In: Proceedings of the LREC-COLING 2024, pp. 11813–11826. ELRA and ICCL, Torino, Italia (2024). https://aclanthology.org/2024.lrec-main.1031
38. Mishra, R., Singh, S., Shah, R., Kumaraguru, P., Bhattacharyya, P.: Indie: a multilingual open information extraction tool for indic languages. In: Findings of the IJCNLP-AACL 2023, pp. 312–326 (2023)
39. Mishra, R., Vennam, S., Shah, R.R., Kumaraguru, P.: Multilingual non-factoid question answering with silver answers (2024). https://arxiv.org/abs/2408.10604
40. Naseem, T., et al.: A grounded preference model for LLM alignment. In: Findings of the ACL 2024, pp. 151–162 (2024)
41. Nassiri, K., Akhloufi, M.: Transformer models used for text-based question answering systems. Appl. Intell. **53**(9), 10602–10635 (2023)
42. Pan, Q., et al.: Human-centered design recommendations for LLM-as-a-judge. In: Proceedings of the 1st Human-Centered Large Language Modeling Workshop, pp. 16–29. ACL (2024). https://doi.org/10.18653/v1/2024.hucllm-1.2
43. Qiu, P., et al.: Towards building multilingual language model for medicine. Nat. Commun. **15**(1), 8384 (2024)
44. Reimers, N., Gurevych, I.: Sentence-bert: sentence embeddings using siamese bert-networks. In: Proceedings of the 2019 Conference on Empirical Methods in Natural Language Processing. Association for Computational Linguistics (2019). http://arxiv.org/abs/1908.10084
45. Ribeiro, M.T., Singh, S., Guestrin, C.: "Why should i trust you?" Explaining the predictions of any classifier. In: Proceedings of the 22nd ACM SIGKDD International Conference on Knowledge Discovery and Data Mining, pp. 1135–1144 (2016)
46. Sagen, M.: Large-context question answering with cross-lingual transfer. Uppsala University (2021)
47. Sen, P., Mavadia, S., Saffari, A.: Knowledge graph-augmented language models for complex question answering. In: Proceedings of the 1st Workshop on Natural Language Reasoning and Structured Explanations (NLRSE), pp. 1–8 (2023)
48. Singh, A.K., Murthy, R., Sen, J., Ramakrishnan, G., et al.: Indic QA benchmark: a multilingual benchmark to evaluate question answering capability of LLMs for indic languages. arXiv preprint arXiv:2407.13522 (2024)
49. Soares, M.A.C., Parreiras, F.S.: A literature review on question answering techniques, paradigms and systems. J. King Saud Univ.-Comput. Inf. Sci. **32**(6), 635–646 (2020)
50. Soleimani, A., Monz, C., Worring, M.: Nlquad: a non-factoid long question answering data set. In: the 16th EACL: Main Volume, pp. 1245–1255 (2021)
51. Surdeanu, M., Ciaramita, M., Zaragoza, H.: Learning to rank answers on large online QA collections. In: Proceedings of ACL-08: HLT, pp. 719–727 (2008)
52. Tandon, J., Sharma, D.M.: Unity in diversity: a unified parsing strategy for major Indian languages. In: Proceedings of the Fourth International Conference on Dependency Linguistics (Depling 2017), pp. 255–265 (2017)
53. Tay, Y., et al.: Long range arena: a benchmark for efficient transformers. In: International Conference on Learning Representations (2021)

54. Mesnard, T., et al.: Gemma: open models based on gemini research and technology. arXiv preprint arXiv:2403.08295 (2024)
55. Costa-jussà, M.R., et al.: No language left behind: scaling human-centered machine translation (2022)
56. Trivedi, H., Jetcheva, J.G., Rojas, C.: LLM-based localization in the context of low-resource languages. In: 2024 Conference on AI, Science, Engineering, and Technology (AIxSET), pp. 276–289. IEEE (2024)
57. Uthus, D., Ontanon, S., Ainslie, J., Guo, M.: mlongt5: a multilingual and efficient text-to-text transformer for longer sequences. In: EMNLP (2023)
58. Wang, S., Xu, F., Thompson, L., Choi, E., Iyyer, M.: Modeling exemplification in long-form question answering via retrieval. In: Proceedings of the 2022 NAACL: Human Language Technologies, pp. 2079–2092 (2022)
59. Wolf, T., et al.: Transformers: state-of-the-art natural language processing. In: Proceedings of the 2020 EMNLP: System Demonstrations, pp. 38–45. ACL, Online (2020). https://www.aclweb.org/anthology/2020.emnlp-demos.6
60. Yang, L., et al.: Beyond factoid QA: effective methods for non-factoid answer sentence retrieval. In: Ferro, N., et al. (eds.) ECIR 2016. LNCS, vol. 9626, pp. 115–128. Springer, Cham (2016). https://doi.org/10.1007/978-3-319-30671-1_9
61. Zhang, T., Kishore, V., Wu, F., Weinberger, K.Q., Artzi, Y.: Bertscore: evaluating text generation with bert. In: International Conference on Learning Representations (2020)

CommuteQA: Visual Question Answering for Bus Transport

Anwar Shaikh[(✉)], Prakrit Garg, Mahisha Ramesh, Ritwik Kashyap, Raghava Mutharaju, and Rajiv Ratn Shah

Indraprastha Institute of Information Technology Delhi, New Delhi, India
{anwars,prakrit20100,mahisha23121,ritwik21485,
raghava.mutharaju,rajivratn}@iiitd.ac.in

Abstract. There are many Visual Question Answering (VQA) datasets like VQA [1], OK-VQA [4] and TextVQA [13] but very few of them focus on public transport domain, especially buses. If you are a bus commuter and standing at a bus stop which is new to you, then you may have many questions such as "When next bus will arrive for the destination" or "What are different buses which you can take to reach the destination". So a commuter generally have questions about the bus routes, timings, frequency and finding alternatives. Current datasets do not include these practical questions. Also the existing datasets does not include the location of the user asking the question and the time when the question was asked. Lack of this information limits the design and development of the applications which can help commuters in their daily travel. To address these issues we created a dataset called "CommuteQA" which focuses on public transport user questions. We also propose method for the Visual Question Answering over this dataset, which considers the image, location, time and bus schedule information to answer the user questions. This dataset also aims to help creation of tools and applications for the visually impaired people who use public transport.

Keywords: VQA · Question Answering · Transport domain · Public transport question answering · LLM

1 Introduction

Public transport systems play a major role in the movement of people across the city. Often, newcomers visit the city or existing residents move to different places within the city and commute using public transport. Many of them have questions about the vehicle they are going to ride for their journey. Visually impaired people also face challenges while using public transport. They too have basic questions about visible features of the vehicles such as color, text written, or whether the vehicle is crowded. Visual Question Answering (VQA) can assist these commuters in their journey by answering their questions. The questions can be related to visual features of the vehicle such as color, route number and

© The Author(s), under exclusive license to Springer Nature Switzerland AG 2026
R. Gupta et al. (Eds.): BDA 2025, LNCS 16041, pp. 19–34, 2026.
https://doi.org/10.1007/978-3-032-15134-6_2

destination; these type of questions can come from visually impaired people or illiterate people who cant read numbers or maybe from a person who is not wearing his eyeglasses. And there are questions which could come from any person; these questions are about source and destination of the vehicle, timing of the vehicle, frequency of the vehicle, route of the vehicle and so on.

The existing VQA systems can handle the questions related to visual features of the vehicle, but lacks ability to answer other questions. Also the existing VQA systems do not consider location from where the user is asking the questions and time at which the question is asked. The location and time parameters plays an important role in getting the correct answer for the users question. If the time and location is not provided along with question then the question about public transport vehicle may not be answered accurately.

There are also another type of solutions which considers time and location of the user but does not include image as part of the question, which makes it difficult for the visually impaired people to provide correct context of the question to the VQA system. In the absence of image or time or location it creates difficulty in understanding the situation of the end user and VQA system provides answer based on some assumptions producing incorrect or partially correct answers.

To enable the development of the VQA system for the public transport domain we introduce a dataset named CommuteQA and a baseline method for answering these questions. This dataset contains images of buses and bus stops, along with the location and time when the image was taken. Images cover different view angles of the buses from user perspective. For each image, we have annotated multiple questions. There are many questions which are relevant to each image such as "what is the bus number?", these questions are covered for each image of the bus.

In this paper we also introduced a baseline approach for solving the visual question answering problem for the Bus transportation system. This includes identification of the bus number display board, followed by detection of the bus number and destination written on the display board. Identifying the destination of the bus is crucial in understanding the route followed by the bus. This information along with image, time and question is used for the fine tuning of the model which can understand the question and generate a function as output which can be further executed over the bus route information database to extract the final answer for the user.

2   Related Work

Many VQA datasets exist today, we describe some of them. VQA dataset [1] contains questions about images which needs language understanding and common-sense knowledge. It contains multiple questions per image and multiple ground truth answers per question. Images in VQA dataset are taken from COCO dataset [14] and abstract scenes. The VQA dataset contains images related to public transport but does not contain questions that require outside knowledge, such as the route or schedule of the vehicle.

OK-VQA [4] is Outside Knowledge Visual Question Answering. The questions in the OK-VQA dataset requires models to incorporate external knowledge beyond the image content to answer questions. The questions in the OK-VQA dataset require external knowledge but are not related to the routes, timings, or stations. Instead, these questions are more oriented towards the features of vehicles that require external knowledge.

VizWiz-VQA [7] is another VQA dataset which consists of images taken by blind people along with questions asked by blind people themselves, making it a unique and important VQA dataset created with help of blind people. VQA task also involves predicting if the question is unanswerable. Goal of this dataset is to educate people about needs of blind people and help them to build technologies that can assist blind people.

In TextVQA [13] focus is on improving Visual Question Answering (VQA) models so they can read and understand text in images. Questions in this dataset require understanding and reasoning about the text in the images, but it is not specifically designed for the public transport domain.

Therefore, there is no VQA dataset specifically focused on public transport. Moreover, these datasets do not consider the time and location of the image. Many questions related to public transport can only be answered if the location and time information about the image is available. For example: 'When will this bus reach the airport?' or 'What is the next stop of this bus?'.

Now, looking at various visual question answering models. MuKEA [15] provides a approach to enhance Visual Question Answering by integrating multi-modal knowledge. To bridge the gap between different modalities, they introduce three objective losses focusing on embedding structure, topological relations, and semantic space. It uses pre-training and fine-tuning strategy to gather both basic and domain-specific multi-modal knowledge, enhancing the system's ability to predict answers accurately.

Another model for VQA - UNITER [16] created with a goal of having a universal representation that can handle various tasks involving both images and text such as visual question answering, image-text retrieval and visual common-sense reasoning. VQA model proposed in UNITER [16] is inspired by BERT [17] and use Transformer to leverage self-attention mechanism designed for learning. This models masks only one modality keeping other modality intact to avoid potential misalignment when masked region happens to be described by masked word.

LXMERT [8] is model designed for learning cross-modality encoder representations from transformers. The model is pre-trained using five diverse tasks: masked language modeling, masked object prediction (feature regression and label classification), cross-modality matching, and image question answering. These pre-training tasks help the model to learn both intra-modality and cross-modality relationships.

Model proposed in TextVQA [13] is designed for letting model learn to read and reason about text within images. This is needed for answering questions that involve textual information present in the images. This model uses an architecture called Look, Read, Reason and Answer (LoRRA). This architecture integrates text reading into the visual question answering process, allowing the model to detect and interpret text in images.

PaLI [9] is designed to handle multiple languages, making it a model for multilingual vision-language tasks including VQA, image captioning and text reasoning. Its main focus is on reusing existing language and vision models and jointly scaling both vision and text components.

3   Dataset

We have collected a comprehensive image dataset of buses and bus stops to ensure that the dataset is emblematically representative of the local commuting environment, capturing the multifaceted nature of public transportation within a city. In this section we discuss geo-tagging of images, various parameters considered for creation of dataset, example of image and question from dataset and finally we discuss the annotation process.

3.1   Geo Tagging of Images

Each image was geotagged through the activation of location services on the capturing device, enabling the embedding of comprehensive metadata within the image file's Exchangeable Image File Format Data (EXIF).

3.2   Dataset Collection Parameters

Various perspectives of the commute were considered while collecting the bus images these are summarized in Table 1.

Table 1. Dataset collection parameters.

Parameter	Variations
Types of buses	AC vs Non-AC, Private vs Public Transport
View angle	Front, side and the back view
Number of buses	One and Many
Time of day	Day and Night
Crowd levels	Crowded and Uncrowded
Occlusion	Occluded and Non occluded

3.3   Dataset Examples

Consider a image from dataset which shows rear view of the bus. Figure 1 shows the questions and answers associated with this image. Some of the questions have direct answers for example 'Is this AC bus?', but there can be questions which are unanswerable such as 'Is there any vacant seat in bus?'. Also we see here many questions which have answer such as 'call function getBusDestination()'. Here *getBusDestination()* is an internal function that is executed on the bus route information database to get the destination of the bus based on bus number extracted from the image.

Considering bus and bus stop images there are total 1074 images in the dataset. With 901 day-time images and 173 night-time images. To cover diverse scenarios in which the user can take images, we selected different parameters for the preparation of the dataset. These parameters are time of the day, view of the bus, color of the bus, number of buses in image, operator of the bus and occlusion of bus. The detailed distribution of images is mentioned in Table 2.

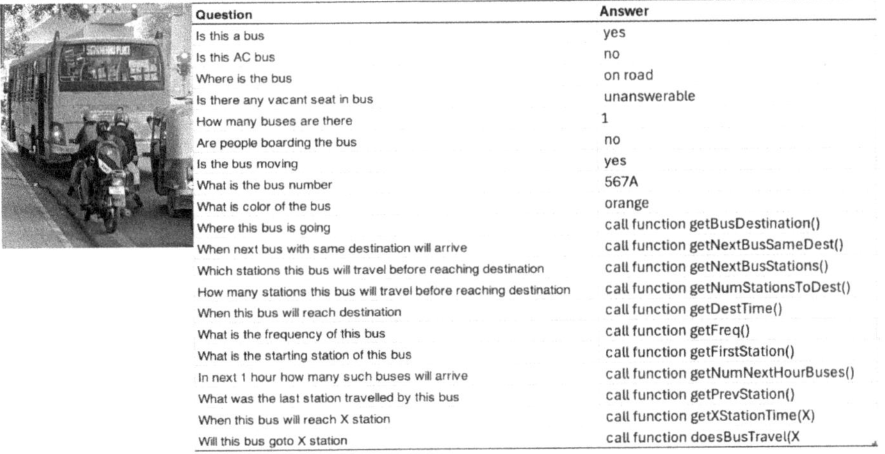

	Question	Answer
	Is this a bus	yes
	Is this AC bus	no
	Where is the bus	on road
	Is there any vacant seat in bus	unanswerable
	How many buses are there	1
	Are people boarding the bus	no
	Is the bus moving	yes
	What is the bus number	567A
	What is color of the bus	orange
	Where this bus is going	call function getBusDestination()
	When next bus with same destination will arrive	call function getNextBusSameDest()
	Which stations this bus will travel before reaching destination	call function getNextBusStations()
	How many stations this bus will travel before reaching destination	call function getNumStationsToDest()
	When this bus will reach destination	call function getDestTime()
	What is the frequency of this bus	call function getFreq()
	What is the starting station of this bus	call function getFirstStation()
	In next 1 hour how many such buses will arrive	call function getNumNextHourBuses()
	What was the last station travelled by this bus	call function getPrevStation()
	When this bus will reach X station	call function getXStationTime(X)
	Will this bus goto X station	call function doesBusTravel(X

Fig. 1. Bus image in question.

3.4   Annotation

Annotation of the images was done using the user interface, which displays the image and questions. There are some common questions that a user can have about the image, those questions are pre-populated on the user interface. Also there is space provided to enter any other question that a user may have about the image. For some of the questions pre-populated options are also provided. We have added one extra option called *Unanswerable* to indicate that the answer can not be determined from the given images. Annotation was done by team of

4 people. Different types of images were equally divided between all annotators. Annotators were advised to at least add one custom question for each image which is not part of the pre-populated list. For each image multiple questions were collected. There are a total of 19857 questions in the dataset.

Table 2. Detailed distribution of images in the dataset

Type	Time of day	View	Blue	Green	Orange	Other color	Red	Total
Bus	Day	Front	146	86	128	10	93	463
		Rear	42	26	41	2	19	130
		Side	83	62	69	1	43	258
	Night	Front	20	23	16	2	23	84
		Rear	5	5	14		5	29
		Side	9	15	8	4	10	46
Bus stop	Day							50
	Night							14
Total			305	217	276	19	193	1074

4 Methodology

Visual Question Answering of the public transport images involves three types of questions. First type is direct questions; questions which can be answered by machine learning models directly from the image. Second type is indirect questions; questions which need additional information provided by the external datasource. Third type is unanswerable questions; questions which can not be answered based on the given image and external information, questions of this type can be found in VizWiz [7]. While training the VQA models, we make the model understand the type of question asked by the user. Following is the process to be followed for indirect questions.

We propose a Visual Question Answering system for the bus images as shown in Fig. 2 which allows us to answer various questions that a passenger may have. As the datasets focus mainly on buses following steps can be followed to answer the questions. The user input is a tuple $(Image - I1, Question - Q1, Time - T1, Location - L1)$. The first step is to identify the coordinates of the display board on the bus where the bus route number and the destination of the bus are written. From the bounding box of display board extract the bus route number as $R1$. Then rephrase the users question using the bus route number. The rephrased question is presented to the VQA model. The VQA Model then outputs either the answer or a string which represents a function call with available parameters. If the output generated is a function call, then execute the function by adding necessary parameter values obtained from image attributes. This function retrieves bus route and schedule information from the bus route databases

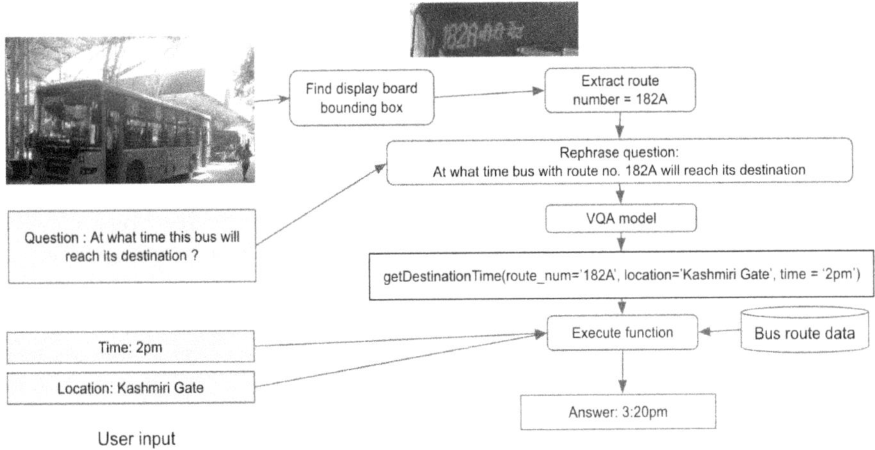

Fig. 2. Visual Question Answering system for Bus images.

such as Open Transit Data [5] or National Transit Map Data [6]. The output of the function is presented to the user as an answer.

Various pre-defined functions are used in the pre-training of the VQA model. The model learns to output these functions when a question is asked. The generated functions are of type $getXStationTime(R_1, L_1, T_1, X)$ where R_1 is the route number, L_1 is location, T_1 is time and X is the station name. This function returns the estimated arrival time at station X for the bus with the given route number based on the given time. This approach can be generalized to different modes of transport and for transport agencies of different cities. The implementation of the above-mentioned functions would differ depending on the underlying datastore used for answering questions.

5 Experiments and Results

The question answering over bus images involves detection of the bounding boxes surrounding the bus destination number, extracting bus number, followed by post processing. So in this section we describe bounding box detection using YOLO [2] model, followed by extraction of bus number using LLaVA model and finally discuss experiments to answer the bus related questions.

5.1 Bounding Box Detection

We have used YOLOv5 [3] for detecting bounding boxes because of its efficient single-stage detection approach that enables real-time bounding box detection with high accuracy. During training, the model learns to predict bounding boxes and classify objects within those boxes based on the annotated dataset. The training process involves optimizing a loss function that penalizes discrepancies between predicted bounding boxes and ground truth annotations. Common loss functions for object detection tasks include the focal loss and the binary cross-entropy loss. The model parameters are updated iteratively using gradient descent optimization algorithms.

Results and Evaluation. We have used 336 images for training and 85 images for validation. At a confidence threshold of 0.001 and an intersection over union (IoU) threshold of 0.6, the model demonstrates an impressive average precision (P) of 82.6% and recall (R) of 71.8%, resulting in mean average precision (mAP50) of 80.3%. Class-specific performance results are shown in Table 3.

Table 3. Results of object detection

Class	Precision(%)	Recall(%)
Overall	82.6	71.8
Bus	91.4	51.7
Bus-stop-board	86.9	88.9
Display board	76.2	79.4
Number plate	75.8	61.2

In Fig. 3, the YOLOv5 model effectively detects and locates a bus. The model's bounding box accurately outlines the bus, digital board, and number plate, separating the bus from its surroundings.

Fig. 3. YOLOV5 model performance on the captured Image.

5.2   Extracting the Bus Numbers

The section describes the methodology used for text extraction from digital displays, such as bus numbers and destinations on buses, as well as destinations indicated on bus stop boards. The LLaVA-1.6-7b-Vicuna model was selected for this task, owing to its superior performance compared to existing optical character recognition (OCR) systems. Traditional OCR technologies demonstrated limitations, particularly in deciphering text from digital boards, where factors such as glare, motion blur, and variable lighting conditions often impede accuracy. In contrast, the LLaVA model showcased enhanced baseline results, outperforming other state-of-the-art OCR solutions. Its advanced algorithms and robust training on diverse datasets enable more reliable recognition of textual information in such challenging scenarios.

Baseline Results. The LLaVA-1.6-7b-Vicuna model proposed in [10] was prompted to identify the numbers and text displayed on a digital board of the bus and to extract the text written on the bus stop. Evaluation of inference results for bus number detection was done based on exact match whereas the evaluation of extraction of bus destination was done based on fuzzy string matching. The model achieved an exact match accuracy of approximately 46% for bus numbers. However, the performance was weaker in extracting bus destinations from the successful bus number matches, with a fuzzy string matching score of only 21%. This lower score was primarily due to difficulties in handling multilingual text (Hindi). In contrast, the evaluation of the model performance on bus stop boards showed promising results. The fuzzy string matching score for extracting bus stop destinations was 79%, suggesting a high success rate in accurately identifying the text on these boards.

Fine Tuning for Detecting Bus Numbers. Owing to LLaVA's challenges with handling digital board text and multilingual (Hindi) content, finetuning was conducted to enhance the baseline results. The fine-tuning of LLaVA utilized LoRA [11] and PEFT [12] techniques on 30% of the positively detected bounding boxes, identified using the YOLO model from the test set.

The evaluation of the inference results after finetuning showed a modest improvement in the exact match accuracy for bus numbers from digital boards, increasing from 46% to 57%. There was also a significant enhancement in the extraction of bus destination text following successful extraction of bus number, with accuracy rising from 23% to 57%. The fine tuned model demonstrated increased capability in handling multilingual text, specifically Hindi bus destination text from digital boards. However, the fine tuned model encountered difficulties in extracting clear bus numbers from some images that the base model had successfully processed, resulting in a loss of accuracy.

Post Processing. To address the limitations of the fine-tuned LLaVA model in accurately extracting bus numbers, destinations, and stop names from images, a post-processing pipeline was implemented using a weighted hybrid string matching algorithm. This approach corrected noisy or partially incorrect outputs from the model by aligning each extracted text string T (e.g., a bus number like "43A" or a destination like "Downtown Terminal") with the closest entry from a known database of correct values $\{C1, C2, \ldots, Cn\}$.

The core of this post-processing lies in a similarity function $S(T, C_i)$ that combines three metrics:

Normalized Levenshtein Distance: The Levenshtein distance $d(T, C_i)$ is the minimum number of single-character edits (insertions, deletions, or substitutions) required to change one string into another. To account for differences in string length, this distance is normalized as

$$dnorm(T, Ci) = \frac{d(T, Ci)}{max(|T|, |Ci|)}$$

where $|T|$ indicates length of the string T.

It is converted to a similarity score as:

$$LevSim(T, Ci) = 1 - dnorm(T, Ci)$$

Jaccard Similarity: Given two strings, this measures the similarity between sets of character n-grams (typically bigrams or trigrams). If $G(T)$ and $G(C_i)$ are the sets of n-grams from T and C_i, respectively, the Jaccard similarity is

$$J(T, C_i) = \frac{|G(T) \cap G(C_i)|}{|G(T) \cup G(C_i)|}$$

where $G(T) \cap G(C_i)$ denotes intersection of the two sets of n-grams, and $G(T) \cup G(C_i)$ denotes union of the two sets of n-grams, $|G(T) \cap G(C_i)|$ represents the number of common n-grams and $|G(T) \cup G(C_i)|$ represents total number of n-grams in both strings.

Cosine Similarity: Here, the strings are converted to vector representations using character count vectors, which are especially well-suited for short, structured text like bus numbers and stop names. The similarity is measured by the cosine of the angle between these vectors:

$$CosSim(T, C_i) = \frac{T \cdot C_i}{||T|| \cdot ||C_i||}$$

where T and C_i are the vector representations of the strings and $||T||$ is the Euclidean magnitude of vector T.

These three scores are combined using a weighted sum to form the final similarity function:

$$S(T, C_i) = w_1 \times (1 - d_{norm}(T, C_i)) + w_2 \times J(T, C_i) + w_3 \times CosSim(T, C_i)$$

with empirically chosen weights $w1 = 0.5$, $w2 = 0.3$, and $w3 = 0.2$. These weights prioritize Levenshtein similarity for precision, Jaccard for structure, and Cosine for redundancy handling, striking a practical balance for short strings.

The best match for each extracted string T is the database entry C_{best} that maximizes this similarity:

$$C_{best} = \arg\max_{C_i} S(T, C_i)$$

This method resulted in significant improvements. For bus number and destination pairs, the exact match rate increased to 66%, while the fuzzy match accuracy for destinations reached 94%, even in cases where the bus number was initially unreadable. For bus stop names, which were processed similarly, the use of the same hybrid matching function against a corpus of all valid stops improved recognition accuracy from 79% to 96%.

The post-processing steps do limit the generalizability of the model but not in a negative way. We encourage a tradeoff between specificity (public transport bus numbers and stops) and generalizability. We prioritize practical accuracy, especially for assisting visually impaired users who depend on reliable transit information.

5.3 Question Answering on CommuteQA Dataset

In this section, we evaluate the performance of various Visual Question Answering (VQA) models using a set of established metrics described in [18]. The models evaluated include llava-1.6-7b-mistral [21], llava-1.6-7b-vicuna [20], llava-1.5-7b (Vicuna) [19], BakLLaVA (llava-1.5-7b-mistral) [22], and paligemma-3b-pt-224 [23]. For each candidate answer generated by the models, we compare it against a single reference answer. Results of evaluation are shown in Table 4.

Table 4. Results of CommuteQA

Model	VQA	Soft VQA	METEOR	CIDER	BERTScore	S-BERTScore
BakLLaVA (llava-1.5-7b-mistral)	33.79	42.76	39.43	24.49	11.33	39.76
llava-1.5-7b (Vicuna)	36.18	43.56	37.71	28.8	19.43	33.29
paligemma-3b-pt-224	43.65	49.98	47.44	50.02	38.66	41.23
llava-1.6-7b-mistral	79.19	83.5	58.73	62.86	79.87	84.03
llava-1.6-7b-vicuna	**83.48**	**86.75**	**58.99**	**65.74**	**81.41**	**85.41**

VQA Accuracy (Exact Match): VQA Accuracy is computed as the proportion of questions for which the model's predicted answer exactly matches the ground truth (reference) answer. Mathematically, it is defined as:

$$\text{VQA Accuracy} = \frac{\text{Number Of Correct Predictions}}{\text{Total Number Of Questions}} \times 100\%$$

Given that each candidate answer is compared against a single reference answer, this metric is particularly stringent, requiring an exact match to achieve a correct prediction. The results show that the 'llava-1.6-7b-vicuna' model achieves the highest exact match accuracy of 83.48%, indicating its superior precision. Conversely, the 'BakLLaVA (llava-1.5-7b-mistral)' model has the lowest accuracy at 33.79

Soft VQA Accuracy (Edit Distance): Soft VQA Accuracy uses Character Error Rate (CER) to measure the similarity between the predicted and reference answers by considering the minimum number of edits (insertions, deletions, or substitutions) needed to transform the predicted answer into the reference answer. It is calculated as:

$$\text{CER} = \frac{\text{Number of Edits}}{\text{Total Number Of Characters In Reference Answer}} \times 100\%$$

This metric is less rigid than exact matching, allowing for partial credit when answers are similar but not identical. With a Soft VQA Accuracy of 86.75%, the 'llava-1.6-7b-vicuna' model again outperforms others, while the 'BakLLaVA (llava-1.5-7b-mistral)' model shows a lower score of 42.76%, reflecting its greater deviation from the reference answers.

METEOR: The METEOR score assesses the quality of the predicted answer by considering not just exact word matches but also stemmed words, synonyms, and word order. The METEOR score is particularly well-suited for single-reference evaluations, as it accounts for linguistic variability while comparing a candidate answer to the reference. It is defined as:

$$\text{METEOR} = 10 \times \frac{Precision \times Recall}{\alpha \times Precision + (1 - \alpha) \times Recall}$$

where α is typically set to 0.85. The 'llava-1.6-7b-vicuna' model records the highest METEOR score of 58.99, indicating a strong performance in generating semantically aligned answers. In contrast, the 'llava-1.5-7b (Vicuna)' model has a METEOR score of 37.71.

CIDEr: The CIDEr (Consensus-based Image Description Evaluation) metric evaluates the consensus between the predicted answer and the reference answer by focusing on n-gram overlaps. While CIDEr is often used in multi-reference scenarios, its application here with a single reference answer still effectively measures the degree of overlap, emphasizing rare n-grams more heavily:

$$\text{CIDEr} = \frac{1}{N} \sum_{n=1}^{N} \sum_{g \in G_n} \log\left(\frac{c(g)}{p(g)}\right) \times \text{IDF}(g)$$

where $c(g)$ is the count of n-grams, $p(g)$ is the probability, and $\text{IDF}(g)$ is the inverse document frequency. The 'llava-1.6-7b-vicuna' model achieves the highest CIDEr score of 65.74, showing the strongest agreement with the reference answers, while 'BakLLaVA (llava-1.5-7b-mistral)' achieves a score of 24.49.

BERTScore: BERTScore evaluates the semantic similarity between the predicted and reference answers by comparing their embeddings using a pre-trained BERT model. This score is particularly effective in single-reference scenarios, as it focuses on capturing the nuanced meanings of words and their contexts:

$$\text{BERTScore} = \frac{1}{N} \sum_{i=1}^{N} \cos(\text{BERT}_{\text{pred}}(i), \text{BERT}_{\text{ref}}(i))$$

The 'llava-1.6-7b-vicuna' model achieves a BERTScore of 81.41, indicating a high level of semantic similarity to the reference answers, whereas the 'Bak-LLaVA (llava-1.5-7b-mistral)' model records a lower score of 11.33.

S-BERTScore: S-BERTScore extends BERTScore by evaluating the overall sentence similarity using sentence embeddings from Sentence-BERT. This method captures the holistic meaning of the predicted answer in relation to the reference answer:

$$\text{S-BERTScore} = \cos(\text{SBERT}_{\text{pred}}, \text{SBERT}_{\text{ref}})$$

The 'llava-1.6-7b-vicuna' model achieves the highest S-BERTScore of 85.41, reflecting its strong ability to generate contextually relevant answers. In comparison, the 'llava-1.5-7b (Vicuna)' model has the lowest S-BERTScore of 33.29.

The evaluation results demonstrate that the 'llava-1.6-7b-vicuna' model consistently outperforms other models across all metrics, making it the most reliable for the VQA task when evaluated against a single reference answer. This model excels in both exact matching and semantic similarity, as indicated by its superior performance in VQA Accuracy, Soft VQA Accuracy, METEOR, CIDEr, BERTScore, and S-BERTScore. On the other hand, the 'BakLLaVA (llava-1.5-7b-mistral)' model lags behind, particularly in semantic similarity metrics such as BERTScore and S-BERTScore, suggesting potential areas for improvement in generating contextually accurate answers.

There was a significant performance gap between LLaVA 1.5 and LLaVA 1.6, primarily due to two key improvements introduced in LLaVA 1.6 such as:

Higher Input Resolution: LLaVA 1.6 supports image inputs at up to 4× higher resolution, which enables better recognition of small or fine-grained visual elements, such as text on digital bus display boards

Enhanced Visual Instruction Tuning: Improved instruction tuning for visual tasks resulted in better multimodal reasoning and OCR capabilities, particularly for real-world scenarios like bus identification.

The PaLI-Gemma model, with 3 billion parameters, underperformed on Hindi text, likely due to limited exposure to Devanagari script during pretraining. To improve its performance in such contexts, the model would require fine-tuning on Hindi-language downstream tasks, ideally using datasets with annotated Hindi transit signage.

In terms of the visual characteristics of the dataset, private buses exhibited greater variability compared to public buses. Public buses typically used standardized LED display boards to show route numbers and destinations, although some were affected by poor maintenance (e.g., missing or dim displays). In contrast, private interstate buses often had clearly visible digital or printed boards. However, private intercity buses frequently lacked consistent signage, making it more difficult to extract structured information from their displays.

Training for all models was enhanced by incorporating additional context specifically tailored for questions involving bus numbers, destinations, and bus stop names. The llava-1.6-7b-vicuna model emerged as the best performer, achieving a 61% accuracy rate in correctly identifying bus numbers, which accounted for approximately 4% of the overall dataset. Despite these improvements, the model continued to face challenges in scenarios where visual perspectives were limited or obstructed, such as when side views were obscured by tinted windows or glare, leading to difficulties in accurately answering related questions.

6   Conclusion

To conclude, we have provided a Visual Question Answering dataset with images and questions specifically designed for the Public Transport domain. We also proposed a baseline architecture which can be used to answer most of the questions that a public transport commuter would have about transport vehicles. We also feel that this dataset would help in building conversational interfaces for the public transport users. The dataset also covers some basic questions which a visually impaired person may have, making it an inclusive dataset.

Limitations and Future Directions: While our system demonstrates strong performance on the CommuteQA dataset, its generalization to other datasets or bus images from different regions may be limited. This is due to variations in the display boards and languages/scripts used. The current OCR and VQA components may require adaptation to handle regional scripts and diverse visual features present in buses around the world.

To scale up the model for multi-region deployment, several steps are necessary such as collecting and annotating a larger and more diverse dataset covering different cities, bus types, and languages; fine-tuning or retraining the OCR and VQA models to handle different scripts, signage styles, and environmental conditions. Future work can explore domain adaptation techniques and active learning to reduce annotation effort and improve model robustness.

Acknowledgments. This work is partly supported by the Infosys Center for AI (CAI) and the Center of Design and New Media (CDNM) at IIIT Delhi, India.

A   Appendix

Different OCR Accuracy
Table 5 and Table 6 details the accuracy of various OCR methods used for detecting bus number and destination, measured as CER - Character Error Rate.

Table 5. OCR accuracy for bus destination extraction

Model	Exact Match	CER
llava-1.6-7b-vicuna	21.39	28.91
llava-1.6-7b-mistral	18.34	24.44
EasyOCR	18.09	29.68
PaddleOCR	12.14	20.43
MMOCR	10.76	18.38
MMOCR-SAM	15.75	23.32

Table 6. OCR accuracy for bus number extraction

Model	Exact Match	CER
llava-1.6-7b-vicuna	46.24	53.98
llava-1.6-7b-mistral	38.65	43.21
EasyOCR	38.72	46.5
PaddleOCR	37.88	47.73
MMOCR	26.34	31.43
MMOCR-SAM	41.9	48.16

References

1. Antol, S., et al.: VQA: visual question answering. In: Proceedings of the IEEE International Conference on Computer Vision, pp. 2425–2433 (2015)
2. Jocher, G., Wightman, R., Bochkovskiy, A.: YOLOv5: An Incremental Improvement. arXiv preprint arXiv:2107.08430 (2021). https://arxiv.org/abs/2107.08430
3. Bochkovskiy, A., Liao, H.Y.M.: YOLOv5-P6: Faster, Better YOLOv5 in Production. arXiv preprint arXiv:2104.13963 (2021). https://arxiv.org/abs/2104.13963
4. Marino, K., Rastegari, M., Farhadi, A., Mottaghi, R.: Ok-vqa: a visual question answering benchmark requiring external knowledge. In: Proceedings of the IEEE/CVF Conference on Computer Vision and Pattern Recognition, pp. 3195–3204 (2019)
5. Open Transit Data. https://otd.delhi.gov.in/
6. National Transit Map Data. https://www.bts.dot.gov/newsroom/national-transit-map-data
7. Gurari, D., et al.: Vizwiz grand challenge: answering visual questions from blind people. In: Proceedings of the IEEE Conference on Computer Vision and Pattern Recognition, pp. 3608–3617 (2018)
8. Tan, H., Bansal, M.: Lxmert: learning cross-modality encoder representations from transformers. arXiv preprint arXiv:1908.07490 (2019)
9. Chen, X., et al.: Pali: a jointly-scaled multilingual language-image model. arXiv preprint arXiv:2209.06794 (2022)
10. Liu, H., Li, C., Wu, Q., Lee, Y.J.: Visual instruction tuning for enhanced multimodal interaction. In: Proceedings of the 37th Conference on Neural Information Processing Systems (NeurIPS 2023) (2023)

11. Huerta, E.A., Kumar, S., Gormley, M.: Low-rank adaptation of large language models (LoRA): streamlining model tuning. In: Proceedings of the International Conference on Learning Representations (ICLR 2021) (2021)
12. Zhang, J., Chen, P., Xiao, L.: Parameter-efficient transfer learning for NLP. In: Proceedings of the 58th Annual Meeting of the Association for Computational Linguistics (ACL 2020) (2020)
13. Singh, A., et al.: Towards VQA models that can read. In: Proceedings of the IEEE/CVF Conference on Computer Vision and Pattern Recognition (CVPR 2019) (2019)
14. Lin, T.Y., et al.: Microsoft coco: common objects in context. In: European Conference on Computer Vision, pp. 740–755. Springer (2014)
15. Ding, Y., Yu, J., Liu, B., Hu, Y., Cui, M., Wu, Q.: Mukea: multimodal knowledge extraction and accumulation for knowledge-based visual question answering. In: Proceedings of the IEEE/CVF Conference on Computer Vision and Pattern Recognition, pp. 5089–5098 (2022)
16. Chen, Y.C., et al.: Uniter: universal image-text representation learning. In: European Conference on Computer Vision, pp. 104–120. Springer (2020)
17. Devlin, J., Chang, MW., Lee, K., Toutanova, K.: BERT: Pre-training of Deep Bidirectional Transformers for Language Understanding. arXiv preprint arXiv:1810.04805 (2018)
18. Mañas, O., Krojer, B., Agrawal, A.: Improving automatic VQA evaluation using large language models. In: Proceedings of the AAAI Conference on Artificial Intelligence, pp. 4171–4179. (2024)
19. llava-1.5-7b (Vicuna) Model. https://huggingface.co/liuhaotian/llava-v1.5-7b. Accessed 18 Aug 2024
20. llava-1.6-7b-vicuna Model. https://huggingface.co/liuhaotian/llava-v1.6-vicuna-7b. Accessed 18 Aug 2024
21. llava-1.6-7b-mistral Model. https://huggingface.co/liuhaotian/llava-v1.6-mistral-7b. Accessed 18 Aug 2024
22. BakLLaVA model. https://huggingface.co/SkunkworksAI/BakLLaVA-1. Accessed 18 Aug 2024
23. paligemma-3b-pt-224 Model. https://huggingface.co/google/paligemma-3b-pt-224. Accessed 18 Aug 2024

IndLaw-QA: Fine-Tuned LLMs with RAG for Indian Legal QA

Aayush Badoni[1], Divyansh Anand Singh[1], Kapil Vuthoo[1],
Shivansh Singh[1], Sonia Khetarpaul[1(✉)], and L. Venkata Subramaniam[2]

[1] Department of Computer Science and Engineering, Shiv Nadar Institution of
Eminence, Delhi NCR, India
{ab681,ds192,kv806,ss398,sonia.khetarpaul}@snu.edu.in
[2] IBM-Research, Gurgaon, India
lvsubram@in.ibm.com

Abstract. Natural Language Processing tasks like information retrieval, question-answering (QA), cross-lingual in legal domain face significant challenges in removing multiple interpretations and ambiguity, contextually complex and jurisdiction-specific legal terminology. This study introduces IndLaw-QA question-answering framework tailored to legal corpora, with a primary focus on the Indian legal system. The proposed architecture leverages latest techniques like Retrieval-Augmented Generation (RAG) with large language models (LLMs), cross-model validation across LLMs and domain-specific fine-tuning, to alleviate limitations in existing legal information retrieval pipelines. The system integrates latest LLMs such as GPT-4o, Llama 3.2, and Claude Haiku 3.5. with prompt optimization and iterative training strategies to enhance semantic contextual precision and output. Further, the framework also incorporates an evaluation protocol to utilize a standardized test corpus comprising 100 legal QA pairs and performance metrics. It includes BERTScore, precision, recall and F1 score, combined with cross-LLM validation with real-time document embeddings. This methodology delivers a scalable, accurate, and semantically adaptive solution for legal information extraction and validation within legal domain specific to Indian law corpora.

Keywords: Legal AI · Legal Question Answering · LLM · Information Retrieval · Cross-validated LLM · RAG · Text embeddings

1 Introduction

The nuances of linguistic heterogeneity inherent in the legal language and more-so in Indian legal framework necessitate an artificial intelligence-driven methodology that is not only legally adept but also linguistically versatile. Given that contemporary language models frequently exhibit deficiencies in contextual profundity essential for legal interpretations and encounter difficulties within the multilingual environment of Indian jurisprudence. In addressing this challenge,

© The Author(s), under exclusive license to Springer Nature Switzerland AG 2026
R. Gupta et al. (Eds.): BDA 2025, LNCS 16041, pp. 35–48, 2026.
https://doi.org/10.1007/978-3-032-15134-6_3

our objective is to create a domain-specific LLM predicated on a comprehensive corpus of Indian legal literature, with the intention of providing precise and contextually pertinent legal insights.

Our methodology encompasses the meticulous processing and refinement of a curated corpus of Indian legal documents - encompassing case law, legislative texts, and scholarly commentary. Through rigorous data preprocessing, annotation, and semantic analysis to encapsulate the unique terminology and procedural subtleties characteristic of Indian legal principles. The model employs Retrieval-Augmented Generation (RAG), facilitating highly precise document retrieval and contextual response formulation. This RAG architecture empowers the LLM to swiftly extract pertinent information from the processed corpus, utilizing it as a foundational knowledge base and to generate coherent, legally robust responses in line with the complexities of Indian law. To enhance efficiency and accessibility, our custom LLM - IndLaw-QA is built on an in-memory database architecture, facilitating rapid retrieval without extensive computational overhead. This architecture supports real-time query responses and scalable deployment, making it well-suited for applications in legal research, case analysis, and document summarization.

The following are the key contributions of the paper:

1. Fine-tuning the language models using legal QA datasets that are tailored specifically to align with the Indian legal framework.
2. Employing Retrieval-Augmented Generation (RAG) with Indian legal documents, which resulted in demonstrably superior outcomes compared to existing studies within the Indian context.
3. Optimized prompt engineering through iterative experimentation with diverse prompts. Thus, contributing to improved response quality.
4. Incorporating a cross-validation methodology between large language models, wherein the responses generated by one model are validated using another. This innovative approach achieved approximately a 15% improvement in the quality of the responses using BERTScore (measure the semantic similarity between a generated text and a reference text by leveraging the power of BERT embeddings. Unlike traditional metrics like BLEU or ROUGE that rely on n-gram overlap).

2   Related Work

Despite the recent advancements like LLMs, their use in legal summarization faces significant challenges. Hallucinations, inconsistencies, and factual inaccuracies are common issues that undermine the reliability of generated summaries. For example, models may produce incorrect dates, names, or relationships between entities, leading to serious errors in the legal context.

Current research efforts are focused on mitigating these issues through domain-specific fine-tuning, improved prompting strategies for LLMs, and incorporating semantic similarity based methods to enhance reliability. Overall, while extractive methods remain robust for factual accuracy, abstractive summarization models and LLMs show potential for generating more natural summaries. But their limitations necessitate a human-in-the-loop approach, where manual oversight ensures the accuracy and consistency of outputs, thus making these models more practical for deployment in real-world legal domain applications.

Fine-tuned and cross-validated LLMs significantly enhance legal QA systems. General-purpose models like BERT perform well, but domain-specific models like LawLLM and LawGPT reach state-of-the-art results. Retrieval augmentation and knowledge frameworks also improve capabilities, although challenges such as legal complexity and ethical concerns remain.

Table 1. Comparison of Key Models in Legal QA Tasks

Model	Description	Citation
BERT	General-purpose transformer-based model with strong performance in legal QA	[3,10]
ALBERT	Lightweight version of BERT with comparable performance	[3]
RoBERTa	Optimized BERT variant with robust performance	[3,10]
LEGAL-BERT	Domain-specific BERT variant	[3]
InternLM-Law	Specialized LLM for Chinese legal tasks	[6]
LawLLM	Multi-task LLM for US legal system	[11]
LawGPT	Open-source Chinese legal LLM	[14]
DISC-LawLLM	Uses legal syllogism prompting and retrieval augmentation	[12]
GPT-3.5, GPT-4	General-purpose models evaluated in legal tasks	[15]

In this paper, we improved Legal QA Tasks as outlined in Table:1 specifically for Indian legal domain by employing fine-tuned language models using Indian legal QA datasets and implemented RAG with Indian legal documents. This resulted in better performance over existing methods. It further enhanced response quality using optimized prompt engineering and a novel cross-LLM validation strategy.

3   Dataset and Tools

3.1   Datasets

- **Indian Supreme Court Judgments Dataset** [8]: Maintained by Legal Information Institute of India, this dataset contains over 42,000 Supreme Court judgments from as early as 1950. Out of this, we used 7,000 judgments for training, which contained the 'Headnote' section, which is the summary of the judgment. This data was used to fine-tune our LLM to the Indian context. This fixes the problem of question-answering in the American context on which the backbone is trained.

– **Kaggle Legal Datasets** [1]: Sourced from Kaggle, this dataset includes various legal datasets such as court rulings, legislative text, and case studies. It provides real-world examples and helps models perform more effectively in legal context understanding and prediction tasks. These datasets are pre-processed, thus reducing the time to cleanup.

– **Custom User PDFs:** These datasets include legal documents like contracts, agreements, and case briefs. These documents allow QA based on the user's specific legal documents. The ability to update and retrieve information from custom files ensures high accuracy during QA sessions. Further, these datasets also contain legal research papers, articles, and case law. It helps in covering legal terminology across various fields of law comprehensively. This data has been scraped by us from the web. It helped in training models to capture complex legal language and produce accurate responses in legal QA tasks as the data covered a range of legal scenarios, from contracts to constitutional law and precedences.

– **Prominent Indian Legal Documents** [2]: This dataset is a collection of key legal documents like the Constitution of India, the Code of Criminal Procedure (CrPC), other codes and acts as shown in Fig. 1. These documents form the backbone of the Indian legal system, thus making it perfect for training models for answering legal queries within the Indian legal context. Their inclusion trains the system to provide better and contextually relevant answers.

3.2 Tools

– **FAISS (Facebook AI Similarity Search):** FAISS [5] is created by Facebook AI. It allows for efficient similarity search and clustering of dense vectors. In this project, FAISS was used for fast document retrieval by indexing the high-dimensional embeddings of legal texts. It allowed for quick identification of relevant documents during QA sessions, thus improving the precision of the system.

– **OpenAI Embeddings:** The OpenAI Embeddings model [7] is used to convert legal documents and queries into high-dimensional vectors aka embeddings that capture semantic meaning. These embeddings enable efficient matching of user queries with relevant documents in the FAISS vector store, thus ensuring contextually accurate retrieval and response.

– **PyPDF2:** PyPDF2[1] is a Python library used to read and extract text from PDFs. This tool is employed to parse legal documents for processing and create embeddings for retrieval purposes. This ensures that custom legal documents are immediately accessible for queries and generating responses.

---

[1] https://pypi.org/project/PyPDF2/.

- **Langchain:** Langchain [4] provides the framework to connect external APIs and data sources. Building the Retrieval-Augmented Generation (RAG) system using this framework makes implementation easier. It integrates document retrieval with language model-based response generation. Further, it enables seamless processing of user queries for retrieval of relevant documents from the vector database (FAISS) and generation of legal responses for enhancing the efficiency and effectiveness of the QA system.
- **BertScore 0.3.13:** The BertScore [13] This Python library was used to compute semantic similarity scores using pre-trained semantic embeddings, BertScore provides a robust metric to compare the generated answers with reference answers. Thus, capturing legal nuances and relevance in QA for legal domain. This tool played a critical role in benchmarking the performance of models at various stages of fine-tuning and validation.
- **OpenAI Fine-Tuning API:** The OpenAI Fine-Tuning API [9] was used to customize the GPT-4o language model with domain-specific legal knowledge. This was accomplished by feeding the model a set of curated legal QA pairs to fine-tune its responses and ensuring that it accurately handles legal questions and provides contextually relevant answers.

4   Methodology

The methodology is designed to address key components of IndLaw-QA legal document-based question-answering model, incorporating a diverse dataset, robust preprocessing, model training, evaluation processes, and advanced strategies for improving contextual relevance and accuracy in the Indian legal domain:

1. **Data Collection:** We curated a diverse dataset by collecting text from various legal documents, articles, and online resources to ensure comprehensive coverage of legal terminology and contexts. Each document was meticulously organized and labeled, facilitating easy retrieval and use during model training and evaluation. The dataset was balanced and representative of different legal scenarios to minimize bias and improve the model's generalization capabilities. Most of the data was sourced from previous research and Kaggle, requiring minimal preprocessing.
2. **Data Preprocessing:** Some preprocessing was needed to optimize performance and reduce reliance on FAISS (our vector database) and the OpenAI Embeddings API. Initially, we removed stop words, punctuation, URLs, mentions, and non-printable characters while lowercasing all text. However, after analyzing early results, we retained only minimal cleaning to preserve semantic richness, improving the performance in subsequent tasks. Non-relevant sections such as appendixes and indexes were also excluded to focus on meaningful content.

   Before processing, legal documents were converted into machine-readable text using PyPDF2 and were cleaned by removing special characters and formatting inconsistencies. This step ensures uniformity in the dataset and optimizes subsequent steps like tokenization and embedding generation.

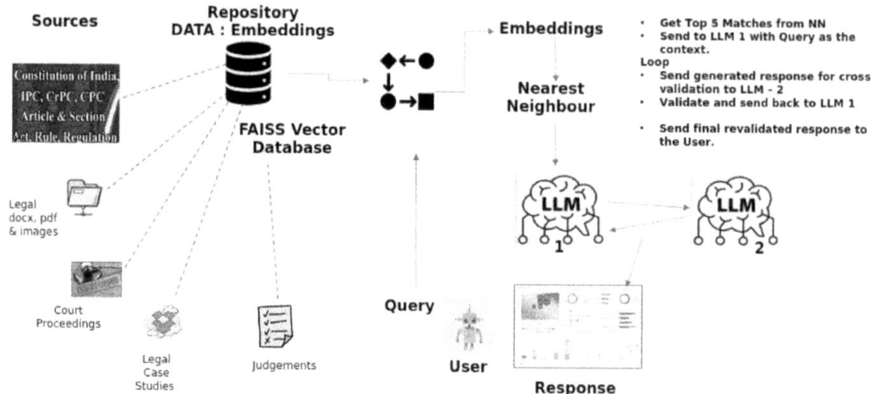

Fig. 1. Proposed Methodology.

3. **Tokenization and Embedding:** We employed the RecursiveCharacterTextSplitter from Langchain to tokenize text into manageable chunks while maintaining contextual integrity. Each chunk was embedded using the OpenAI Embeddings model (text-embedding-3-large), which captures semantic nuances and relationships effectively. We measured the quality of embeddings using cosine similarity metrics to ensure that similar concepts were accurately clustered in the embedding space, leading to better retrieval and response generation.

   The extracted text was tokenized using Hugging Face's tokenizers, breaking down the text into smaller units for granular analysis. This was followed by generating high-dimensional embeddings using OpenAI's Embedding API, ensuring that both the user's query and legal documents are transformed into comparable vector representations.

4. **Model Training and Fine-Tuning:**
   We used the pre-trained GPT-4o language model, leveraging its robust text generation capabilities for legal contexts. The model was fine-tuned on domain-specific data to enhance its understanding of legal tasks and terminology. Using the OpenAI Fine-Tuning API, we provided a dataset of a few hundred QA pairs in JSON format, refining the model's performance for specific legal queries. We also optimized hyper-parameters like temperature and max tokens to balance creativity and coherence.

   To enhance the model's performance in the Indian legal context, we introduced additional fine-tuning using Indian Supreme Court judgements dataset[2]. This focused training enabled the model to better understand region-specific legal terminology, precedents, and nuances.

5. **Retrieval-Augmented Generation (RAG):** This framework was implemented to enhance the model's ability to generate the correct answers. The system works by pre-processing user queries, embedding them using the Ope-

---

nAI Embeddings model and finally retrieving relevant documents from the vector store (FAISS). The top k similar document vectors were selected using Nearest-Neighbor (kNN) algorithm based on cosine similarity. Users could also upload custom PDFs. These pdfs are parsed, split, and embedded in real-time, thus ensuring immediate retrieval during QA sessions.

6. **Model Selection:**
The selection of the optimal model was a critical step in our research, aimed at identifying the most effective large language model (LLM) for legal question-answering in the Indian context. To achieve this, we adopted a rigorous approach involving the training and evaluation of multiple models, including Llama 3.2, Claude Haiku 3.5, and GPT-4o. Each model was evaluated at three stages: untrained (baseline performance), fine-tuned on our custom LLM Fine-Tuning Dataset of Indian Legal Texts from Kaggle, and validated using outputs from another LLM. This three-step enhancement process allowed us to systematically assess the impact of domain-specific fine tuning and cross-validation on performance.

7. **Advanced Prompt Engineering:** To improve the model's contextual relevance and sensitivity, we integrated advanced prompt engineering techniques. By crafting prompts that guide the model to generate answers specifically tailored to the Indian context, including cultural and legal nuances, we improved the quality of responses. Prompts also ensured that the system remained sensitive to user tone and intent, enhancing the overall user experience.

8. **Cross-LLM Validation:** Cross-LLM validation strategy was introduced to boost the accuracy and reliability of responses. In this approach, user queries are answered by our primary LLM (LLM-1) and the generated response is sent to a secondary LLM (LLM 2) along with the relevant context. The secondary LLM validates the response and provides suggestions/ corrections. The primary LLM then iteratively refines its answer based on this feedback, ensuring higher-quality responses.

9. **Evaluation and Testing:**
The model's performance was evaluated on a benchmark test set of 100 legal QA pairs - derived from the LLM Fine-Tuning Dataset of Indian Legal Texts. The responses generated by the models were compared against expert-verified answers to assess their contextual accuracy and relevance. Metrics such as precision, recall and F1 score were calculated to quantify performance. Moreover, we employed BERTScore as the primary evaluation metric, enabling a nuanced comparison of semantic similarity between the generated and reference answers. Further for evaluation, a user interface was implemented featuring authentication, chat history, and session persistence. This ensured consistent interactions during testing and allowed for qualitative feedback on the model's usability and relevance. These combined efforts provided a comprehensive assessment of the model's performance in addressing legal queries.

10. **Iterative Development and Optimization:**
The entire development process followed an iterative approach, where we continuously refined the model based on evaluation feedback and real-world testing. By assessing the model's performance at each stage and incorporating

user feedback, we made continuous improvements, ensuring that the final system meets the requirements for real-world legal application.

5   Performance Evaluation and Results

The use of LLMs within the context of legal domain assesses logical reasoning abilities and accuracy and also guides the trajectory of future developments in the field of legal NLP. Performance evaluation metrics such as BERTScore, benchmarks the extent of generated text aligns with reference text in terms of semantic content. Further, extensive benchmarks like LawBench and Legal-Bench offer systematic methodologies for evaluating LLMs across various dimensions of legal knowledge. They cover areas of practical implementation in the domain with curated tasks by legal professionals. These evaluative benchmarks help researchers and practitioners to understand the strengths and limitations of LLMs. It helps pave way to the enhancement of more efficient and dependable NLP systems tailored for legal tasks.

5.1   Dataset and Evaluation Process

The dataset used for LLM Fine-tuning is from Kaggle - Dataset of Indian Legal Texts. It contains one-line legal questions and corresponding answers. This dataset is tailored for fine-tuning language models in the context of Indian legal texts. The dataset was split into following parts:

- **Fine-tuning Set:** The fine-tuning set was used to train the last layers of frozen model. While the testing set was reserved exclusively for evaluation to ensure an unbiased assessment and generalization of model performance. The fine-tuning set consisted of a substantial portion of the dataset, while the final testing phase was conducted on 100 questions selected from the dataset. These questions were carefully curated to ensure diversity and relevance to the Indian legal context.
- **Testing set:** Each question in the testing set was followed by a corresponding legal answer, and the models were tasked with generating answers based on the provided queries. The models' outputs were compared with the ground-truth answers from the dataset.

5.2   Model Evaluation

The evaluation of the models was performed at various stages of enhancement, including untrained, fine-tuned, and validated configurations. For each stage, we computed the Precision, Recall, and F1 scores, providing a comprehensive understanding of how the models improved with each enhancement step.

- **Base Model:** The initial evaluation was performed on the untrained versions of each model, where the models were evaluated based on their raw performance without fine-tuning or validation.
- **Fine-tuned Model:** The models were then fine-tuned on the LLM Fine-tuning Dataset of Indian Legal Texts to adapt their responses specifically to the Indian legal context. We recalculated the evaluation metrics at this stage to determine the impact of fine-tuning on model performance.
- **Cross-validated Model Untrained:** Lastly, we validated the fine-tuned models using Claude Haiku for cross-LLM validation. This validation step aimed to further enhance the accuracy and relevance of the answers by incorporating a second model's feedback and refinement, and we evaluated the models at this stage as well.

5.3 Performance Metrics

The scores derived from the evaluation process, specifically - Precision, Recall and the F1 score were meticulously calculated employing the BERTScore methodology. This serves to rigorously analyze and assess the semantic congruence and alignment that exists between the responses that were generated by various models and the reference responses that were utilized as benchmarks, encompassing all stages of enhancement and development of these models. The employment of these particular metrics facilitated a comprehensive and nuanced understanding of the performance exhibited by each model, effectively capturing not only the aspect of accuracy represented by Precision but also encompassing the breadth and comprehensiveness of the generated outputs as reflected by Recall, alongside providing insights into their harmonic mean, commonly referred to as the F1 score, which integrates these two vital dimensions of performance evaluation into a singular, coherent measure. Ultimately, the diligent application of these assessment techniques allowed for a deeper exploration and elucidation of the intricate dynamics that characterize model performance, thereby contributing valuable insights into the effectiveness and reliability of the generated responses in relation to established reference standards.

5.4 Results

In our evaluation, we compared the performance of several models, including Llama 3.2, Claude Haiku 3.5, and GPT-4o, in both untrained and fine-tuned configurations, as well as with the addition of cross-LLM validation using Claude Haiku. Our goal was to assess how these models performed in a legal question-answering task, measured through Precision (P), Recall (R), and F1 score metrics.

As depicted in Table 2, the GPT-4o model, both in its untrained and fine-tuned forms, consistently outperformed the other models across all evaluation metrics. Initially, GPT-4o, in its untrained form, achieved a Precision of 0.619, Recall of 0.612, and an F1 score of 0.624. Upon fine-tuning the GPT-4o model, we observed significant improvements, with the Precision, Recall, and F1 scores reaching 0.672, 0.670, and 0.677, respectively.

The most notable enhancement came when we integrated Claude Haiku validation into the GPT-4o framework. This approach not only incorporated cross-LLM validation, where GPT-4o's generated responses were validated and refined by Claude Haiku, but also helped correct errors and adjust context, leading to more accurate and reliable outputs. The Table 2 and Fig. 2 shows that GPT-4o with Claude Haiku validation achieved the highest scores: Precision of 0.702, Recall of 0.699, and F1 of 0.705. GPT-4o with Claude Haiku validation outperformed the other models, including the Llama 3.2 and Claude Haiku 3.5 variants. While Llama 3.2 and Claude Haiku models demonstrated improvements with fine-tuning and validation, their scores remained significantly lower than those of GPT-4o, indicating that GPT-4o's architecture, coupled with cross-validation, provided superior performance for the legal question-answering task.

The improvements in F1 score, rising from an initial 0.624 to 0.705 with GPT-4o's cross-validation, reflect a significant leap in model performance. This result highlights the effectiveness of combining fine-tuning on legal domain data with the validation mechanism provided by Claude Haiku, ensuring not only more precise answers but also enhancing recall to improve the model's ability to cover diverse legal contexts.

As the final comparative study showcases : GPT-4o model, augmented with Claude Haiku validation, emerged as the most effective solution, underscoring its suitability for legal text processing in the Indian context.

6   Limitations

This research is specifically designed for the Indian legal framework, concentrating on the creation of a question-and-answer model tailored for Indian legal texts and processes. Consequently, its direct relevance to other legal systems is restricted due to notable differences in legal terminology, formats, and practices across various nations. The training and assessment of the model depend on a Kaggle dataset that consists of concise legal questions and their corresponding answers, which may not adequately capture the intricacies or scope of authentic legal discussions. Moreover, while employing robust models such as GPT-4o and implementing a cross-validation strategy with Claude Haiku improves the system's dependability, it simultaneously raises computational demands, possibly impeding scalability in resource-limited environments.

The model undergoes assessment with a fairly compact test set of 100 question-answer pairs, utilizing traditional evaluation metrics such as BERTScore, precision, recall, and F1 score. Although the metrics here give relevant information, the small test set size triggers inquiries about the extent to which the results can be generalized. The methodology also incorporates sophisticated elements like Retrieval-Augmented Generation and prompt engineering, which, while effective, may complicate implementation in practical scenarios.

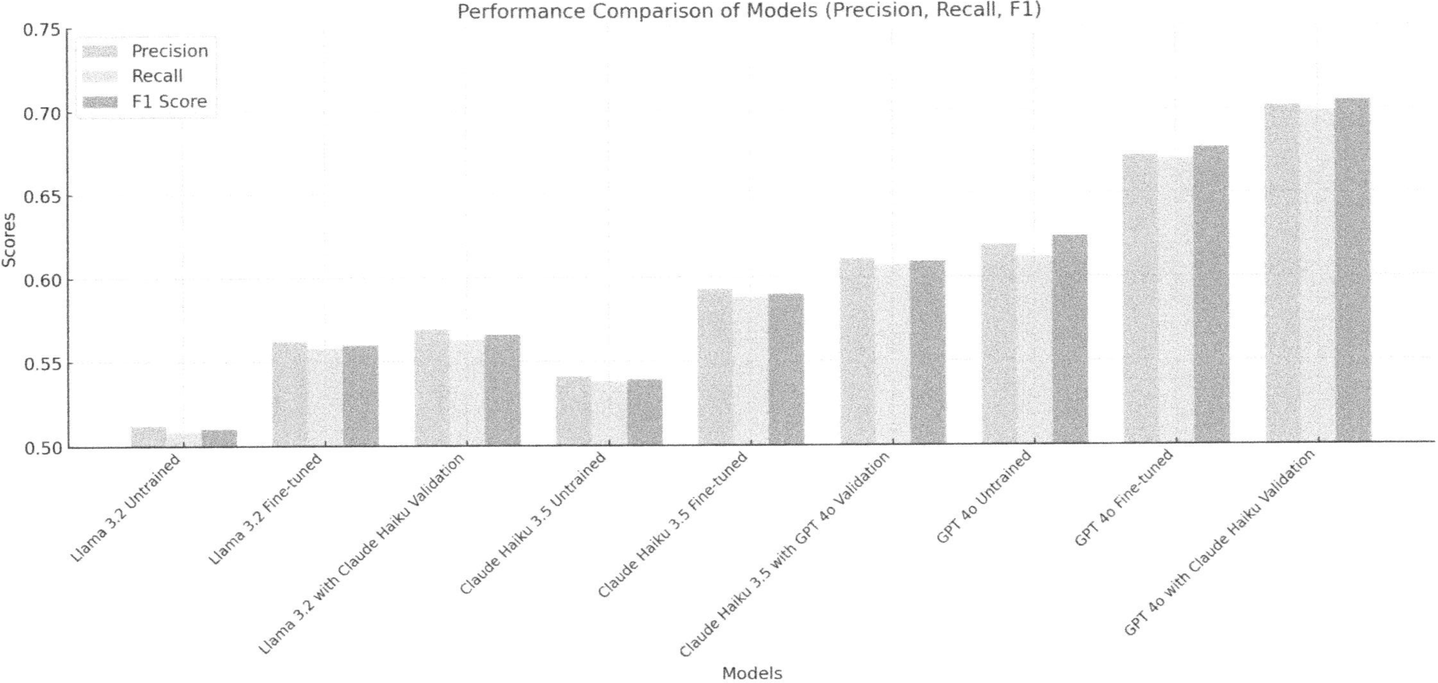

Fig. 2. BertScores of Various Trained and Untrained Models.

Table 2. BERTScore - Precision, Recall and F1 Scores of LLMs

Model	Precision	Recall	F1 Score
Llama 3.2 Untrained	0.512	0.508	0.510
Llama 3.2 Fine-tuned	0.562	0.558	0.560
Llama 3.2 with Claude Haiku Validation	0.569	0.563	0.566
Claude Haiku 3.5 Untrained	0.541	0.538	0.539
Claude Haiku 3.5 Fine-tuned	0.593	0.588	0.590
Claude Haiku 3.5 with GPT 4o Validation	0.611	0.607	0.609
GPT 4o Untrained	0.619	0.612	0.624
GPT 4o Fine-tuned	0.672	0.670	0.677
GPT 4o with Claude Haiku Validation	**0.702**	**0.699**	**0.705**

Furthermore, despite attempts to mitigate factual inaccuracies through domain-specific fine-tuning, challenges such as hallucinations persist in legal summarization. The research also fails to thoroughly address the ethical and legal ramifications of implementing such a system, particularly regarding user trust and accountability, highlighting the necessity for further exploration into the responsible application of AI in legal contexts.

7   Conclusion

In conclusion, our findings support that the integration of cross-LLM validation, particularly using Claude Haiku alongside GPT-4o, substantially boosts performance. The F1 score improvement from 0.6 to 0.7 demonstrates the robustness of this approach in legal question-answering tasks, marking GPT-4o with Claude Haiku validation as the most effective configuration in our experiments. We evaluated the efficacy and applicability of fine tuned LLMs with RAG for legal QA tasks within the complex domain of the Indian legal framework. Our methodology entailed a comprehensive assessment of multiple LLMs, encompassing both pre-trained and fine-tuned variants, specifically Llama 3.2, Claude Haiku 3.5, and GPT-4o. Employing an feedback pipeline of domain-specific fine-tuning and cross-model validation using secondary LLM leading to enhancements in model performance. Notably, GPT-4o achieved the highest performance, attaining an F1 score of 0.677 post-fine-tuning and validation with Claude Haiku, a marked improvement over its baseline score of 0.624. These results highlight the critical role of targeted fine-tuning and cross-LLM validation in optimizing LLMs for specialized tasks of legal question answering in the Indian legal ecosystem. These impressive results underscore the fundamental significance of fine-tuning and validation strategies in advancing the performance of large language models when engaged in specialized tasks, which inherently call for answers that are not merely accurate but also contextually aligned and pertinent to the unique legal contexts in question.

The evidence we gathered makes it clear that, even though untrained models perform reasonably well, the act of fine-tuning these models using domain-specific datasets, combined with validation via external large language models, can greatly boost the accuracy and pertinence of the produced answers. Furthermore, this work serves to highlight the promising potential that advanced validation techniques hold for the further enhancement of model performance, particularly within the realm of complex legal domains that often involve intricate and nuanced considerations. Looking ahead, future research endeavors could explore the promising integration of additional domain-specific data, as well as the further refinement of validation strategies, in order to optimize the overall applicability and effectiveness of large language models for legal practitioners and the development of automated legal systems. In summary, developing a domain-specific Indian legal LLM using RAG and a curated legal corpus is a validated approach that addresses the contextual and linguistic demands of Indian law, with current research showing marked improvements in precision and relevance for legal tasks.

References

1. Indian Supreme Court Judgments—kaggle.com. https://www.kaggle.com/datasets/vangap/indian-supreme-court-judgments. Accessed 25 Apr 2025
2. Papers with Code - ILDC Dataset—paperswithcode.com. https://paperswithcode.com/dataset/ildc. Accessed 25 Apr 2025
3. Caballero, E.Q., et al.: Study of question answering on legal software document using bert based models. In: Proceedings of LatinX in AI Workshop at ICML (2022)
4. Chase, H., et al.: Langchain: Building applications with llms through composability (2022). https://github.com/langchain-ai/langchain. Accessed 25 Apr 2025
5. Douze, M., et al.: The faiss library. arXiv preprint arXiv:2401.08281 (2024)
6. Fei, Z., et al.: Internlm-law: an open source Chinese legal large language model. arXiv preprint arXiv:2406.14887 (2024)
7. Korade, N.B., Salunke, M.B., Bhosle, A.A., Kumbharkar, P.B., Asalkar, G.G., Khedkar, R.G.: Strengthening sentence similarity identification through openai embeddings and deep learning. Int. J. Adv. Comput. Sci. Appl. **15**(4) (2024)
8. LII of India: Indian supreme court cases. http://www.liiofindia.org/in/cases/cen/INSC/. Accessed 26 Nov 2024
9. OpenAI: Openai fine-tuning api (2023). https://platform.openai.com/docs/guides/fine-tuning. Accessed 25 Apr 2025
10. Shah, N., Thakkar, H.K., Mewada, H.: On the analysis of a bert-based domain-specific question answering models for Indian legal system. In: Proceedings of IEEE International Symposium on Advanced Electrical Communication Technology (ISAECT) (2024)
11. Shu, D., et al.: Lawllm: law large language model for the us legal system. In: Proceedings of ACM Conference on Fairness, Accountability, and Transparency (FAccT) (2024)
12. Yue, S., et al.: Disc-lawllm: fine-tuning large language models for intelligent legal services. arXiv preprint arXiv:2309.11325 (2023)

13. Zhang, T., Kishore, V., Wu, F., Weinberger, K.Q., Artzi, Y.: Bertscore: evaluating text generation with bert. In: International Conference on Learning Representations (ICLR) (2020). https://github.com/Tiiiger/bert_score, version 0.3.13. Accessed 25 Apr 2025
14. Zhou, Z., et al.: Lawgpt: a Chinese legal knowledge-enhanced large language model. arXiv preprint arXiv:2406.04614 (2024)
15. Zubaer, A.A., Granitzer, M., Mitrovic, J.: Performance analysis of large language models in the domain of legal argument mining. Front. Artif. Intell. (2023)

Generative AI-Driven Continuous Monitoring and Management of Diabetes

Snehali Biswas, Aman Sagar, Suchi Kumari$^{(\boxtimes)}$, and Abhishek Soni

Department of Computer Science and Engineering, Shiv Nadar Institute of Eminence, Delhi-NCR 201314, U.P., India
{sb855,as624,as630}@snu.edu.in, suchi.singh24@gmail.com

Abstract. Purpose: An AI-driven solution is provided for predicting diabetes risk and managing future glucose levels using advanced machine learning models. By analyzing features such as age, gender, hypertension, BMI, and blood glucose levels, we employ models like Logistic Regression and Random Forest to assess diabetes risk.

Methods: We leverage Random Forest model and Long Short-Term Memory (LSTM) networks on CGM data to forecast future glucose levels based on historical readings, insulin doses, and carbohydrate intake. The models are evaluated using performance metrics such as accuracy, loss, confusion matrix, ROC curve, MAE, and RMSE.

Results: The best-performing model is integrated into a web application, offering real-time glucose predictions and personalized recommendations. Generative AI further enhances the solution by providing patients with actionable suggestions for improving their health and supporting self-management of diabetes.

Conclusion: This innovative tool has the potential to enhance long-term diabetes management by offering data-driven insights and actionable suggestions, particularly benefiting patients who require personalized care to maintain stable glucose levels.

Keywords: Diabetes monitoring · CGM · Gen-AI · Diabetes management

1 Introduction

Diabetes mellitus is a chronic metabolic disorder that poses a significant global health challenge. According to a survey conducted by International Diabetes Federation in 2021 [1], over 537 million adults worldwide, with projections estimating this number to rise to 643 million by 2030 and 783 million by 2045 will be affected by diabetes. The condition is characterized by the body's inability to regulate blood glucose levels due to either insufficient insulin production (Type 1 diabetes) or insulin resistance (Type 2 diabetes), leading to serious complications such as cardiovascular diseases, kidney failure, and neuropathy if not properly managed. Effective diabetes management requires regular monitoring

R. Gupta et al. (Eds.): BDA 2025, LNCS 16041, pp. 49–64, 2026.
https://doi.org/10.1007/978-3-032-15134-6_4

of blood glucose levels, adherence to tailored treatment regimens, and timely interventions. Continuous Glucose Monitoring (CGM) systems have emerged as a transformative technology in diabetes care, offering real-time glucose data to patients and clinicians. Butt *et al.* [2] showed that CGM systems improve glycemic control, reduce the frequency of hypoglycemia, and enhance patients' quality of life by promoting informed decision-making regarding insulin administration and dietary adjustments. However, despite their benefits, CGM devices face critical barriers to widespread adoption due to their high costs and limited monitoring duration, typically lasting only 14 days.

The motivation for this project stems from two critical issues i.e., cost and limited monitoring duration, in current diabetes management. The CGM tools are prohibitively expensive, making them inaccessible to many patients, especially in developing countries. Current CGM devices typically track glucose levels for only 14 days. Once this monitoring period ends, patients often revert to untracked routines, which can lead to poor diabetes management and health deterioration. To address these limitations, this project aims to develop a cost-effective, AI-driven solution that extends the benefits of CGM data for diabetes management. The tool will focus on both self-management of diabetes through tracking predicted future glucose levels and predictive analytics for assessing the risk of developing diabetes. The primary objectives of this Generative AI-Driven Continuous Diabetes Management Tool are:

- Develop AI models using LSTM and Random Forest to predict glucose levels based on historical CGM data, insulin doses, carbohydrate intake, and other factors.
- Implement a predictive algorithm to assess a patient's likelihood of developing diabetes based on historical glucose levels, lifestyle factors, and relevant health metrics.
- Allow patients to input their glucometer readings and receive real-time predictions and insights on their future glucose levels.
- Use generative AI to offer customized advice on maintaining glucose control, including dietary suggestions, insulin dosing recommendations, and lifestyle adjustments.
- Design the tool to be accessible and cost-effective, particularly for patients in low-resource settings.
- Measure the tool's effectiveness in predicting and managing blood glucose levels, with comparisons to traditional diabetes management methods.

The manuscript is organized as follows: Sect. 2 reviews existing research related to this field. In Sect. 3, we present the proposed strategies, which are organized into several phases. Section 4 presents the results along with a detailed analysis. Lastly, Sect. 5 offers the conclusions of the study and discusses potential future directions.

2   Literature Review

Extensive research has been conducted in the field of diabetes prediction using machine learning techniques, aiming to improve early diagnosis and risk assessment. Mujumdar *et al.* [1] proposed a diabetes prediction model utilizing key features such as glucose levels and BMI. They demonstrated the effectiveness of logistic regression, achieving high classification accuracy and emphasizing its potential for large-scale health screenings. Butt *et al.* [2] employed a hybrid machine learning system combining Random Forest and Multilayer Perceptron (MLP). They used diverse datasets for classification, achieving an accuracy of 87.26%, proving the robustness of ensemble methods in identifying diabetes risk. Khanam *et al.* [3] conducted a comparative study of multiple machine learning algorithms, such as Logistic Regression and Support Vector Machines (SVM), concluding that these methods provided superior accuracy in early diabetes detection. Ahmed *et al.* [4] further enhanced predictive models by combining SVM and artificial neural networks with fuzzy logic. This integrated system achieved an impressive 94.87% accuracy, demonstrating the potential of combining machine learning with soft computing techniques for real-time diabetes risk prediction. Similarly, Shahriare *et al.* [5] proposed a hybrid machine learning model leveraging ensemble techniques for diabetes classification. These studies showcase the advancements in predictive modeling and emphasize the role of machine learning in enabling early and accurate diabetes diagnosis.

In addition to prediction, significant efforts have been directed toward improving diabetes management through the use of generative AI and personalized tools. Dey [6] explored the role of ChatGPT in augmenting diabetes care. The study highlighted ChatGPT's ability to provide personalized medical guidance, answer patient queries, and engage users with tailored recommendations. Zheng *et al.* [7] analyzed the potential of ChatGPT in enhancing self-management and diabetes education. However, the study also noted the limitations of generative AI, such as restricted medical expertise and contextual understanding. Mashatian *et al.* [8] developed a retriever-augmented generative AI model for delivering accurate medical information about diabetes care, achieving an accuracy of 98%. This work demonstrated the reliability of AI systems in providing trusted health advice while emphasizing the importance of content validation in medical applications. Jadon *et al.* [9] addressed the challenges of synthetic data generation in healthcare using generative AI models. All these studies underline the transformative potential of generative AI in providing proactive, patient-centered solutions for managing diabetes, thereby addressing a critical gap in healthcare accessibility and engagement.

The development of advanced methods for glucose monitoring and forecasting has been pivotal in managing diabetes more effectively. Pappada *et al.* [10] utilized neural networks for real-time glucose level prediction, specifically targeting insulin-dependent diabetes patients. Zhu *et al.* [11] proposed an innovative framework based on signal decomposition methods for blood glucose prediction. These approaches significantly improved prediction accuracy and was particularly effective in handling irregularities in glucose data. Kavakiotis *et al.* [12] provided an extensive review of machine learning and data mining techniques in diabetes research, emphasizing the role of CGM systems in improving glycemic control. Their findings highlighted the potential of these systems to provide real-time insights, enabling clinicians and patients to make timely interventions. Zhu *et al.* [13] conducted a systematic review of deep learning applications in diabetes, focusing on their ability to process large-scale CGM datasets. The review concluded that deep learning models are highly effective in identifying complex patterns and trends in glucose variability, offering superior diagnostic and predictive capabilities.

3 Proposed Methodology

The study is divided into various phases; prediction of diabetic risk based on constant glucose, future glucose level using CGM data and personalized AI-based insights for patients.

3.1 Prediction of Diabetes Risk Based on Constant Glucose

Our focus is on predicting the risk of diabetes using various machine learning models. By analyzing features such as age, gender, hypertension, BMI, and blood glucose levels, we utilize models like Logistic Regression, Random Forest, and Support Vector Machine (SVM). These models assess the relationship between these attributes and the probability of a diabetes diagnosis. The architectural diagram in Fig. 1 illustrates the process, where the dataset is trained on multiple models, and the best-performing model is selected for predicting the risk of diabetes.

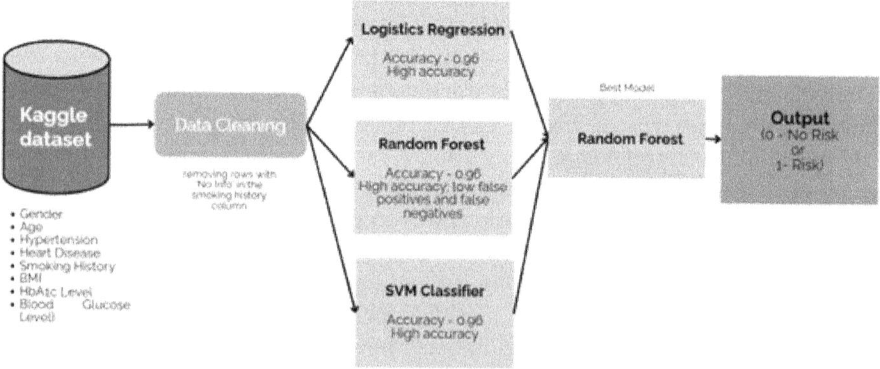

Fig. 1. The architectural diagram for prediction of diabetic risk on constant glucose level.

3.2   Future Glucose Level Prediction Using CGM Data

We utilize Continuous Glucose Monitoring (CGM) data to predict future glucose levels. By employing Long Short-Term Memory (LSTM) networks and Random Forest, we analyze sequential data, including historical glucose levels, insulin dosages, and carbohydrate (CHO) intake. Our model forecasts glucose levels for the upcoming week, enabling individuals to manage their diabetes more effectively by anticipating fluctuations and adjusting insulin and dietary intake accordingly. The methodology for predicting future glucose levels is illustrated in Fig. 2, where preprocessed data is fed into both the LSTM and Random Forest models, with the best-performing model selected for glucose level prediction using CGM data.

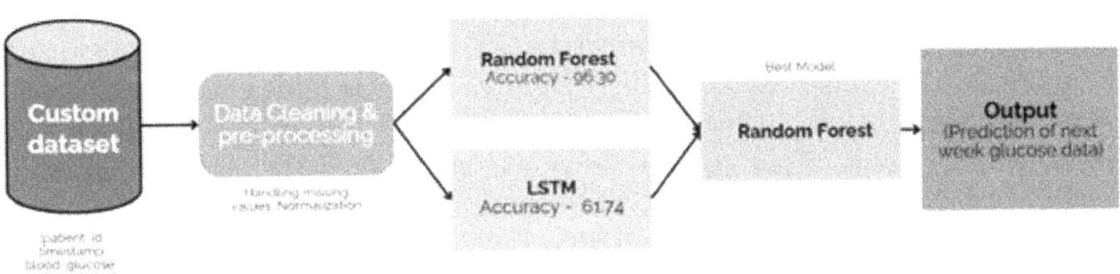

Fig. 2. The architectural diagram for prediction of future glucose level considering CGM data

3.3   Dataset and Preprocessing

The Diabetes Prediction Dataset [14] was utilized to predict the risk of diabetes. Table 1 displays the attributes of the data set.

Table 1. Attributes of Kaggle Dataset for Diabetes Risk Prediction

Attribute	Description
Gender	Biological sex (Male, Female, Other)
Age	Ranges from 0 to 80
Hypertension	0 (No), 1 (Yes)
Heart Disease	0 (No), 1 (Yes)
Smoking History	Not current, Former, No Info, Current, Never, Ever
BMI	Ranges from 10.16 to 71.55
HbA1c Level	Indicates average blood sugar; ≥ 6.5
Blood Glucose Level	Measures glucose in the bloodstream
Diabetes	Target variable; 0 (No diabetes), 1 (Diabetes)

In the data cleaning phase, rows with 'No Info' in the smoking history column, are removed. Categorical variables were converted to numeric values: 'Female' was mapped to 0 and 'Male' to 1 for the gender variable, while 'never' was mapped to 0, 'former' to 1, and 'current' to 2 for smoking history. Rows with missing values were removed. For feature selection, all columns except for the'diabetes' variable were included as features (X), while the'diabetes' column was used as the target variable (y). The dataset was split into training (80%) and testing (20%) sets. Finally, feature standardization was performed using the StandardScaler to normalize the feature scales.

The dataset for predicting future blood glucose levels is generated using the Simglucose simulator [15] for Type-1 Diabetes research. The dataset includes simulated data from 30 virtual patients. To generate data using a simulator we used a Python script [15,16] for basal-bolus insulin management by simulating glucose levels for different patients (adults, adolescents, and children). Table 2 displays the input and output parameters. The dataset was processed to ensure the model's consistency and usability. Data cleaning included handling missing values, outlier removal, and sorting. Min-max scaling was applied to prepare the features for training. This scaling technique normalized the data, bringing all features; CGM, insulin, and carbohydrate intake (CHO), into the range [0, 1]. Two types of features were created from the raw data and feature engineering was done considering lag features, feature selection and train-test split.

Table 2. Input Attributed for Future Glucose Prediction

Input Parameter	Description
Duration of the Simulation	24 h
Number of Samples	1000
Glucose Levels	Randomly generated between 40 and 400 mg/dL
Meal Intake	Random meal amounts (g/min)
Output Parameter	Description
patient_id	Identifier for each virtual patient
timestamp	Time of each glucose measurement
blood_glucose	Blood glucose level at each measurement
insulin_dose	Amount of insulin administered
risk_index	Risk index at each time point

3.4 Model Employed

Some models of machine learning and deep learning were employed for the prediction and monitoring of diabetes.

- **Logistic regression:** It is a widely used statistical method for binary classification, particularly useful in predicting the likelihood of an outcome based on one or more predictor variables. The logistic regression model is trained using the training data with a maximum iteration of 100 to ensure sufficient convergence.

- **Random Forest Model:** It is an ensemble learning technique that utilizes multiple decision trees to enhance predictive accuracy and reduce the risk of overfitting. The Model was employed for the Prediction of Diabetes Risk as well as the Future Blood Glucose Level Prediction. For the prediction of diabetic risk, We trained a Random Forest classifier with 100 estimators to make predictions. For future glucose level prediction, a sliding window approach created sequences of past observations, which served as input features for the model.

- **Long Short-Term Memory (LSTM):** It is a specialized type of recurrent neural network (RNN) designed to process sequential and time series data. Unlike traditional RNNs, LSTMs are equipped with memory cells that can capture long-term dependencies within the data, making them particularly suitable for time-dependent tasks. This capability is crucial for predicting glucose levels, as LSTMs can effectively analyze historical glucose measurements, insulin doses, and carbohydrate intake over time.

We implemented an LSTM model to predict glucose levels based on CGM data, insulin intake, and carbohydrate (CHO) consumption. A custom PyTorch dataset class was created to facilitate the handling of sequential data. The LSTM model was defined with two layers and trained using the Adam optimizer until convergence, monitored by the mean squared error loss.

4   Results and Analysis

This section highlights the performance of the models used for predicting diabetes risk based on constant glucose measurements and forecasting future blood glucose levels using CGM data. We begin by discussing the parameters used to evaluate the models, followed by the results obtained, and conclude with a summary of the final conclusions.

4.1   Evaluation Parameters

The prediction performance of the models is measured by considering key evaluation metrics such as accuracy, precision, recall, F1-score, MAE and RMSE.

- **Validation and training accuracy** measures the percentage of correct predictions made by the model on the training and validation datasets.
- **Validation and training loss** represents the error or loss incurred by the model during training and validation; lower values indicate better performance.
- The **Receiver Operating Characteristic (ROC)** curve illustrates the true positive rate against the false positive rate at various threshold settings, while the **Area Under the Curve (AUC)** quantifies the overall performance of

the model, with values closer to 1 indicating better discrimination between classes.

- **Confusion Matrix** is a table that summarizes the performance of a classification model by showing the true positive, true negative, false positive, and false negative counts. It helps visualize the performance of the model and identify misclassification.
- **Mean Absolute Error (MAE)** measures the average absolute difference between predicted and actual values.
- **Root Mean Squared Error (RMSE)** measures square root of the average of squared differences.

4.2   Visualization

In Figs. 3 (a) and (b), AUC is 0.96 which shows excellent distinction between diabetic and non-diabetic cases.

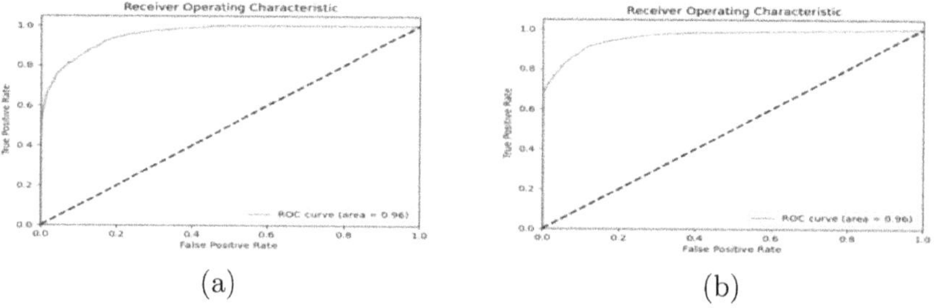

(a)   (b)

Fig. 3. ROC Curve for (a) Logistics Regression Model and (b) Random Forest Model

In Fig. 4(a), true positive cases are 9428, true negatives are 767, false positives are 424, and false negatives are 129. While in Fig. 4(b), true positive cases are 9500, true negatives are 826, false positives are 365, and false negatives are 55.

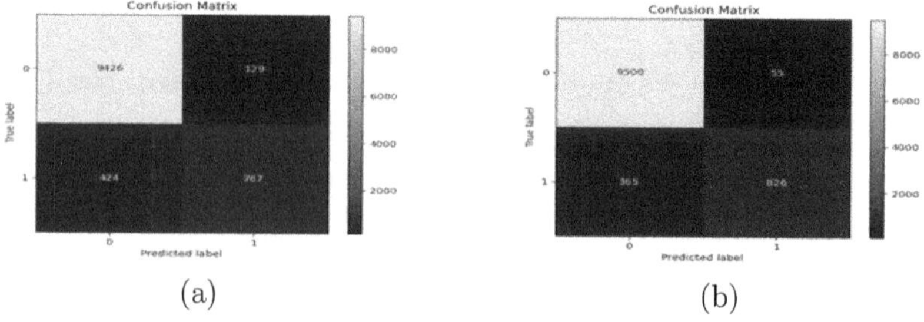

(a)   (b)

Fig. 4. Confusion Matrix for (a) Logistics Regression Model and (b) Random Forest Model

In Figs. 5 and 6, accuracy for both training and validation sets improves steadily, reaching about 96% by the end of training. While, decreasing trend for both training and validation loss, stabilizing at the end for both the models.

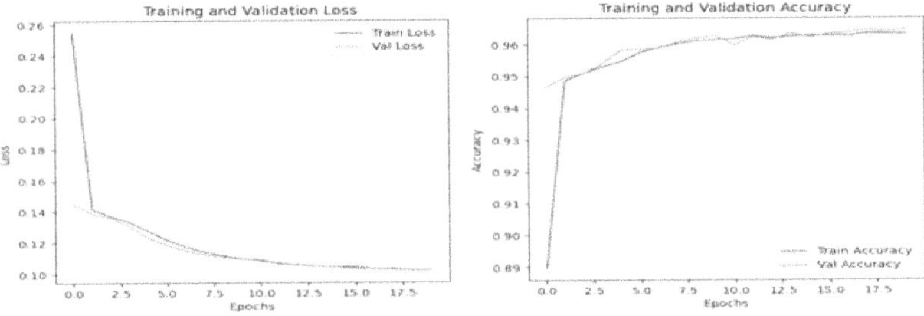

Fig. 5. Accuracy and loss curve plot for Logistics Regression model on the diabetes dataset.

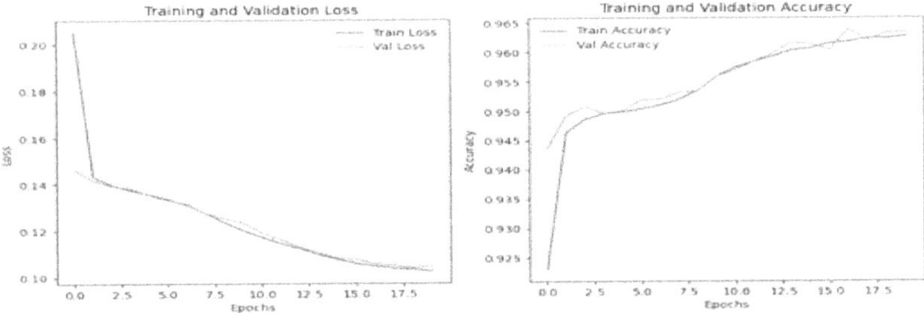

Fig. 6. Accuracy and loss curve plot for Random Forest model on the diabetes dataset.

Figs. 7(a) and (b) illustrate the Mean Absolute Error (MAE), RMSE, and accuracy results for nine patient datasets, showing the model's consistency across varying conditions. In Fig. 7(a), MAE is in the range $(11.86 - 12.63)$ which is higher in comparison to Random Forest. RMSE is in the range $(16.28 - 16.82)$ which indicates significant error. Hence, future prediction accuracy is very low in range $(59.97\% - 61.74\%)$. In Fig. 7(b), the MAE ranges in $(0.86 - 1.44)$ which indicates better close predictions. RMSE is also significantly small i.e., 3.02. There the prediction accuracy is in the range $(96.30\% - 99.49\%)$.

4.3   User Interface and Generative AI Insights

The top-performing model from all those tested is integrated into the backend to guarantee seamless functionality. A user-friendly frontend is created to facilitate easy interaction and visualization of glucose trends. Additionally, generative AI models are employed to deliver personalized insights, offering users actionable recommendations for effectively monitoring and managing their glucose levels.

4.3.1   Backend Architecture

The backend serves as the core of the system, managing predictions, user inter-actions, and secure data flow. It enables real-time inference by processing input data and delivering predictions within 1015 seconds, ensuring timely insights for users. For authentication and security, the backend uses JWT (JSON Web Token) for secure user verification, while passwords are hashed with Bcrypt, following industry standards for data protection. The system also handles CSV file uploads efficiently, validating and processing them with accuracy. Through RESTful API endpoints, the backend ensures seamless communication with the frontend, managing file uploads, predictions, and downloads. The predicted results are exported as downloadable CSV files, making them easily compatible with analytical tools like Excel for further user analysis.

4.3.2   Frontend Framework

The frontend, built with React and styled using Bootstrap, delivers a responsive, modern, and interactive user interface. Key features include a file upload func-tionality, allowing users to easily upload CGM data for analysis. A "Download Predicted CSV" button provides straightforward access to the prediction results. Validation mechanisms are in place to prevent erroneous uploads, ensuring that only valid data is processed. Additionally, dynamic visualizations, such as time-series plots, are incorporated to showcase glucose trends, providing users with clear and actionable insights. The clean, intuitive design of the interface improves usability, making the system both accessible and user-friendly. The development of the web application significantly enhances the research work by integrating advanced backend and frontend technologies. The system efficiently combines predictive analytics with an intuitive user interface, ensuring accessibility and practicality for end-users. End to end workflow is illustrated in Fig. 8.

Figure 9 showcases the prediction of diabetes risk based on user-provided inputs such as age, gender, BMI, HbA1c levels, and blood glucose levels. It visualizes the risk percentage generated by the prediction models, offering a clear, user-friendly interpretation of individual risk factors.

Figure 10 represents the average glucose level predictions derived from the 14-day CGM data. It provides a comparison of time-series glucose trends, empha-sizing the predicted glucose levels and enabling users to analyze patterns and variability. The visualization serves as a foundation for further insights, such as comparing actual glucose levels from glucometer readings with the predicted trends.

```
Person 0: MAE = 11.862015129915758, RMSE = 16.287033324423152, Accuracy = 60.16114592658908%
Person 1: MAE = 10.469909439494112, RMSE = 13.632748563065281, Accuracy = 58.93763055804237%
Person 2: MAE = 12.561935218118194, RMSE = 16.979285441925402, Accuracy = 63.14532975231274%
Person 3: MAE = 16.484691873382875, RMSE = 22.74476328682043, Accuracy = 51.44732915547598%
Person 4: MAE = 12.380289923196724, RMSE = 15.930287389708228, Accuracy = 58.01253357206804%
Person 5: MAE = 11.81183593385326, RMSE = 14.592196858042291, Accuracy = 59.65383467621606%
Person 6: MAE = 11.009888771943087, RMSE = 14.122157478636872, Accuracy = 62.2202327663384%
Person 7: MAE = 10.395626172412756, RMSE = 13.177236411915993, Accuracy = 61.65323783945092%
Person 8: MAE = 14.542776392886054, RMSE = 18.82471833836296, Accuracy = 47.00089525514772%
Person 9: MAE = 12.627949270360697, RMSE = 16.82475922284834, Accuracy = 61.74276335422262%
```

(a)

```
Person 0: MAE = 1.07, RMSE = 1.66, Accuracy = 98.54%
Person 1: MAE = 0.92, RMSE = 1.49, Accuracy = 98.87%
Person 2: MAE = 1.02, RMSE = 2.25, Accuracy = 98.90%
Person 3: MAE = 1.09, RMSE = 1.59, Accuracy = 98.54%
Person 4: MAE = 0.98, RMSE = 1.42, Accuracy = 99.05%
Person 5: MAE = 1.44, RMSE = 3.02, Accuracy = 96.30%
Person 6: MAE = 1.05, RMSE = 2.15, Accuracy = 98.39%
Person 7: MAE = 0.86, RMSE = 1.21, Accuracy = 99.46%
Person 8: MAE = 1.11, RMSE = 1.86, Accuracy = 98.27%
Person 9: MAE = 1.02, RMSE = 1.40, Accuracy = 99.49%
```

(b)

Fig. 7. MAE, RMSE, Accuracy Results for 9 Patients for (a) LSTM Model and (b) Random Forest Model

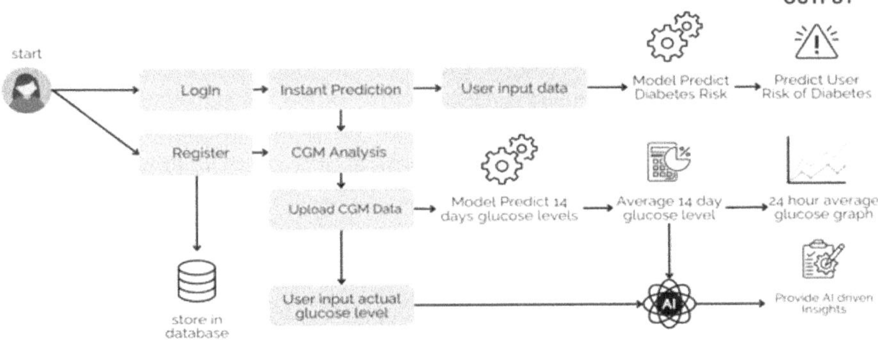

Fig. 8. End-to-End Workflow of the Diabetes Management Tool

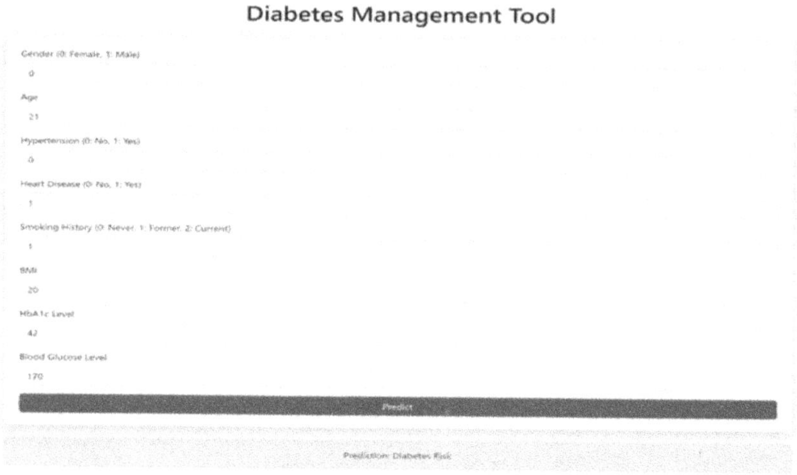

Fig. 9. Predicted Diabetes Risk based on user input

4.4 Integrating Generative AI

Generative AI has emerged as a transformative technology in healthcare, enabling personalized insights and data-driven recommendations. Here, OpenAI's GPT-3.5 Turbo model is integrated to enhance the self-management module by providing actionable insights based on glucose level data analysis. This combination of predictive analytics and generative intelligence bridges the gap between complex data and actionable user outcomes, setting a new standard in healthcare innovation. Figure 11 showcases the personalized recommendations provided by the generative AI model. Based on the analysis of deviations between predicted and actual glucose levels, the AI offers tailored insights such as lifestyle adjustments, dietary changes, and timing for physical activity to improve glucose control. These recommendations are presented to users in an actionable and easy-to-understand format.

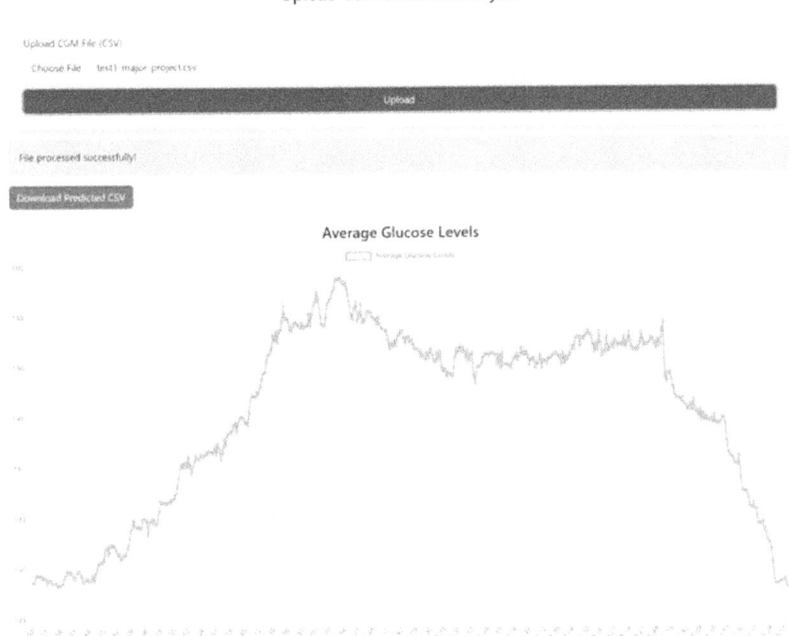

Fig. 10. Average Glucose Level Predicted based on 14 days Predicted Glucose Levels

The AI model interprets differences between predicted glucose levels, derived as an average of 14 days of CGM data, and actual readings from glucometers. Deviations are categorized into controlled, moderately controlled, and uncontrolled based on predefined thresholds, providing users with a clear understanding of their glucose control status. GPT identifies recurring trends, such as specific times when glucose spikes or persistent discrepancies occur. These patterns may highlight underlying lifestyle factors or physiological changes affecting glucose levels. Tailored suggestions are generated based on the identified patterns, including dietary changes, adjustments in physical activity, or modifications to medication schedules. The recommendations are designed to be practical, actionable, and user-friendly, fostering proactive diabetes management.

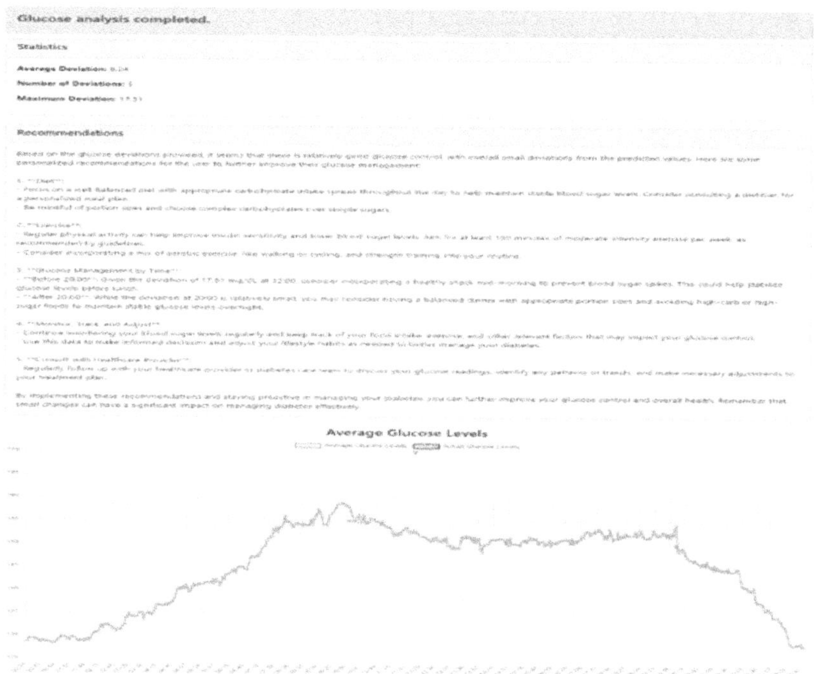

Fig. 11. Personalized Recommendations and Insights using Generative AI

5   Conclusion and Future Scope

This work offers transformative benefits for diabetes management, making it more accessible, cost-effective, and user-centric. Traditionally, Continuous Glucose Monitoring devices cost over 5000 INR for just 14 days of monitoring, with no continued feedback or guidance. This work significantly reduces costs by 65% by limiting user expenses to only CGM device purchases while providing continuous monitoring and tailored recommendations, extending the utility of CGM data. By integrating machine learning and generative AI models, the platform provides affordable and effective diabetes management solutions to a broader population, reducing reliance on expensive, proprietary tools. The approach enables users to monitor glucose levels beyond CGM device durations through predictive modeling, offering continuous support and actionable recommendations tailored to individual needs. With real-time analysis of glucose patterns and personalized insights, users can make informed decisions to better control glucose levels, reducing the risk of complications and improving overall health outcomes.

The current work can be utilized for enhancement in the impact and usability. Some additional data sources such including physical activity levels, sleep patterns, and dietary intake, can be integrated to provide more comprehensive and holistic glucose management insights. A real time monitoring system can be developed to send instant alerts to users during critical glucose deviations, enhancing timely intervention. Long-term data analysis can be provided by enabling longitudinal tracking of glucose trends to predict future health risks, such as potential diabetes complications, and recommend preventive measures. More advanced generative AI models can be used for even deeper and more nuanced insights, including psychological support for better adherence to recommendations.

References

1. Mujumdar, A., Vaidehi, V.: Diabetes prediction using machine learning algorithms. Procedia Comput. Sci. **165**, 292–299 (2019)
2. Butt, U.M., Letchmunan, S., Ali, M., Hassan, F.H., Baqir, A., Sherazi, H.H.R.: Machine learning based diabetes classification and prediction for healthcare applications. J. Healthcare Eng. **2021**(1), 9930985 (2021)
3. Khanam, J.J., Foo, S.Y.: A comparison of machine learning algorithms for diabetes prediction. ICT Express **7**(4), 432–439 (2021)
4. Ahmed, U., et al.: Prediction of diabetes empowered with fused machine learning. IEEE Access **10**, 8529–8538 (2022)
5. Shahriare Satu, M., Atik, S.T., Moni, M.A.: A novel hybrid machine learning model to predict diabetes mellitus. In: Uddin, M.S., Bansal, J.C. (eds.) Proceedings of International Joint Conference on Computational Intelligence. AIS, pp. 453–465. Springer, Singapore (2020). https://doi.org/10.1007/978-981-15-3607-6_36
6. Dey, A.K.: Chatgpt in diabetes care: an overview of the evolution and potential of generative artificial intelligence model like chatgpt in augmenting clinical and patient outcomes in the management of diabetes. Int. J. Diab. Technol. **2**(2), 66–72 (2023)
7. Zheng, Y., Wu, Y., Feng, B., Wang, L., Kang, K., Zhao, A.: Enhancing diabetes self-management and education: a critical analysis of chatgpt's role. Ann. Biomed. Eng. **52**(4), 741–744 (2024)
8. Mashatian, S., et al.: Building trustworthy generative artificial intelligence for diabetes care and limb preservation: a medical knowledge extraction case. J. Diab. Sci. Technol. 19322968241253568 (2024)
9. Jadon, A., Kumar, S.: Leveraging generative ai models for synthetic data generation in healthcare: Balancing research and privacy. In: 2023 International Conference on Smart Applications, Communications and Networking (SmartNets), pp. 1–4. IEEE (2023)
10. Pappada, S.M., Cameron, B.D., Rosman, P.M., Bourey, R.E., Papadimos, T.J., Olorunto, W., Borst, M.J.: Neural network-based real-time prediction of glucose in patients with insulin-dependent diabetes. Diab. Technol. Therapeut. **13**(2), 135–141 (2011)
11. Zhu, T., Wang, W., Yu, M.: A novel blood glucose time series prediction framework based on a novel signal decomposition method. Chaos Solitons Fract. **164**, 112673 (2022)
12. Kavakiotis, I., Tsave, O., Salifoglou, A., Maglaveras, N., Vlahavas, I., Chouvarda, I.: Machine learning and data mining methods in diabetes research. Comput. Struct. Biotechnol. J. **15**, 104–116 (2017)
13. Zhu, T., Li, K., Herrero, P., Georgiou, P.: Deep learning for diabetes: a systematic review. IEEE J. Biomed. Health Inf. **25**(7), 2744–2757 (2020)

14. Diabetes prediction dataset—kaggle.com. https://www.kaggle.com/datasets/iammustafatz/diabetes-prediction-dataset. Accessed 20 Dec 2024
15. GitHub - jxx123/simglucose: A Type-1 Diabetes simulator implemented in Python for Reinforcement Learning purpose—github.com. https://github.com/jxx123/simglucose. Accessed 20 Dec 2024
16. Katbin—katb.in. https://katb.in/viverutoxob. Accessed 20 Dec 2024

Learning Paradigms and Optimisations

Enhancing Credit Risk Prediction Using Stacked Ensemble and Quantum Approaches on PySpark

Robin Chauhan, Aditya Bhardwaj[✉], and Ajay Kumar

School of Computer Science Engineering and Technology (SCSET), Bennett University, Greater Noida, India
`{e23soep0041,aditya.bhardwaj,ajay.kumar1}@bennett.edu.in`

Abstract. Objective: Accurate prediction of credit risk is essential for financial organizations to handle defaults efficiently. As data complexity increases, the integration of traditional and novel AI models becomes imperative. This study improves the prediction of credit defaults with a hybrid quantum-classical stacking methodology. **Methods**: This research utilizes an interdisciplinary approach that integrates quantum machine learning with conventional ensemble methodologies with the Pyspark framework. The classical layer consists of Light Gradient Boosting Machine (LGBM), Extreme Gradient Boosting (XGBoost), Categorical Boosting (CatBoost), and HistGradientBoosting (HGB) classifiers optimized using Optuna. An ensemble is augmented with a variational quantum classifier developed using PennyLane. Stacking is executed via LGBM as a meta-learner. Principal Component Analysis (PCA) is used for dimensionality reduction, while SMOTETomek tackles class imbalance. **Results**: In the credit data set, the mixed-stacked model attained an accuracy 94.47%, with a precision 93.13%, a recall 97.60%, F1-score 95.31% and an AUC 98.33%. The incorporation of quantum characteristics improved the variety and resilience of the model, resulting in more refined decision limits. **Conclusion**: The findings indicate that including quantum classifiers in standard ensemble frameworks can significantly enhance the effectiveness of credit risk modelling, particularly in unbalanced and high-dimensional financial datasets. **Significance**: This adapted methodology presents an innovative direction in credit scoring by integrating classical elucidative models with quantum AI. Improve predictive accuracy and model interpretability and facilitate the development of advanced financial risk assessment tools.

Keywords: Credit risk prediction · Pyspark · Quantuam AI · Ensemble · Credit scoring

1 Introduction

In urban financial ecosystems, there has been notable growth in individual credit demand, influenced by a post-pandemic increase in consumption, housing investments, and the accessibility of digital lending. As of September, 2024, the total

R. Gupta et al. (Eds.): BDA 2025, LNCS 16041, pp. 67–83, 2026.
https://doi.org/10.1007/978-3-032-15134-6_5

outstanding individual housing loans in Delhi amounted to Rs. 33.53 lakh crore, reflecting a 14% increase year-on-year, with Rs. 9.07 lakh crore disbursed in FY 2023–24 [17]. In Q3 FY24, Delhi NCR was the leading region for loan origination in India, disbursing Rs. 61,482 crore across various categories, including Rs. 18,215 crore in personal loans alone [18].

The expansion of unsecured lending has revealed significant vulnerabilities. Major banks have reported significant growth in unsecured loan portfolios: According to the Reserve Bank of India (RBI), several banks experienced a 10% increase in unsecured credit, whereas some banks noted a 17% rise in personal loans and a 28% increase in credit card lending [19]. By late 2023, personal loans constituted 34% of total outstanding bank credit, exceeding Rs. 50 lakh crore [20]. The increasing volume of credit has coincided with a rise in defaults. In the first half of the fiscal year 2025 (FY25), long-term non-payment rates, which are more than 90 days overdue, are increased across borrower segments in Delhi, rising from 2.5% to 3% [21]. This trend indicates increased risk levels for retail portfolios, particularly within unsecured and subprime segments.

The urgency for precise and scalable credit risk prediction models has intensified in light of these developments. The complexity of contemporary credit behaviour, driven by digital lending, BNPL schemes, and microcredit, renders conventional risk assessment methods inadequate. In Delhi, the intersection of rapid urbanisation and diverse borrower profiles required the use of advanced machine learning techniques, including ensemble models, explainable boosting machines, and quantum-classical hybrid models, to enhance the precision of default predictions [27].

In the current credit environment, an individual machine learning model frequently struggles to precisely assess credit risk due to the growing complexity and variability in borrower profiles. Urban cities such as Delhi demonstrate diverse financial habits shaped by freelance employment, informal earnings, and digital financing, which models like logistic regression or decision trees may inadequately represent. Conventional models like Support Vector Machines (SVMS) and k-Nearest Neighbours (k-NN) have difficulties in accommodating high-dimensional, imbalanced data characteristic of credit failures. Furthermore, these models frequently lack the capability to analyse dynamic behavioural data from sources such as BNPL schemes or fintech sites. Consequently, they often exhibit suboptimal performance in practical situations, resulting in inadequate generalisation and elevated misclassification rates. To alleviate these restrictions, ensemble techniques like XGBoost and LGBM and interpretable models like Explainable Boosting Machines (EBMS) are gaining preference alongside novel quantum-classical hybrid methodologies.

1.1 Motivation for the Study

This research establishes a hybrid quantum-classical framework for credit risk prediction, designed explicitly for default categorisation, by combining quantum computing methodologies with conventional machine learning models. The classical component comprises optimised ensemble models like XGBoost, LGBM, and CatBoost, each fine-tuned with Optuna for hyperparameter optimisation. These models use a quantum classifier developed with PennyLane, utilising variational quantum circuits to elucidate intricate non-linear correlations in borrower data. The ensemble outputs are integrated using a meta-learner, a LGBM model, to improve prediction efficacy. The pipeline incorporates Synthetic Minority Over-sampling Technique (SMOTE) to rectify class imbalance and utilizes SHAP values for explainability, hence ensuring transparency in predictions. This methodology enhances credit risk forecasting by including both classical and quantum characteristics, minimizing false positives, and facilitating improved credit distribution to high-risk clients. The system is engineered for scalability and interpretability, catering to financial institutions' dynamic requirements in varied and high-volume loan markets.

1.2 Key Contributions

The main contributions of this study are as follows:

- We proposed quantum-classical hybrid pipeline for credit risk prediction that integrates a variational quantum classifier with classical ensemble models such as XGBoost, LGBM, and CatBoost. These classical models are fine-tuned using Optuna for hyperparameter optimization, enabling enhanced predictive accuracy and automation in credit scoring tasks.
- The hybrid stacking framework combines quantum and classical predictions using a LGBM meta-learner, effectively capturing complex patterns in borrower behaviour. The pipeline incorporates SMOTETomek to handle class imbalances and SHAP for interpretability. This allows the model to address the limitations of traditional classifiers by taking advantage of the ability of quantum computing to explore high-dimensional feature spaces, making it suitable for high-risk and non-traditional credit segments.
- We validated the proposed model against individual classical models and demonstrate that it achieves superior performance in accuracy, recall, and F1 score while maintaining computational feasibility. Integrating explainable AI and quantum-enhanced learning significantly improves default detection, reduces false positives, and facilitates more informed and equitable credit decision-making in rapidly evolving financial environments.

Article Structure. The subsequent sections of the paper are structured in the following manner. Section 2 examines the related work, while Sect. 3 outlines the proposed system model. Section 4 provides an evaluation of the performance, Sect. 5 represents the conclusion of the paper, and finally, Sect. 6 discusses future work.

2 Related Work

The domain of credit risk prediction has seen substantial advancements due to the application of Machine Learning (ML) and Deep Learning (DL) models.

Various research methodologies are examined to enhance predictive accuracy, interpretability, and robustness. This section analyzes current research that has progressed the development of credit risk prediction models. Blended classifiers, frequently employing ensemble methodologies, have been extensively utilized in credit risk assessment, delivering enhanced performance as compared to individual models. A machine learning-based system proposed by [5] utilizes a combined approach that incorporates Random Forest (RF), Gradient Boosting (GB), and Extreme Gradient Boosting (XGBoost). The proposed ademonstrated significant performance improvements, achieving Area Under Curve (AUC) scores of 93.4%, 94.4%, and 87.0% on the finance datasets, respectively. Similarly, a hybrid credit risk evaluation model proposed by [3] which integrated Three-Way decisions with a stacking ensemble learning approach, utilizing LGBM for assessing default probabilities. The model achieved a balanced accuracy of 75.30%, AUC of 88.44%, F-measure of 56.25%, and G-mean of 74.32%, while minimizing decision costs to 5.08 million. In another, the approach proposed by [7] in which the stacking ensemble approach focused on the ensemble classifiers with model explanation, combining the SMOTE-ENN resampling technique to address class imbalance. In this study, XGBoost attained a recall of 93.0% and specificity of 84.6% on the German dataset. For the Australian dataset, Random Forest achieved a recall of 90.7% and specificity of 92.2%, reflecting improved robustness and balance. Deep learning has been effectively combined with ensemble learning in two-stage hybrid models. One such approach proposed by [6] employed XGBoost to generate high-dimensional sparse feature matrices, followed by a graph-based deep neural network (forgeNet). This hybrid model achieved 87.52% accu- racy, 93.13% F1-score, and 85.59% G-mean on Lending Club data spanning 2007–2016, outperforming other benchmark methods. In another study [8] the authors proposed a hybrid deep learning model integrated sand cat swarm optimization (SC-SOFS) for feature selection and a Deep LSTM Supervised Autoencoder Neural Network (DLSTM-SANN) for classification. Leveraging the political optimizer for hyperparameter tuning, their model achieved accuracies of 96.49% on the German dataset and 96.12% on the Australian dataset. Advanced feature selection and interaction modeling have further refined credit risk prediction. The Adaptive Feature Cross-Compression (AFCC) method, which incorporates SENET gating mechanisms and cross compression units, improved feature representation and interaction proposed by [4]. Evaluated on datasets from Tianchi and Lending Club, AFCC increased AUC by 1–2%, KS by 0.5–4.5%, and G-mean by 1–1.5% over 11 competing models. Additionally, [9] proposed a single default discrimination model combined chi-square and RFECV for feature selection and Local Outlier Factor (LOF) for instance selection, deploying six classifiers (LR, KNN, NB, DT, LDA, MLP). On the Chinese listed companies dataset, the KNN classifier achieved an accuracy of 97.4%, a precision of 96.5%, recall of 98.4%, F-measure of 97.5%, and G-mean of 97.4%. On the German dataset, the MLP classifier attained 85.8% accuracy, 91.3% AUC, 81.2% F-measure, 88.9% precision, and recall of 84.8% and Gmean (GM) of 86.0%.

Table 1. Comparative analysis of existing credit risk prediction-related work and our contributions.

Authors	Year	Problem/Solution	Advantages	Disadvantages
[3]	2022	Credit risk prediction by proposing a two-stage hybrid model using XGBoost for feature transformation and a graph-based deep neural network to handle complex credit data	The model improves prediction accuracy and handles complex feature relationships effectively	High computational overhead for large-scale deployment, limited focus on scalability in dynamic networks
[4]	2023	Credit default prediction models in capturing feature interactions and handling information interference lead to inaccurate predictions	Effectively capture feature interactions and reduce information interference, resulting in better AUC, KS, and G-mean values than traditional models	Facing challenge to accurately predicting credit default risk, which is critical for minimizing economic losses and ensuring the stability of financial institutions
[1]	2024	Mitigates class imbalance and high-dimensional data complexity by implementing a stacked classifier and filter-based feature selection	The proposed model enhances the accuracy of credit risk prediction and addresses class imbalance through the implementation of a stacked classifier combined with effective filter-based feature selection. It provides enhanced performance and computational efficiency across various datasets, ensuring reliable and scalable implementation for financial institutions	The proposed model requires significant computational resources and adds complexity, making interpretation challenging. Its generalization, scalability, and reliance on optimal feature selection need further validation, especially for highly imbalanced or large-scale, real-time datasets
[11]	2024	High risk of misjudgment in traditional two-way credit risk evaluation models	The hybrid model improves classification accuracy and reduces decision errors by using three-way decisions and stacking ensemble learning. It effectively handles high decision cost samples, minimizing financial risks	The proposed model increases misclassification of non-default samples as default, leading to potential overestimation of risk
Proposed Work	2025	Addresses the problem of credit risk prediction by accurately identifying potential loan defaulters in imbalanced and high-dimensional financial datasets	It utilizes a hybrid quantum-classical approach by combining classical models like XGBoost, LGBM, CatBoost, and HistGradientBoosting with a quantum classifier to enhance prediction accuracy and feature representation	Requires significant computational resources due to its complex ensemble structure and hyperparameter tuning

2.1 Summary of Gaps in Literature

Overall, recent developments in credit risk prediction have been driven by ensemble learning, explainable AI, and hybrid methodologies that aim to improve model interpretability and predictive accuracy. As demonstrated in Table 1, these methods have significantly contributed to the evolution of scalable and intelligent credit scoring systems. However, despite the growing interest in quantum machine learning, there remains limited research on integrating quantum classifiers with classical ensemble models for financial applications. Furthermore, the application of hybrid quantum-classical pipelines especially those incorporating explainability techniques like SHAP and scalable tuning frameworks like Optuna remains underexplored in the credit risk domain. This gap highlights a promising direction for future work, especially improving the precision, interpretability, and robustness of credit risk models in complex, data-intensive financial environments.

3 Proposed System Framework

This section defines the proposed framework, which articulates a quantum-classical credit risk categorization pipeline. The procedure encompasses data preparation, which comprises addressing class imbalance with SMOTETomek and creating interpretable features utilizing SHAP. Several classical base models XGBoost, LGBM, and CatBoost are refined using Optuna and trained in conjunction with a variational quantum classifier developed with PennyLane. Their forecasts are consolidated into a layered feature set, subsequently input into a LGBM meta-model to enhance final predictions. This hybrid stacking method improves model precision and resilience by utilizing both classical and quantum learning abilities. The model's performance is thoroughly assessed using measures like accuracy, precision, recall, F1-score, AUC, and a confusion matrix, providing a detailed evaluation of financial risk in intricate borrower profiles.

3.1 System Model

The proposed quantum-classical architecture, depicted in Algorithm 1, offers a systematic pipeline that combines advanced preprocessing methods, ensemble learning, and quantum machine learning to enhance credit risk prediction. The pipeline begins with data cleaning and preprocessing. Missing numerical values are imputed using the mean, while missing categorical features are imputed using the mode. Class imbalance is addressed using SMOTETomek, which applies oversampling to the minority class and cleans the dataset by removing noise, thus augmenting the model's ability to generalize across imbalanced distributions.

After preprocessing, SHAP values (Shapley Additive Explanations) are computed to evaluate feature significance, thereby improving the model's interpretability and explainability. Additionally, feature normalisation techniques are employed to standardise the features, thereby ensuring consistency among all input variables. The dataset is subsequently partitioned into training and testing subsets to guarantee an unbiased evaluation of the model.

The model utilises several base learners that illustrate feature interactions and encapsulate the complexity of the dataset. The base models consist of XGBoost, LGBM, CatBoost, and a QuantumClassifier, which is configured using the PennyLane framework. Each model undergoes optimisation through the application of Optuna's Bayesian optimisation method, aimed at enhancing the

---

Algorithm 1. Quantum-classical hybrid credit risk prediction model

---

Input: $D = \{x, y\}$, where x denotes the features and y denotes the labels.

Output: M, the trained hybrid ensemble model, and E, the evaluation metrics.

1: Preprocess D using SMOTETomek to balance classes.

2: Apply SHAP for explainability and scale features.

3: Split dataset D into training and test sets:

$$(x_{\text{train}}, x_{\text{test}}, y_{\text{train}}, y_{\text{test}}) \leftarrow \text{train_test_split}(x, y, \text{size} = 0.3)$$

4: **for** each base model $M_i \in \{\text{XGBoost, LGBM, CatBoost, QuantumClassifier}\}$ **do**

5:     Train model M_i on $(x_{\text{train}}, y_{\text{train}})$ using Optuna

6:     $\hat{y}_{\text{train}}^{(i)} \leftarrow M_i.\text{predict_proba}(x_{\text{train}})$

7:     $\hat{y}_{\text{test}}^{(i)} \leftarrow M_i.\text{predict_proba}(x_{\text{test}})$

8: **end for**

9: Combine predicted probabilities to create stacked training and testing sets:

$$\hat{y}_{\text{stacked}}^{\text{train}} \leftarrow \left[\hat{y}_{\text{train}}^{(1)}, \hat{y}_{\text{train}}^{(2)}, \hat{y}_{\text{train}}^{(3)}, \hat{y}_{\text{train}}^{(4)}\right]$$

$$\hat{y}_{\text{stacked}}^{\text{test}} \leftarrow \left[\hat{y}_{\text{test}}^{(1)}, \hat{y}_{\text{test}}^{(2)}, \hat{y}_{\text{test}}^{(3)}, \hat{y}_{\text{test}}^{(4)}\right]$$

10: Train the final estimator M_{final} on $\hat{y}_{\text{stacked}}^{\text{train}}$

11: $\hat{y}_{\text{pred}} \leftarrow M_{\text{final}}.\text{predict}(\hat{y}_{\text{stacked}}^{\text{test}})$

12: Calculate evaluation metrics

$$E = \{\text{Accuracy, Precision, Recall, F1-score, AUC, Confusion Matrix}\}$$

13: **return** M: Trained hybrid stacked ensemble model

14: **return** E: Evaluation metrics

15: **End**

AUC-ROC (Area Under Curve – Receiver Operating Characteristics) on the training dataset.

Upon completion of the base model training, the resulting output probabilities are aggregated to form new feature representations, subsequently utilised for layered generalisation. A final meta-learner, LGBM, is employed to integrate the predictions generated by the base models. This ensemble method contributes to the reduction of model bias, minimises volatility, and improves the overall reliability of predictions.

Threshold optimization is conducted subsequent to stacking through the application of scipy otimize and minmize function, enhancing the decision boundaries' precision. This establishes a fair equilibrium among accuracy, recall and F1 score, enhancing the performance of the model on these essential metrics. The final model undergoes evaluation through various performance metrics, which encompass accuracy, precision, recall, F1-score, AUC-ROC, and a confusion matrix. This assessment facilitates the identification of false positives and false negatives, thereby offering detailed insights into the model's performance.

The architecture of this quantum-classical system leads to the development of a credit risk prediction model that is interpretable, adaptable, and highly precise. This model is designed for practical applications in financial decision-making, providing reliable predictions for high-risk borrowers.

3.2   Proposed Scheme

The proposed credit risk prediction framework, as in Fig. 1, integrates distributed data processing, dimensionality reduction, quantum-classical modelling, and ensemble learning to provide a resilient and scalable solution for practical financial datasets.

The framework begins with an imbalanced dataset, which is divided into training and testing subsets. This division guarantees proper validation and evaluation of the model. The data is subsequently transmitted to the Data Processing Phase, conducted within a distributed computing framework utilising Apache Spark. A central driver program and Spark context manage many worker nodes via a cluster manager, facilitating parallel processing, caching, and task execution throughout the dataset.

Upon incorporation of the raw data, preprocessing procedures are executed to sanitise and organise the data. The procedures encompass managing absent values, implementing SMOTETomek to rectify class imbalance, and utilising SHAP for interpretability.

Role of Quantum classifier : The PennyLane-powered Variational Quantum Classifier (VQC) in our proposed hybrid ensemble framework captures complex nonlinear patterns in borrower data using a four-qubit quantum circuit with parameterized Ry rotations and entangling CNOT gates. The model is optimized using the Adam optimizer and executed on a qubit-based backend. Its probabilistic outputs are integrated with predictions from classical base learners and subsequently aggregated by a LGBM meta-learner. Incorporating the quantum component improved the model's robustness on imbalanced datasets, yielding a 4.5% increase in AUC and a 3% improvement in F1-score on the German credit dataset.

In our proposed framework, Principal Component Analysis (PCA) was employed for dimensionality reduction due to its strong interpretability, computational efficiency, and compatibility with Spark-based distributed computing. PCA transforms the original features into a lower-dimensional orthogonal space, helping to reduce noise and improve generalization while retaining most of the variance. Although we evaluated alternative methods such as Least Absolute Shrinkage and Selection Operator (LASSO) and Autoencoders, they were excluded from this version due to challenges in adaptability and maintaining interpretability—an essential requirement for credit risk modeling. Truncated SVD and PolynomialFeatures were also explored as part of feature engineering, but their impact on model performance was found to be less effective than PCA. A quantum classifier is subsequently trained on the PCA-transformed subset, using quantum-enhanced learning capabilities for intricate, non-linear feature patterns.

During the modeling phase, the predictions from the quantum classifier are integrated with outputfrom several traditional gradient boosting models, such as XGBoost, LGBM (LGBM), CatBoost, and HistGradientBoost. Each model is optimised with Optuna to enhance AUC-ROC performance. Their outputs create probability vectors, which are further integrated into a stacking ensemble architecture.

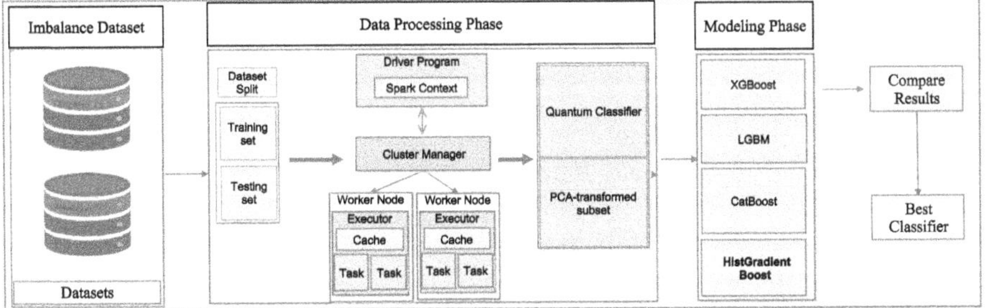

Fig. 1. System architecture of proposed credit risk prediction framework.

A meta-classifier, usually LGBM, is trained on these aggregated outputs to generate final predictions. The ensemble method guarantees enhanced generalisation and less variance. All models are evaluated, and the optimal classifier is chosen based on a wide range of assessment criteria, including precision, recall, F1 score, and AUC-ROC.

This modular and scalable system facilitates effective credit risk prediction on extensive, high-dimensional financial datasets, utilising both classical and quantum paradigms for improved accuracy and interpretability.

4   Performance Evaluation

This section assesses the performance of the proposed stacking ensemble model using key performance measures to determine its usefulness in credit risk prediction. A confusion matrix is utilised to illustrate the model's capacity to distinguish between high-risk and low-risk people in relation to previous methodologies.

4.1   Experimental Setup

The experimental configuration includes a high-performance system designed for machine learning and deep learning applications. The system has eight NVIDIA Tesla V100 GPUS, each with 16 GB, interconnected using NVLink Hybrid Cube Mesh to facilitate high-bandwidth, low-latency multi-GPU computing. Rapid data transmission is facilitated by two 10 GbE and four InfiniBand 100 Gbps

EDR connections. The system employs Intel Xeon E5-2698 v4 processors, 512 GB DDR4 LRDIMM memory, and four 1.92 TB SSDs to ensure rapid data access and storage dependability. The redundant 1600w power supply provides continuous high-performance operation, rendering the configuration optimal for extensive AI research and deep learning endeavours.

4.2 Results and Discussion

This section establishes the performance of the proposed hybrid model is evaluated using three benchmark credit risk datasets: Australian, German, and Taiwan. These datasets offer diverse characteristics and class distributions, making them ideal for assessing model robustness. Evaluation metrics include accuracy, precision, recall, F1-score, and AUC to provide broad understanding of predictive performance. The model is implemented in Python and employs SMOTE-Tomek for handling class imbalance, StandardScaler for feature scaling, and PolynomialFeatures with PCA for feature engineering. Hyperparameter tuning is conducted using Optuna, and models are validated via stratified K-fold cross-validation. StackingClassifier is used as the ensemble method, with threshold optimization applied to balance precision, recall, and F1-score. The complete configuration of simulation tools and techniques is summarized in Table 2.

Table 2. Simulation parameters.

Parameter	Description
Language	Python
SMOTE Method	SMOTETomek (oversampling and noise cleaning)
Scaling Method	StandardScaler (applied to numerical features)
Feature Engineering	PolynomialFeatures, PCA, Transformer
Ensemble Method	StackingClassifier
Model Hyperparameters	max_depth, learning_rate, subsample, colsample_bytree
Performance Metrics	Accuracy, Precision, Recall, F1-Score, AUC
Optimization Method	Optuna used for hyperparameter tuning
Cross-Validation	Stratified K-Fold Cross-Validation
Threshold Optimization	Loss function balancing precision, recall, and F1-Score

4.3 Analysis of Model Performance on the Australian Credit Risk Dataset

Taking into consideration the given dataset, proposed model outperforms others with a accuarcy of 90%, precision of 87.72%, recall of 91.74% as shown in Fig. 2a, and an F1-score of 89.69%. The model achieved an AUC of 96.58% as in Fig. 2b and a log loss of 0.29.

These results suggest that proposed model effectively distinguishes between high-risk and low-risk individuals while maintaining strong balance between false positives and false negatives, making it a reliable tool for assessment of risk.

In comparison, the performance of other models such as LGBM, XGBoost, Logistic Regression, SVM, and AdaBoost show varying degrees of success in classifying the dataset. LGBM, with an accuracy of 87.92%, and XGBoost with a slightly lower recall, still demonstrated competitive results. Despite showing promising performance, Logistic Regression had a higher AUC 92.11% but lower recall compared to the proposed model, while SVM effectively detected defaulters but misclassified a higher number of non-defaulter individuals. AdaBoost, although competitive, struggled with poorer accuracy 83.57% and a lower recall of 74.07%, highlighting its limitations in capturing risky cases.

The confusion matrix Fig. 2c further illustrates the model's ability to identify 100 true positives and only 9 false negatives, while having 14 false positives. This demonstrates the model's strength in reducing false negatives, crucial for identifying true risk cases. The proposed model outperforms others by minimizing false negatives while controlling false positives.

4.4 Analysis of Model Performance on the German Credit Risk Dataset

Credit risk classification is challenging, especially with the imbalanced dataset like german dataset. The proposed model surpasses others with a remarkable performance as showin in Fig. 3a , achieving 94.47% accuracy, 93.13% precision, and 97.60% recall. As shown in the Fig. 3b , the model maintains an F1-score of 95.31% and an AUC of 98.33%, efficiently balancing precision and recall. It outperforms models like LGBM, which showed 75.33% accuracy, 79.74% precision, and 86.60% recall, and XGBoost, which exhibited similar performance with 74.67% accuracy and 79.04% precision.

While Logistic Regression demonstrated good recall of 90.43%, it had a high false positive rate and 72.33% accuracy, misclassifying non-risk individuals as high-risk. Support Vector Machine (SVM) achieved 75% accuracy and 94.26% recall, but its precision 75.77% was lower, leading to more misclassified non-risk

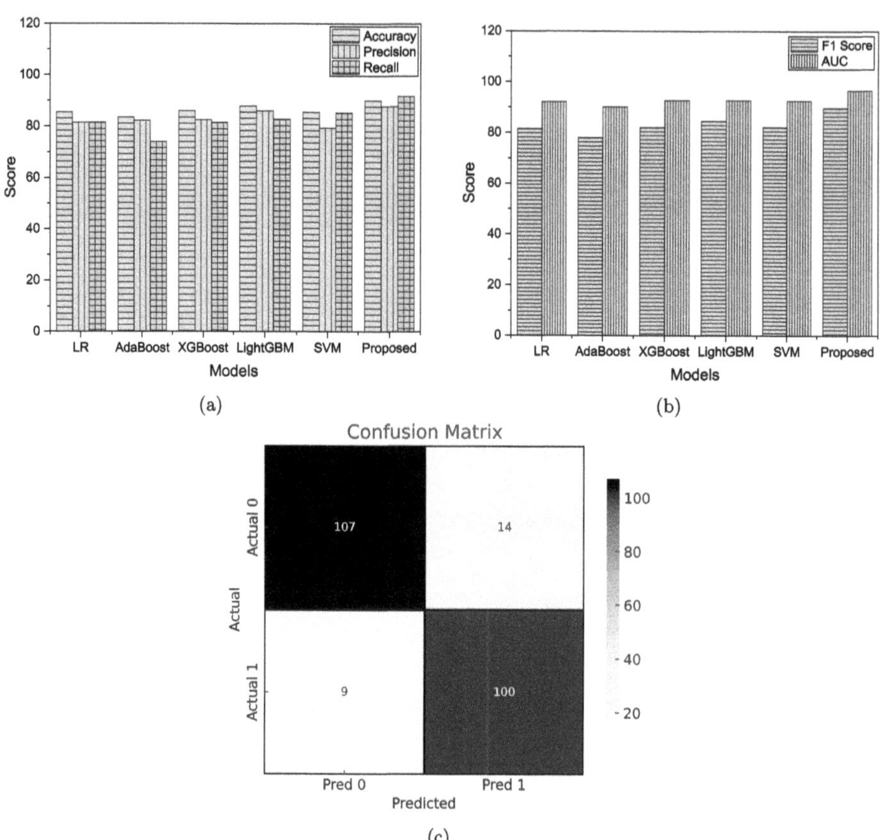

Fig. 2. Performance evaluation on the Australian dataset: a) Comparative analysis of accuracy, precision, and recall; b) F1-Score vs. Area Under the Curve (AUC) analysis; c) Confusion matrix representation.

individuals. AdaBoost also performed decently with 72.67% accuracy and 89% recall but did not match the performance of ensemble models. The confusion matrix Fig. 3c highlights that the proposed model correctly identified 122 true positives and only 3 false negatives, with 9 false positives and 83 true negatives. This shows the model's strong capability to detect high-risk individuals while minimizing false positives. The proposed model, with its impressive AUC of 98.33%, effectively differentiates between high-risk and low-risk borrowers, making it a highly reliable option for credit risk evaluation in financial institutions.

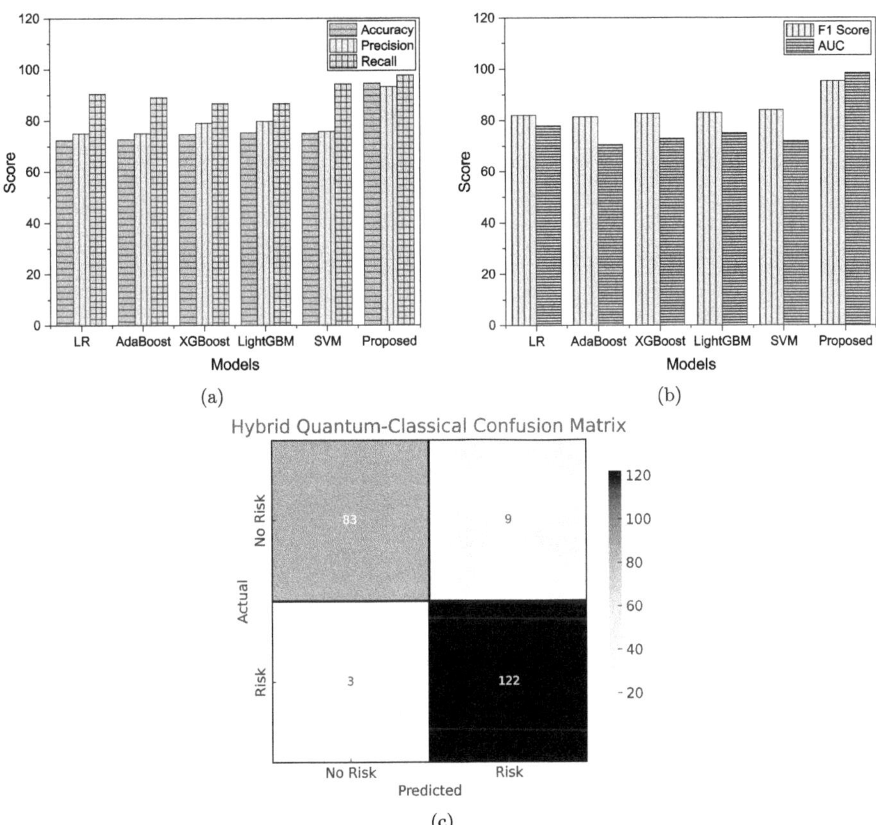

Fig. 3. Performance evaluation on the German dataset: a) Comparative analysis of accuracy, precision, and recall; b) F1-Score vs. Area Under the Curve (AUC) analysis; c) Confusion matrix representation.

4.5   Analysis of Model Performance on the Taiwan Credit Risk Dataset

The Taiwan dataset's diverse credit risk data posed a significant challenge for prediction algorithms. As shown in the Fig. 4a the proposed model outperformed all others with an accuracy of 86.67%, precision of 84.44%, recall of 48.68%, an F1-score of 61.76%, and AUC of 93.92%, as shown in the Fig. 4b. This model excels in its ability to detect defaulters while maintaining lower rate of false positives, making it highly suitable for financial applications in Taiwan.

In comparison, LGBM showed an accuracy of 82.12%, precision of 66.51%, and recall of 36.07%, while XGBoost had an accuracy of 81.27%, precision of 61.61%, and recall of 37.09%. Both models exhibited good accuracy but struggled with lower recall, indicating a reduced ability to identify defaulters. The pro-

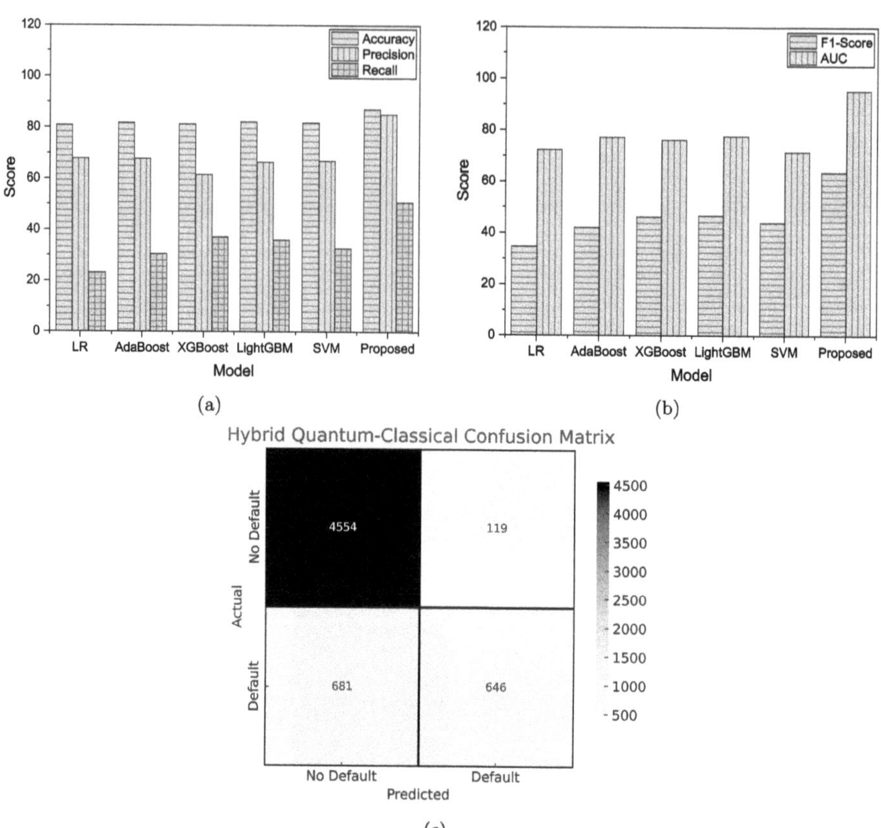

Fig. 4. Performance evaluation on the Taiwan dataset: a) Comparative analysis of accuracy, precision, and recall; b) F1-Score vs. Area Under the Curve (AUC) analysis; c) Confusion matrix representation.

posed SVM model showed 81.84% accuracy, 66.98% precision, and 32.81% recall, demonstrating its proficiency in detecting defaulters despite the class imbalance.

The confusion matrix in Fig. 4c reveals that the system accurately detected 653 true positives, 117 false positives, 674 false negatives, and 4,500 true negatives. The proposed model significantly reduces false negatives, compared to traditional models like Logistic Regression (LR), SVM, and AdaBoost, which often prioritize precision over recall and miss a significant number of defaulters.

Although competitive, LGBM and XGBoost struggled with finding defaulters, leading to misclassifications. The proposed model, with its AUC of 93.92%, strikes a balance between accuracy and recall, making it the best option for credit risk classification in this dataset. The proposed financial risk assessment approach is highly accurate, with the ability to identify defaulters and non-defaulters with minimal misclassification risks. Its high AUC and efficiency make it an excellent choice for financial institutions to enhance credit risk evaluation and financing decisions.

5 Conclusion

This study proposes a quantum-classical hybrid model for credit risk prediction, combining optimized ensemble learners with a variational quantum classifier. The model employs robust machine learning methods such as XGBoost, LGBM, CatBoost, and HistGradientBoosting inside a stacking ensemble framework, augmented by a meta-classifier, to tackle class imbalance using SMOTE-Tomek. Combining polynomial feature generation with Truncated SVD improves feature space and improves computing efficiency, resulting in superior predictive performance. By employing hyperparameter tweaking with Optuna and optimizing thresholds, the model achieves an equitable balance between precision and recall, surpassing conventional classifiers and typical machine learning techniques.

6 Future Work

The proposed methodology effectively identifies high-risk borrowers early, enabling financial institutions to apply risk-adjusted pricing, restructure loans, and boost financial inclusion, especially for micro, small and medium enterprises (MSMEs) and rural borrowers. This method improves the credit risk evaluation, ensuring more accurate and reliable loan decisions. Future efforts will focus on integrating real-time data processing to enhance predictive accuracy. Incorporating explainable AI (XAI) will improve transparency and trust in model predictions [22]. Advanced techniques such as Recurrent Neural Networks (RNNs) and Long Short-Term Memory (LSTM) networks can capture temporal patterns in transaction data for more accurate risk predictions. Transformer models like BERT and GPT could help interpret unstructured data in borrower profiles, such as text from loan applications. Incorporating alternative data sources, including social media and transaction records, alongside ensemble learning methods like XGBoost and LGBM, will further enhance predictive power and model robustness [24]. Autoencoders and Generative Adversarial Networks (GANS) could be used for anomaly detection, while reinforcement learning (RL) can adapt the model to changing market conditions in real time.

Furthermore, our proposed credit risk prediction framework can be explored in a low computing resource environment with classical ensemble models that function autonomously while preserving robust predictive performance.

References

1. Emmanuel, I., Sun, Y., Wang, Z.: A machine learning-based credit risk prediction engine system using a stacked classifier and a filter-based feature selection method. J. Big Data **11**(1), 23 (2024)
2. Shi, Y., Qu, Y., Chen, Z., Mi, Y., Wang, Y.: Improved credit risk prediction based on an integrated graph representation learning approach with graph transformation. Eur. J. Oper. Res. **315**(2), 786–801 (2024)
3. Liu, J., Zhang, S., Fan, H.: A two-stage hybrid credit risk prediction model based on XGBoost and graph-based deep neural network. Expert Syst. Appl. **195**, 116624 (2022)
4. Zhang, L., Yu, Q., Zhang, Y., Zhou, C.: Adaptive feature cross-compression for credit default prediction. IEEE Access **11**, 94322–94334 (2023)
5. Yıldırım, M., Okay, F., Yıldırım, F., Özdemir, S.: Big data analytics for default prediction using graph theory. Expert Syst. Appl. **176**, 114840 (2021)
6. Bulut, C., Arslan, E.: Comparison of the impact of dimensionality reduction and data splitting on classification performance in credit risk assessment. Artif. Intell. Rev. **57**(9), 252 (2024)
7. Ramesh, R., Jeyakarthic, M.: Enhancing credit risk prediction with hybrid deep learning and sand cat swarm feature selection. Multimedia Tools Appl. **83**(21), 60243–60263 (2024)
8. Bone-Winkel, G.F., Reichenbach, F.: Improving credit risk assessment in P2P lending with explainable machine learning survival analysis. Dig. Finan. **6**(3), 501–542 (2024)
9. Chang, V., Xu, Q. A., Akinloye, S. H., Benson, V., Hall, K.: Prediction of bank credit worthiness through credit risk analysis: an explainable machine learning study. Ann. Oper. Res. 1–25 (2024)
10. Guan, C., Suryanto, H., Mahidadia, A., Bain, M., Compton, P.: Responsible credit risk assessment with machine learning and knowledge acquisition. Human-Centric Intell. Syst. **3**(3), 232–243 (2023)
11. Li, Y., Zhao, R., Sha, M.: A hybrid credit risk evaluation model based on three-way decisions and stacking ensemble approach. Comput. Econ. 1–24 (2024)
12. Mandour, M. A., Chi, G., Amran, G. A., Alsalman, H.: A single default discrimination model based on the selection of multiple single models. IEEE Access (2024)
13. Aruleba, I., Sun, Y.: Effective credit risk prediction using ensemble classifiers with model explanation. IEEE Access (2024)
14. Miljkovic, T., Wang, P.: A dimension reduction assisted credit scoring method for big data with categorical features. Finan. Innov. **11**(1), 29 (2025)
15. Hafez, I.Y., Hafez, A.Y., Saleh, A., Abd El-Mageed, A.A., Abohany, A.A.: A systematic review of AI-enhanced techniques in credit card fraud detection. J. Big Data **12**(1), 6 (2025)
16. Zhang, X., Yu, L., Yin, H.: Domain adaptation-based multistage ensemble learning paradigm for credit risk evaluation. Finan. Innov. **11**(1), 27 (2025)
17. Economic Times Realty: Individual Housing Loans in Delhi Surge 14%. Economic Times (2025)
18. CRIF High Mark Report: Delhi NCR leads in loan origination. CRIF High Mark (2025)
19. New Indian Express: Unsecured loans grow at double-digit rates. New Indian Express (2025)

20. Economic Times BFSI: Personal loans cross Rs. 50 lakh crore. Economic Times (2025)
21. Economic Times BFSI: Delinquency rates rising in H1FY25. Economic Times (2025)
22. Wahab, F., Khan, I., Sabada, S.: Credit card default prediction using ML and DL techniques. Internet Things Cyber-Phys. Syst. **4**, 293–306 (2024)
23. Gasmi, I., Neji, S., Smiti, S., Soui, M.: Features selection for credit risk prediction problem. Inf. Syst. Front. 1–20 (2025)
24. Montevechi, A.A., de Carvalho Miranda, R., Medeiros, A.L., Montevechi, J.A.B.: Advancing credit risk modelling with machine learning: a comprehensive review of the state-of-the-art. Eng. Appl. Artif. Intell. **137**, 109082 (2024)
25. Dua, D., Graff, C.: UCI Machine Learning Repository: Credit Card Datasets. https://archive.ics.uci.edu/datasets?search=credit+card. Accessed 10 Mar 2025
26. Bhardwaj, A., Singh, V., Narayan, Y., et al.: Analyzing bigdata with hadoop cluster in hdinsight azure cloud. In: 2015 Annual IEEE India Conference (INDICON), pp. 1–5 (2015)
27. Aslam, A., Bhardwaj, A., Chaudhary, R.: Quantum-resilient blockchain-enabled secure communication framework for connected autonomous vehicles using post-quantum cryptography. Veh. Commun. **52**, 100880 (2025)

Label, Learn, Enhance: Multimodal Large Language Models-Assisted Annotation and Optimized Classifier Training for Retail Outlet Image Quality Assessment

Prithviraj Purushottam Naik[(✉)] [iD] and Rohit Agarwal [iD]

Bizom (Mobisy Technologies Private Limited), Bengaluru, India
`{prithviraj.naik,rohit}@mobisy.com`

Abstract. In the retail industry, acquiring high-quality image data is essential for accurate audits, stock analysis, and compliance verification. However, image collection is often hindered by inconsistencies, as sales professionals, driven by incentive-based targets, frequently capture low-quality photographs that fail to adhere to established guidelines. Such noisy inputs degrade downstream performance in automated systems. Traditional manual quality assurance methods are labour-intensive, inconsistent, and lack scalability, particularly given the vast diversity of India's 12–13 million retail outlets. While machine learning-based image classifiers present a scalable solution, their performance is intrinsically tied to the availability of large-scale, meticulously labelled training datasets which are often difficult, time-consuming, and expensive to obtain. To address these challenges, we propose a novel framework for automated retail image quality assessment using Multimodal Large Language Models (MLLMs). Our approach begins with domain-adapted prompt engineering to generate supervision signals from multimodal models such as BLIP, LLaVA, and Qwen2-VL. We then employ an ensemble-based consensus mechanism to self-label images with improved reliability. The resulting high-confidence pseudo-labelled dataset with weak manual supervision is used to train and evaluate image classifiers, specifically ConvNeXt, ResNet-50, and Vision Transformers, using a carefully optimized training strategy. Experimental results demonstrate the effectiveness of this approach in evaluating multiple quality dimensions, including blurriness, composition, content clarity, and contextual relevance. By drastically reducing reliance on manual labelling, our pipeline enables scalable, accurate, and cost-effective retail image validation. This innovation ensures that only high-quality images contribute to business operations, leading to better decision-making and more efficient retail management.

Keywords: Retail Image Quality Assessment · Multimodal Large Language Models · Image Classification · Self-labelling Techniques

R. Gupta et al. (Eds.): BDA 2025, LNCS 16041, pp. 84–99, 2026.
https://doi.org/10.1007/978-3-032-15134-6_6

1   Introduction

In India, the retail sector contributes nearly 10% to the country's GDP, making it a cornerstone of the national economy [1]. At the heart of this sector lies salesforce management, a crucial driver of retail success particularly in regions like India and Southeast Asia, where businesses navigate a highly fragmented and dynamic ecosystem. As globalization accelerates, economies become more service-oriented, and technology reshapes consumer behaviour, competition in the retail space has grown fiercer than ever [2]. In this evolving landscape, small independent retail stores locally known as "Kirana" shops or "mom-and-pop" stores remain dominant players. These neighbourhood outlets form the backbone of commerce in the region, serving as vital touchpoints between global brands and everyday consumers. Leading FMCG companies like Unilever and P&G rely on vast networks of field sales representatives who visit these outlets regularly to manage relationships and ensure product availability. For example, a P&G salesman might stop by Anmol Provision Store to restock shelves with Head & Shoulders shampoo, ensuring visibility and availability at the point of sale. A key mission for these on-ground teams is to expand the "retail universe" by identifying, engaging, and onboarding new outlets to increase reach and market penetration. In a landscape shaped by hyperlocal needs and intense competition, smart, agile salesforce strategies are not just an advantage they're essential.

Traditionally, sales representatives follow structured journey plans, using Sales Force Automation (SFA) applications like Bizom to streamline order placement and ensure inventory dispatch within 48–72 h [3]. The efficiency of these systems critically depends on the availability of high-quality outlet data, including accurate store names, addresses, geolocations, and contact details, as these elements form the foundation for reliable downstream tasks such as accurate outlet mapping, efficient route planning, personalized sales strategies and effective field force planning. In addition to textual and tabular information, image data offers rich visual context and deep insights into retail outlets capturing storefront visibility, product placements, branding compliance, and overall store conditions. Such visual cues not only aid in better outlet understanding but also support decision-making in areas like marketing, merchandising, and field audits. The process of image collection is often compromised by inconsistencies sales personnel, motivated by incentive-driven goals, frequently capture subpar images that do not conform to prescribed guidelines. Common issues such as blurriness, poor framing and irrelevant content introduce significant noise, thereby undermining the performance of downstream processes.

Inadequate data quality leads to suboptimal routing, inaccurate demand forecasting, and inefficient resource allocation, ultimately increasing costs and reducing operational effectiveness. Additionally, to counter this, companies have implemented measures such as mandatory photographic evidence of shop exteriors to verify store authenticity and reduce the risk of fabricated data. Despite these efforts, existing retail software solutions lack automated image authentication capabilities, enabling sales representatives to submit irrelevant or fraudulent photos, which compromises data integrity. Given the vast number of retail outlets over 12–13 million in India alone manual verification of images is impractical, making automation a necessity. The challenge is further compounded by the architectural diversity of Kirana stores, which lack a standardized appearance, making it difficult for machine learning models to generalize effectively. To address this issue, machine learning models such as ConvNext [11], ResNet-50 [12] and Vision Transformers (ViT) [13] have been proposed for classifying and validating retail outlet images. However, these models require extensive labelled training datasets, which are both costly and time-consuming to acquire at scale. Traditional manual annotation methods are prone to errors and inefficiencies, further hampering model performance.

At Bizom, we propose an innovative approach utilizing MLLM-Based Generative AI to generate labelled training data in an unsupervised manner using domain-specific prompts. By analyzing image features such as signboards, products, and store entrances, we determine Kirana store authenticity, overcoming the limitations of existing MLLMs, which are constrained by non-Indian datasets and a lack of familiarity with local nuances. Our prompt engineering approach facilitates the generation of high-quality labelled datasets by combining pseudo-labelling with weak supervision and limited manual validation. With a robust dataset in place, we fine-tune advanced image classification models such as ConvNext, ResNet-50, and Vision Transformer(ViT) to accurately differentiate between valid and invalid retail outlet images. To further optimize the pipeline, we replace MLLMs with the trained classifiers in subsequent stages. This not only enhances inference speed and scalability but also addresses the resource-intensive nature of MLLMs, which are best suited for the cold-start stage. In contrast, our trained classifiers are lightweight enough to run on CPUs, making them ideal for large-scale production deployments. This cyclical enhancement starting with MLLMs and evolving into lean classifier-based inference offers a sustainable and scalable solution for intelligent retail image quality assessment. Our findings highlight the effectiveness and reliability of our approach in streamlining image verification, improving salesforce productivity, and ensuring data integrity in retail operations.

2   Related Work

Automatic image tagging and annotation have seen significant advancements with the rise of deep learning and multimodal models. These innovations are particularly relevant for retail image analysis, where accurate and scalable labelling is essential for training high-performance classifiers. However, challenges such as noisy data, the need for large annotated datasets, limited domain generalization, and computational inefficiencies continue to hinder fully automated solutions in real-world retail environments. In this section, we review several influential approaches ranging from open-vocabulary recognition models to semi-automatic labelling frameworks and evaluate their potential and limitations in the context of retail outlet image quality assessment.

The Recognize Anything Model (RAM) [4] enhances multi-label image tagging with open-vocabulary recognition, addressing limitations in models like

Tag2Text and ML-Decoder. Inspired by CLIP and ALIGN, RAM employs annotation free tagging, a data engine for label refinement, and fine-grained alignment for accurate region-based recognition. It surpasses fully supervised methods and commercial APIs, recognizing over 6,400 categories with open-set capabilities, though it faces challenges with noisy web data, limited label scope, efficiency trade-offs, open-vocabulary recognition, and high computational requirements.

"Tag2Text: Guiding Vision-Language Model via Image Tagging" [5] proposes a deep learning approach that leverages both textual and visual features for image tagging. While effective in handling noisy and ambiguous labels, its reliance on large-scale annotated datasets and computationally expensive deep networks makes it less suitable for real-time retail applications. Additionally, the method struggles with domain-specific generalization, limiting its adaptability to diverse retail outlet conditions.

"Semi-automatic image labelling using depth information" [6] introduces a method that enhances image labelling by incorporating depth data to improve object boundary detection. While this approach reduces manual annotation efforts and improves label accuracy, it depends on the availability of high-quality depth sensors, which may not be feasible for large-scale retail datasets. Additionally, its performance is limited in scenarios with poor depth estimation, making it less reliable for complex retail environments.

"WebLabel: OpenLABEL-compliant Multi-Sensor Labelling" [7] introduces a web-based tool for annotating multi-sensor data while adhering to the OpenLABEL standard. While it improves annotation consistency and supports multiple data modalities, its reliance on web infrastructure may introduce latency issues for large-scale retail datasets. Additionally, manual intervention is still required for fine-grained labelling, limiting its scalability in fully automated retail applications.

3   Background

With the rapid digitization of retail operations, the quality of outlet images has become crucial for tasks such as automated auditing, visual merchandising, and inventory verification. However, the manual validation of these images remains time-consuming and inconsistent, creating a bottleneck in scaling retail analytics. To address this, recent advancements in Multimodal Large Language Models (MLLMs) and vision-based image classifiers offer promising solutions for automating and optimizing image quality assessment at scale.

3.1   Multimodal Large Language Models (MLLMs)

More intricate comprehension and creation capabilities are now possible thanks to recent developments in Multimodal Large Language Models (MLLMs), which greatly improve the integration of visual and textual input. Popular models include:

- **BLIP (Bootstrapping Language-Image Pre-training):** BLIP introduces a unified vision-language pre-training framework that excels in both understanding and generation tasks. By employing a bootstrapping strategy with a captioner and a noise filter, BLIP effectively leverages noisy web data, achieving state-of-the-art performance across various benchmarks, including image-text retrieval and visual question answering [8].
- **LLaVA (Large Language and Vision Assistant):** LLaVA combines a vision encoder with a large language model, trained end-to-end on instruction-following datasets generated by GPT-4. This integration results in strong multimodal conversational capabilities, achieving high accuracy on benchmarks like Science QA and demonstrating robust performance on unseen image-text tasks [9].
- **Qwen2-VL:** Alibaba Group's Qwen2-VL employs the Naive Dynamic Resolution method to transform images with varying pixel densities into distinct visual tokens. This method mimics human perception, improving visual representation efficiency and accuracy. Qwen2-VL also uses Multimodal Rotary Position Embedding (M-RoPE) to unify text, images, and videos for image and video analysis. The model scales across parameters (2B, 8B, and 72B) and datasets, achieving competitive results with leading models like GPT-4o on multimodal benchmarks [10].

3.2 Vision-Based Classification Models

In parallel, vision-based classification models have evolved to address complex image recognition tasks:

- **ConvNeXt:** Enhanced convolutional network architecture using Layer Normalization and inverted bottlenecks as design concepts borrowed from Vision Transformers. The efficiency of convolutional networks is maintained by ConvNeXt, which delivers competitive performance with transformer-based models [11].
- **ResNet-50:** An advanced convolutional neural network that allowed for the training of extremely deep networks by introducing residual learning, which mitigated the vanishing gradient problem. Several image classification benchmarks have shown that ResNet-50 performs quite well [12].
- **Vision Transformer (ViT):** An approach for image classification that makes use of the transformer architecture, which was first developed for NLP. With the help of self-attention techniques and image patch tokenization, ViT successfully captures long-range dependencies and achieves excellent accuracy on large-scale image recognition benchmarks [13].

By leveraging the strengths of MLLMs and advanced vision-based classifiers, our approach aims to automate the assessment of retail outlet image quality, addressing challenges such as inconsistent image capture and the need for scalable, accurate validation mechanisms.

4   Methodology

This section presents our methodology, which involves leveraging an ensemble of pre-trained vision-language models BLIP, LLaVA, and Qwen2-VL for automated image annotation. The pseudo-labels generated through this ensemble are minimally validated using weak supervision. Visual inspection by human annotators played a key role in validating model predictions and refining labels, especially in borderline or ambiguous cases. While we did not explicitly define numerical thresholds, the human-in-the-loop feedback process helped maintain label quality and mitigate noise. These labels are then used to train image classifiers through an optimized training strategy. Figure 2 presents an overview of the full methodology, combining the pseudo-labelling strategy with the optimized image classifier training workflow. The overarching goal is to develop a robust and scalable framework for retail image quality assessment.

4.1   Dataset

The dataset used in this study is sourced from Bizom, a leading retail intelligence platform. Figure 1 illustrates a sample set of retail outlet images captured by salespersons during routine retail outlet visits and reflect real-world variations in quality, lighting, framing, and content. These images are paired with standardized textual prompts designed to evaluate the quality and relevance of each image for downstream retail analytics tasks. We utilized a dataset comprising 10,000 retail outlet images as input for the pseudo-labelling process and subsequently evaluated the effectiveness of the generated labels through comprehensive testing. Dataset composition for image classifiers with pseudo-labels after manual validation with weak supervision:

- **Training Set**: 8,000 images
- **Validation Set**: 1,000 images
- **Test Set**: 1,000 images

The dataset is carefully curated to ensure diversity in retail environments and to reflect typical challenges encountered in field-level data acquisition. All samples are subjected to the same preprocessing and annotation pipeline described in subsequent sections.

4.2   Data Input and Preprocessing

Each sample in the dataset comprises an image paired with a standardized textual prompt. A uniform preprocessing pipeline ensures consistency across models:

Fig. 1. Sample of Retail Images captured by the salesman.

- **Ensemble models**: Images are resized and normalized according to each model's specific input requirements.
- **Image classifiers**: Images are resized to 224×224 and normalized using ImageNet statistics, aligning them with the training distributions of the classification models.

Textual prompts are tokenized and formatted as per each model's architecture to ensure consistency in input representation across annotation and classification.

4.3   Ensemble Inference for Annotation

To generate pseudo-labels, the preprocessed image-prompt pairs are simultaneously fed into three pre-trained MLLMs: BLIP, LLaVA and Qwen2-VL.

Each model is provided with the same set of images and prompts. For example, prompts include: "Is the retail outlet fully visible with its boundaries?" or "Is the store board clearly visible and not occluded?". Each model independently evaluates the alignment between the image and the prompt, producing a binary decision:

- **Positive** – The image aligns with the prompt.
- **Negative** – The image does not align with the prompt.

4.4   Pseudo-label Generation via Unanimous Voting

A unanimous voting mechanism is used to ensure high annotation precision and reduce model-specific biases. Three MLLMs: BLIP, LLaVA, and Qwen2-VL independently evaluate each image. An image is labelled positive only if all three models agree it is of good quality. If even one model returns a negative decision, the image is labelled negative. This conservative strategy prioritizes precision over recall, producing clean and reliable labels for training image classifiers. This approach significantly minimizes noisy labels that could degrade classifier performance, especially in real-world retail scenarios. By leveraging the consensus of diverse MLLMs, the annotation process becomes more robust and trustworthy, laying a strong foundation for training high-accuracy image quality assessment models.

4.5   Training Image Classifiers with Pseudo-labels

The generated pseudo-labels serve as supervisory signals, supplemented by minimal manual validation, for training image classifiers capable of assessing image quality independently of textual prompts. Three state-of-the-art architectures: ConvNeXt, ResNet-50, and Vision Transformer (ViT) are employed for classification.

- **Hardware**: NVIDIA A6000 GPU
- **Batch Size**: 64

4.6   Optimized Training Strategy and One-Cycle Policy

A structured fine-tuning approach is employed to optimize the classifier's performance while preventing overfitting. The training process follows a staged learning strategy, incorporating the one-cycle policy, which dynamically adjusts the learning rate and momentum to achieve faster convergence and improved generalization. The training phases include:

- **Initial warm-up phase:** Training begins with only the final classification layers (typically the head) being updated. This phase allows the model to quickly adapt to the pseudo-labelled dataset without disrupting the pre-trained feature extractor.
- **Progressive unfreezing:** Deeper layers of the model are incrementally unfrozen, allowing the network to fine-tune higher-level representations. A gradually decreasing learning rate is applied to these layers to retain pre-trained knowledge while adapting to the new domain.
- **One-cycle learning policy**: The core of the optimization strategy is the One-Cycle Policy, which schedules the learning rate and momentum in a controlled, cyclical pattern throughout training [14].
  - **Learning Rate Schedule:** The learning rate starts at a lower bound, increases linearly to a maximum value (known as the peak or "valley point"), and then decreases gradually toward the end of training.
  - **Momentum Schedule:** Inversely, momentum decreases during the warm-up (when the learning rate is increasing), and increases when the learning rate decays ensuring model stability and improved generalization.

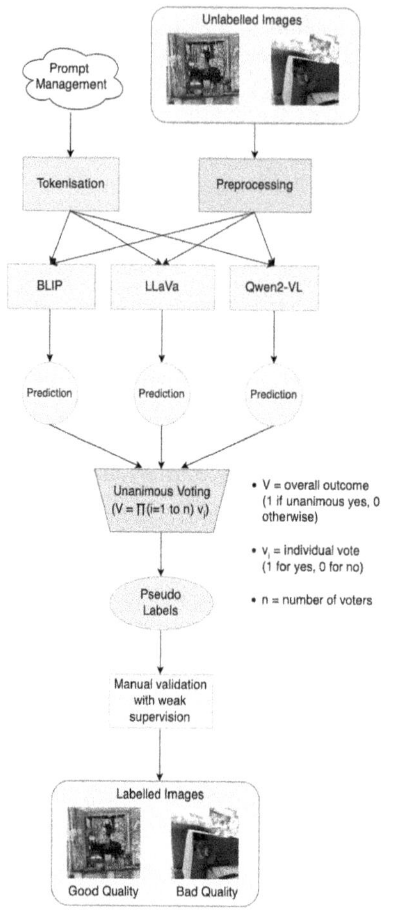
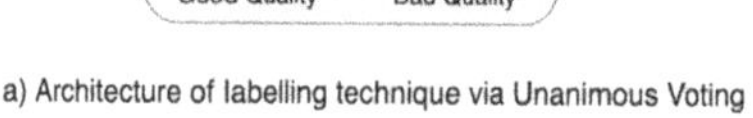
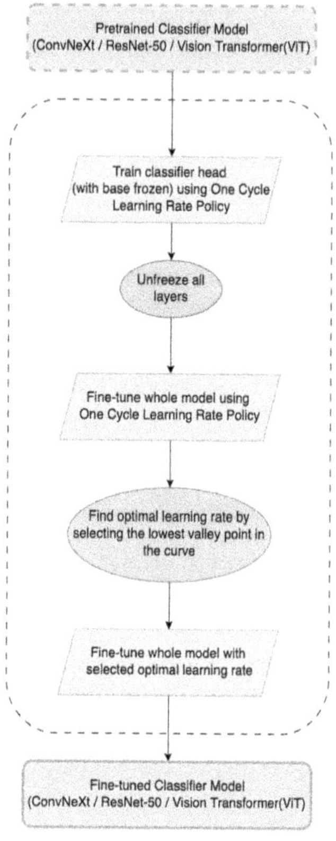

Fig. 2. Overview of the Labelling Technique and Optimized Classifier Training: a) Pseudo-labels are generated by aggregating predictions from multiple Multimodal Large Language Models (BLIP, LLaVA, and Qwen2-VL). Only unanimously agreed-upon labels are retained, ensuring high-confidence, low-noise annotations for weak supervision. b) Pseudo-labelled data is used to train image classifiers (ConvNeXt, ResNet-50, and ViT) using staged fine-tuning with warm-up, progressive unfreezing, and a One-Cycle learning rate policy to enhance convergence and generalization.

- **Valley Point:** The valley point refers to the mid-training peak of the learning rate. Reaching this point encourages the optimizer to explore a broader parameter space initially (reducing chances of getting stuck in local minima), followed by a gentle refinement as the learning rate tapers off. Empirically, this has been shown to result in models that generalize better and converge faster.

- **Final Refinement Phase**:
  Once the one-cycle schedule concludes, the model undergoes additional fine-tuning at the optimal learning rate, determined via an empirical learning rate finder. This final phase stabilizes performance and aligns the model closely with the target distribution. By leveraging the one-cycle policy, the model efficiently learns task-specific representations with minimal overfitting, enabling stable convergence.

The classifiers, such as ConvNeXt, ResNet-50 and Vision Transformer(ViT) learn to assess image quality based on the consensus labels from the ensemble. By leveraging these pseudo-labels, the classifiers develop an understanding of visual quality independent of textual prompts, enabling a more scalable and efficient approach for retail image quality assessment.

5 Evaluation Metrics

To assess the performance of the pseudo-labelling approaches and image classification models, we use standard evaluation metrics including **Accuracy**, **Precision**, **Recall**, and **F1-Score**. These metrics are defined based on the following terms:

- **TP (True Positives)**: Correctly predicted positive samples.
- **TN (True Negatives)**: Correctly predicted negative samples.
- **FP (False Positives)**: Incorrectly predicted positive samples.
- **FN (False Negatives)**: Incorrectly predicted negative samples.

5.1 Accuracy

Accuracy measures the overall correctness of the model and is defined as:

$$\text{Accuracy} = \frac{TP + TN}{TP + TN + FP + FN} \tag{1}$$

5.2 Precision

Precision quantifies the number of correct positive predictions made out of all positive predictions:

$$\text{Precision} = \frac{TP}{TP + FP} \tag{2}$$

5.3 Recall

Recall measures the model's ability to correctly identify all actual positive samples:

$$\text{Recall} = \frac{TP}{TP + FN} \tag{3}$$

5.4 F1-Score

The F1-Score achieves a balance between recall and precision using a harmonic mean:

$$\text{F1-Score} = 2 \times \frac{\text{Precision} \times \text{Recall}}{\text{Precision} + \text{Recall}} \tag{4}$$

6 Evaluation Results and Discussion

This section presents and analyzes the results of our experiments across two main stages: pseudo-labelling quality and final classifier performance. We evaluate using standard metrics: Accuracy, Precision, Recall, and F1-score to provide a comprehensive assessment.

Table 1. Comparative Evaluation of Pseudo-labelling Approaches

Model	Accuracy	Precision	Recall	F1-score
BLIP	0.4869	0.3569	0.9045	0.5118
LLaVA	0.3420	0.3085	**0.9771**	0.4690
Qwen2-VL	0.7635	0.5587	0.9744	0.7102
Our Approach	**0.8066**	**0.6236**	0.8816	**0.7305**

Table 1 summarizes the performance of different pseudo-labelling approaches: BLIP, LLaVA, Qwen2-VL and our proposed ensemble-based method. Among the individual Multimodal Large Language Models (MLLMs), Qwen2-VL performs the best with an Accuracy of 76.35% and a high Recall of 97.44%, indicating its effectiveness in capturing positive instances. However, its Precision (55.87%) and F1-score (71.02%) suggest that it tends to over-predict positives, leading to some false positives. BLIP achieves strong Recall (90.45%) but suffers from low Precision (35.69%), resulting in a moderate F1-score of 51.18%. LLaVA, despite its high Recall (97.71%), demonstrates the weakest overall performance with the lowest Accuracy (34.20%) and F1-score (46.90%), primarily due to poor Precision (30.85%).

In contrast, our approach, which uses unanimous voting and ensemble agreement across MLLMs, outperforms all individual models. It achieves an Accuracy of 80.66%, Precision of 62.36%, and the highest F1-score of 73.05%, indicating a well-balanced trade-off between Precision and Recall. This validates our hypothesis that leveraging multiple MLLMs via ensemble methods can produce more reliable pseudo-labels for training downstream classifiers.

Table 2 presents the final classification performance using three different models trained on the pseudo-labelled data: ConvNeXt, ResNet-50 and Vision Transformer (ViT). Among the classifiers, ConvNeXt achieves the best overall results,

Table 2. Comparative Evaluation of different classification models

Model	Accuracy	Precision	Recall	F1-score
ConvNext	**0.9150**	**0.9238**	**0.9150**	**0.9170**
ResNet-50	0.9030	0.9117	0.9030	0.9052
ViT	0.8410	0.8370	0.8410	0.8316

with an Accuracy of 91.50% and an F1-score of 91.70%, indicating superior generalization. ResNet-50 closely follows with an Accuracy of 90.30% and F1-score of 90.52%, making it a strong alternative, especially in resource-constrained settings due to its lighter architecture. ViT, while demonstrating reasonable performance (Accuracy: 84.10%, F1-score: 83.16%), lags behind the other two, suggesting that it may require more data or longer training to fully leverage its transformer-based architecture [15].

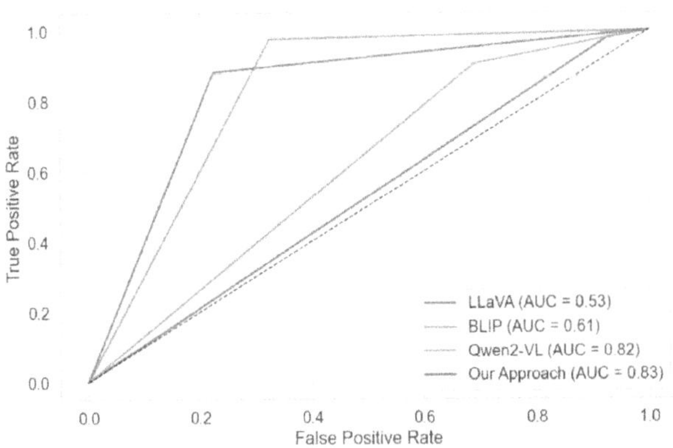

Fig. 3. Receiver Operating Characteristic (ROC) curves of Pseudo-labelling Approaches.

Figure 3 compares the ROC performance of various MLLM-based pseudo-labelling methods. LLaVA exhibits the lowest AUC (0.53), indicating poor discriminatory power, while BLIP performs moderately (AUC = 0.61). Qwen2-VL significantly improves pseudo-label quality with an AUC of 0.82. Our ensemble-based annotation approach slightly surpasses Qwen2-VL, achieving the highest AUC of 0.83.

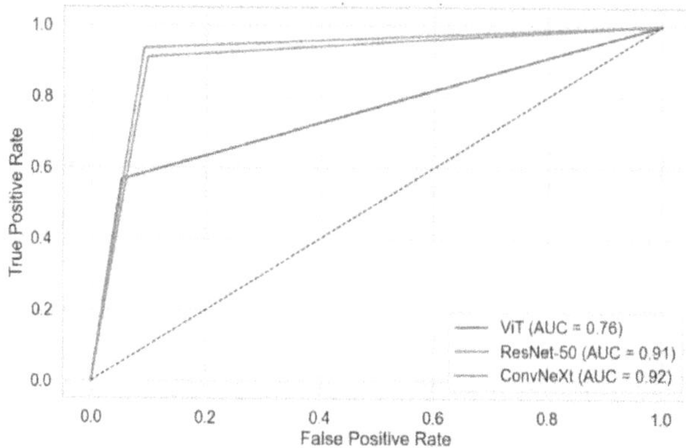

Fig. 4. Receiver Operating Characteristic (ROC) curves of Classification Models.

Figure 4 presents the ROC curves of different vision classifiers trained using the pseudo-labelled dataset. Vision Transformer (ViT) achieves an AUC of 0.76. ResNet-50 and ConvNeXt demonstrate superior performance with AUC scores of 0.91 and 0.92, respectively, highlighting their effectiveness in retail outlet image quality assessment.

7   Key Takeaways

- The quality of pseudo-labels significantly impacts downstream classifier performance.
- Ensemble-based pseudo-labelling using multiple MLLMs leads to more robust supervision than relying on any single model.
- ConvNeXt yields the best classification performance but comes with higher computational requirements.
- ResNet-50 offers a strong balance between performance and efficiency, aligning with real-world deployment constraints.

These findings reinforce the effectiveness of our two-stage approach: first improving annotation quality through multimodal ensemble agreement, and then optimizing classifier training for retail outlet image quality assessment.

8   Conclusion

In this study, we proposed a scalable and efficient framework for retail outlet image quality assessment by integrating the strengths of Multimodal Large Language Models (MLLMs) and advanced vision-based classifiers. Training robust image classifiers typically requires a large volume of accurately labelled data a process that is often time-consuming, costly, and subject to inconsistencies in manual annotations. To overcome these challenges, we leverage an ensemble of state-of-the-art vision-language models: BLIP, LLaVA, and Qwen2-VL to automatically generate high-confidence pseudo-labels for retail outlet images. These MLLMs analyze the visual content and produce labels, which are then aggregated through a unanimous voting mechanism to ensure high precision and reliability. This significantly reduces dependency on manual labelling efforts. To further enhance label trustworthiness, we incorporate weak supervision techniques along with minimal manual validation, ensuring quality without incurring excessive annotation costs. The validated pseudo-labels are then used to train lightweight yet powerful image classifiers such as ConvNeXt, ResNet-50, and Vision Transformer (ViT) using an optimized fine-tuning strategy based on the One-Cycle Policy. By using Generative AI (GenAI) to train predictive AI models, we not only enable smarter learning but also reduce dependency on traditional labelled datasets. To ensure scalability and speed in real-world deployments, we replace the computationally intensive MLLMs with the trained classifiers in production, as they are more resource-efficient, CPU-friendly, and capable of real-time inference. This cyclical pipeline initially using MLLMs for annotation and subsequently transitioning to efficient classifiers enables cost-effective, scalable deployment for automated image quality validation. This approach allows us to use expensive foundation models to train inexpensive models, thereby offering affordable AI for practical deployments.

Experimental results show that our ensemble-based annotation method produces high-quality training labels by outperforming individual MLLMs in accuracy, precision, and F1-score. Thus, classifiers trained on these labels predict image quality well, enabling scaled real-time validation. Our generalizable, resource-efficient end-to-end pipeline solves retail analytics problems like inconsistent image capture, human validation overhead, and lack of scalable labelling. This work opens the door to multimodal annotation systems and automated retail data pipelines, making retail operations smarter and more scalable.

9   Limitations and Future Works

While our framework shows promising results, certain limitations remain. The unanimous voting mechanism, though effective in maintaining high label precision, may exclude a substantial number of images due to model disagreement. Exploring soft-labelling [16] strategies or confidence-based thresholds could allow for the inclusion of more samples without compromising label quality. Additionally, while we utilize multiple MLLMs to capture diverse semantic perspectives, we assume that these models are equally calibrated and reliable across domains, which may not always be the case.

In future work, we plan to explore adaptive voting mechanisms and integrate human-in-the-loop feedback to further improve label quality. We also aim to extend our framework to handle multi-label classification and fine-grained quality attributes [17], as well as evaluate its generalizability across other computer vision tasks in retail such as planogram compliance and on-shelf availability detection. Finally, investigating low-resource or edge-device deployment strategies for real-time inference will be a key focus to make this system production-ready at scale. Future work can also explore refining the ensemble strategy by incorporating confidence-based weighting schemes to mitigate potential false negatives.

Acknowledgments. We thank the anonymous reviewers and participants for their valuable feedback. We extend our gratitude to Lalit Bhise, Founder and CEO of Bizom (Mobisy Technologies Private Limited), for his unwavering support throughout this initiative.

References

1. Vaz, J.: Theoretical Insights and Value Orientation on Growth and Challenges of Retailing Prospects in India. Available at SSRN 4701067 (2022)
2. Kumar, V., Sharma, A., Shah, R., Rajan, B.: Establishing profitable customer loyalty for multinational companies in the emerging economies: a conceptual framework. J. Int. Mark. **21**(1), 57–80 (2013)
3. Sales Automation. The Future of Marketing and Sales Automation. Marketing and Sales Automation 431 (2023)
4. Zhang, Y., et al.: Recognize anything: a strong image tagging model. In: Proceedings of the IEEE/CVF Conference on Computer Vision and Pattern Recognition, pp. 1724–1732 (2024)
5. Huang, X., et al.: Tag2text: guiding vision-language model via image tagging. arXiv preprint arXiv:2303.05657 (2023)
6. Wu, W., Yang, J.: Semi-automatically labelling objects in images. IEEE Trans. Image Process. **18**(6), 1340–1349 (2009)
7. Urbieta, I., et al.: Weblabel: openlabel-compliant multi-sensor labelling. Multimed. Tools Appl. **83**(9), 26505–26524 (2024)
8. Li, J., Li, D., Xiong, C., Hoi, S.: BLIP: bootstrapping language-image pre-training for unified vision-language understanding and generation. In: International Conference on Machine Learning, pp. 12888–12900. PMLR (2022)
9. Liu, H., Li, C., Wu, Q., Lee, Y.J.: Visual instruction tuning. In: Advances in Neural Information Processing Systems, vol. 36, pp. 34892–34916 (2023)
10. Wang, P., et al.: Qwen2-VL: enhancing vision-language model's perception of the world at any resolution. arXiv preprint arXiv:2409.12191 (2024)
11. Liu, Z., Mao, H., Wu, C.-Y., Feichtenhofer, C., Darrell, T., Xie, S.: A convnet for the 2020s. In: Proceedings of the IEEE/CVF Conference on Computer Vision and Pattern Recognition, pp. 11976–11986 (2022)
12. He, K., Zhang, X., Ren, S., Sun, J.: Deep residual learning for image recognition. In: Proceedings of the IEEE Conference on Computer Vision and Pattern Recognition, pp. 770–778 (2016)
13. Dosovitskiy, A., et al.: An image is worth 16 × 16 words: transformers for image recognition at scale. arXiv preprint arXiv:2010.11929 (2020)
14. Smith, L.N.: A disciplined approach to neural network hyper-parameters: Part 1—Learning rate, batch size, momentum, and weight decay. arXiv 2018. arXiv preprint arXiv:1803.09820 (1803)
15. Zhai, X., Kolesnikov, A., Houlsby, N., Beyer, L.: Scaling vision transformers. In: Proceedings of the IEEE/CVF conference on Computer Vision and Pattern Recognition, pp. 12104–12113 (2022)

16. Yin, D., Liu, X., Wu, X., Chang, B.: A soft label strategy for target-level sentiment classification. In: Proceedings of the Tenth Workshop on Computational Approaches to Subjectivity, Sentiment and Social Media Analysis, pp. 6–15 (2019)
17. Wang, Y., et al.: Fine-grained autoaugmentation for multi-label classification. arXiv preprint arXiv:2107.05384 (2021)

Fusion-Based Reduced Variation Representations for Acoustic Scene Classification

Akansha Tyagi[✉] and Padmanabhan Rajan

School of Computing and Electrical Engineering, Indian Institute of Technology Mandi, Mandi, India
`d19030@students.iitmandi.ac.in`, `padman@iitmandi.ac.in`

Abstract. Different sources of variation affect real-world acoustic scene (environmental sounds) data, including different recording locations and devices. This work proposes a fusion-based framework that uses local acoustic scene information from convolutional neural network (CNN)-based features and global acoustic scene information from transformer-based features to construct fused acoustic scene representations that contain the best of both worlds. Additionally, this work uses orthogonal projection loss for acoustic scene classification and demonstrates its effectiveness in reducing variations due to differing recording conditions (devices, locations). We evaluate the performance of the proposed fusion-based and reduced variation acoustic scene representations for the detection and classification of acoustic scenes and events (DCASE) 2020 dataset. The fusion framework enhances the scene classification performance by 7% compared to the non-fusion method, with an additional 2% improvement achieved by incorporating orthogonal projection loss. This incorporation also reduces variations due to different recording conditions, as confirmed by evaluating the proposed system on two auxiliary tasks: device classification and location classification.

Keywords: Acoustic scene classification · Recording-location variation · Recording-device variation · Orthogonal projection loss · Local acoustic information · Global acoustic information · Fusion-based acoustic scene representations

1 Introduction

Automating a machine to recognize the environmental/surrounding sounds termed as 'acoustic scenes' such as a park, market, etc. is called acoustic scene classification (ASC) [2]. It forms an important component for computational auditory scene analysis (CASA) and has been a well-researched problem for more than a decade. ASC is useful in applications where 'making sense of sound' is desired [26], like acoustic monitoring, mobile devices with context awareness, smart wheel chairs, etc.

R. Gupta et al. (Eds.): BDA 2025, LNCS 16041, pp. 100–110, 2026.
https://doi.org/10.1007/978-3-032-15134-6_7

The rise of deep learning, along with the introduction of Detection and Classification of Acoustic Scenes and Events (DCASE) challenge, has motivated the development of many convolutional neural network (CNN)-based acoustic scene classification (ASC) systems, with architectures mainly inspired by ResNet-18 [22] and EfficientNet [13]. Additionally, some methods use embeddings extracted from pre-trained CNN networks like VGGish [11], SoundNet [1], and L3Net [4].

Acoustic scenes consist of multiple overlapping acoustic events, where individual acoustic events represent local acoustic scene patterns, and the relationships between these events correspond to global acoustic scene patterns. CNNs use convolutional filters that operate on small regions of the input data, effectively capturing information at the local level. However, they struggle to comprehend the broader context of an acoustic scene, which requires insight into the relationships and patterns among acoustic events over time. This limitation has been addressed in recent years by the introduction of transformer-based frameworks [9,15]. Considering that acoustic scene representations should contain both local and global contextual information, it is logical to construct a framework that incorporates both CNN-based and transformer-based representations, as demonstrated in [17] and [16].

Real-world ASC systems are affected by different sources of variations including different recording conditions [12] such as different recording locations and recording devices. Recording locations introduce indirect variation due to cultural and regional factors, such as native spoken languages [3,10], while recording devices cause direct variation, introducing device-specific characteristics in the recorded acoustic scene data [19]. These recording conditions can be regarded as a source of intra-scene variation, introducing disparities in two audio signals corresponding to the same acoustic scene class. Furthermore, the presence of similar sound events in two different acoustic scenes, causes inter-scene variation. The existence of intra-scene and inter-scene variations in the acoustic scene data is undesirable.

To address these variations, most of the ASC methods primarily used data augmentation methods like random temporal cropping [18], mixup [27], spectrum correction [14], and specaugment [20]. In contrast, some methods used domain adaptation [7] and multi-view learning [24] to address recording-device and recording-location variation respectively. Authors in [23] addressed both sources of variation, while authors in [25] and [6] addressed different sources of variations in general.

In this work, we aim to construct acoustic scene representations that capture both local and global patterns while minimizing intra-scene variation and maximizing inter-scene variation for improved effectiveness. The first objective is achieved by using a fusion-based framework that consists of three main components: *Local-Subnetwork* (captures local acoustic scene patterns), *Global-Subnetwork* (captures global acoustic scene patterns), and *Fusion-Subnetwork* (combines local and global acoustic patterns to construct fused acoustic scene representations). The second objective is achieved by training the above framework using orthogonal projection loss (OPL) [21] along with categorical cross-entropy. We chose OPL over other margin-based loss functions [25] because it enhances inter-class separation and intra-class compactness. Additionally, OPL can be combined with a categorical cross-entropy loss without the need to learn any additional parameters [21]. Further details about the proposed fusion-based framework for classifying different acoustic scenes are discussed in Sect. 2.

The main contributions of this work can be summarized as follows:

1. We proposed a fusion-based acoustic scene classification system that combines both local and global acoustic scene patterns.
2. Orthogonal projection loss (OPL) is integrated as an additional component in the fusion-based ASC system to reduce variations.
3. To validate the effectiveness of OPL, we perform two auxiliary tasks: recording location classification and recording device classification; in addition to the main task of classifying different acoustic scenes.

2   Methodology

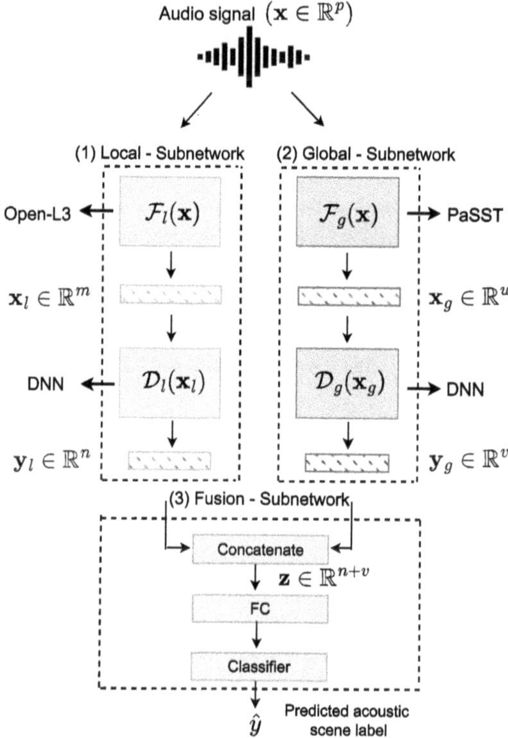

Fig. 1. Overall pipeline for fusion-based acoustic scene classification system. It consists of three main components: (1) Local-Subnetwork (2) Global-Subnetwork and (3) Fusion-Subnetwork.

To begin, we define $\mathbf{x} \in \mathbb{R}^p$ as the time domain representation of an acoustic scene signal recorded at a particular location using a particular device. The

Local-Subnetwork extracts local features from $\mathbf{x}$, while the Global-Subnetwork captures global patterns. The Fusion-Subnetwork combines the outputs of both to create a unified representation that incorporates both characteristics. Figure 1 illustrates an overview of the fusion-based ASC system that comprises of three key components listed as follows:

2.1 Fusion-Based ASC System

1. **Local-Subnetwork**: It consists of Open-L3 [4] pre-trained network with a convolutional neural network (CNN)-based architecture, denoted as $\mathcal{F}_l$ and a shallow deep neural network (DNN) represented by $\mathcal{D}_l$.
   Initially, $\mathbf{x} \in \mathbb{R}^p$ is processed by $\mathcal{F}_l$ which transforms it to $\mathbf{x}_l \in \mathbb{R}^m$. We assume that, after this transformation, $\mathbf{x}_l$ contains local acoustic scene information. Since Open-L3 is used as a frozen feature extractor, thus fine-tuning the obtained representations is crucial. Therefore, $\mathbf{x}_l$ is then processed by $\mathcal{D}_l$, which transforms it into $\mathbf{y}_l \in \mathbb{R}^n$, further reducing its dimension from m to n.
   The overall transformation can be mathematically represented as:

$$\mathbf{y}_l = \mathcal{D}_l(\mathcal{F}_l(\mathbf{x})) \tag{1}$$

2. **Global-Subnetwork**: This component includes PaSST [15] pre-trained network that has a transformer-based architecture, referred as $\mathcal{F}_g$ and a shallow deep neural network (DNN), denoted as $\mathcal{D}_g$.
   In parallel to $\mathcal{F}_l$, $\mathbf{x}$ is also processed by $\mathcal{F}_g$, converting it into $\mathbf{x}_g \in \mathbb{R}^u$. We assume that after this transformation, $\mathbf{x}_g$ captures global acoustic scene patterns. For fine-tuning, $\mathbf{x}_g$ is further processed by $\mathcal{D}_g$, which transforms it into $\mathbf{y}_g \in \mathbb{R}^v$, thereby reducing its dimension from u to v.
   Thus, the overall transformation can be mathematically represented as:

$$\mathbf{y}_g = \mathcal{D}_g(\mathcal{F}_g(\mathbf{x})) \tag{2}$$

   It should be noted that $\mathbf{x}$ is processed in parallel by both Local and Global-Subnetworks.
3. **Fusion-Subnetwork**: This module combines the outputs from the Local and Global-Subnetworks to create fused acoustic scene representations. It consists of a concatenation layer followed by two fully connected layers, with the second layer acting as the classifier. $\mathbf{y}_l$ and $\mathbf{y}_g$ obtained as the outputs of the Local and Global-Subnetworks, are concatenated to form $\mathbf{z} \in \mathbb{R}^{n+v}$, which is further processed by a fully connected layer and a classifier to predict the acoustic scene label $\hat{y}$ corresponding to the input $\mathbf{x}$.

The framework described above learns acoustic scene representations by fusion; however, to further enhance these representations' effectiveness, it is necessary to obtain invariance to different recording conditions. To meet this objective, we train this system using orthogonal projection and standard categorical cross-entropy losses.

2.2 Orthogonal Projection Loss (OPL)

Categorical cross-entropy (CE) facilitates the learning discriminative representations but it does not explicitly reduces inter-class similarity and increases intra-class similarity, which are desired attributes of any classification system. In contrast, OPL ensures the aforementioned key characteristics by enforcing class-wise orthogonality in the feature space learned in the penultimate layer [21] of a neural-network-based framework.

For understanding the mathematics of OPL, we consider the pairs of acoustic scene examples and their corresponding acoustic scene labels in the form $(\mathbf{x}_i, y_i)$, where i is the example index and N is the total number of examples across all acoustic scene classes. OPL operates at the second last layer, thus it will function on the layer labeled 'FC' (Fig. 1). Let $\mathbf{W}$ and $f_i = \mathbf{W}(\mathbf{z}_i)$, denote the weights and output of FC, with $\mathbf{z}$ representing the fused representation. To achieve intra-class compactness and inter-class separation, intra-class similarity (S_{intra}) and inter-class similarity (S_{inter}) are calculated using the features produced as the output of the FC layer as follows:

$$S_{intra} = \frac{\sum_{i,i \in B} < f_i, f_i >}{C_{intra}} \tag{3}$$

$$S_{inter} = \frac{\sum_{i,j \in B} < f_i, f_j >}{C_{inter}} \tag{4}$$

where $< f_i, f_i >$ and $< f_i, f_j >$ denote the normalized cosine similarities between examples from the same and different classes respectively. C_{intra} represents the number examples with same class and C_{inter} corresponds to the number of examples with different classes within batch B.

OPL is calculated as the sum of intra-class and inter-class similarity as per the following equation:

$$\mathcal{L}_{\text{OPL}} = (1 - S_{intra}) + \lambda * |S_{inter}| \tag{5}$$

where λ controls the contribution of inter-class similarity. Intra-class similarity is increased by pushing S_{intra} towards one, and inter-class similarity is decreased by pushing S_{inter} towards zero, achieved by minimizing $\mathcal{L}_{\text{OPL}}$.

The total loss used to train the overall system (refer Fig. 1) is represented as:

$$\mathcal{L} = \mathcal{L}_{\text{CE}} + \gamma * \mathcal{L}_{\text{OPL}} \tag{6}$$

where CE represents categorical cross entropy and γ controls the contribution of OPL in the total loss. For further details about OPL, we direct the readers to reference [21].

3 Experimental Evaluation

This section describes the experimental methods and findings for the fusion-based and reduced variation ASC system. First, we provide a brief overview of the dataset used and a detailed explanation of the feature extraction process. Next, we describe the baseline and proposed systems, discuss the obtained results, and compare them with similar systems.

3.1 Dataset Description

We conducted experiments using DCASE 2020 task 1A: 'acoustic scene classification using multiple devices' development dataset[1]. The dataset contains audio recordings from ten acoustic scene classes ('airport', 'metro', 'metro-station', 'bus', 'park', 'public square', 'shopping mall', 'street pedestrian', 'tram', 'street traffic'), captured using three real devices (A, B, C) and six simulated devices (S1-S6) and recorded across ten European cities namely: 'barcelona', 'helsinki', 'london', 'stockholm', 'vienna', 'lyon', 'milan', 'paris' and 'prague'. Devices B, C, and S1S6 contain randomly selected segments from simultaneous recordings, resulting in partial overlap with device A's data, though not necessarily among themselves. The total duration of audio in the development set is 64 h. The dataset includes both recording location and device labels for each audio sample and is provided with a predefined train/test split, where 70% of the data from each device is used for training and 30% for testing. Additionally, some devices are present only in the test subset.

3.2 Feature Extraction

Referring to Fig. 1, $\mathcal{F}_l$ and $\mathcal{F}_g$ represent the feature extractors for the Local and Global-subnetworks, respectively. For the former, we used Open-L3 [4] network, while for the latter, we used PaSST [15]. Further details about both architectures are provided as follows:

3.2.1 Open-L3

To obtain CNN-based features for all audio recordings, we have used torchopenl3 library[2] which contains Open-L3 network's[3] 'env' version trained on Audioset dataset [8]. The network inputs a raw audio signal and converts it into a mel-spectrogram time-frequency representation with 256 mel bands. It then outputs a 512-dimensional embedding from a hidden layer within the fourth convolutional block of the audio subnetwork. Thus, the Local Subnetwork's feature ($\mathbf{x}_l$) dimension (m) is 512.

3.2.2 PaSST

To obtain the transformer-based features, we have used Patchout faSt Spectrogram Transformer (PaSST)[4] package for Holistic Evaluation of Audio Representations (HEAR) 2021 NeurIPS challenge, which is an efficient version of

---

[1] https://dcase.community/challenge2020/task-acoustic-scene-classification.
[2] https://github.com/torchopenl3/torchopenl3.
[3] https://github.com/marl/openl3.
[4] https://github.com/kkoutini/PaSST.

audio spectrogram transformer (AST) [9]. It is pre-trained on ImageNet [5] and fine-tuned on Audioset [8] dataset. The network's architecture comprises of self-attention blocks and fully connected layers. The network takes a raw audio signal as input and produces output from the 'pre-logits' layer, which returns an embedding of dimension 768. Thus, for the Global-Subnetwork feature $(\mathbf{x}_g)$ dimension (u) is 768.

3.3   Baseline Systems

We implemented two baseline models. The first, referred to as *Local-Baseline*, includes only the Local-Subnetwork and a fully connected layer as the classifier, while the second, *Global-Baseline*, integrates the Global-Subnetwork with the classifier, as presented in Fig. 1. The DNN $(\mathcal{D}_l)$ for the former baseline consists of an input layer with 512 neurons, two fully connected layers with 256 and 128 neurons, and a classification layer with 10 neurons. Similarly, the DNN $(\mathcal{D}_g)$ for the latter baseline consists of an input layer with 768 neurons, two fully connected layers with 384 and 128 neurons, and a classification layer with 10 neurons.

3.4   Proposed Systems

We have proposed four systems. The first two systems, namely: *Local + OPL* and *Global + OPL*, enhance their respective baseline systems by using OPL loss. The third system, termed as *Fusion*, includes a Fusion-Subnetwork in addition to the Local and Global-Subnetworks. Finally, the fourth system, referred as *Fusion + OPL* extends the Fusion-Subnetwork by incorporating OPL loss. The Fusion-Subnetwork consists of a concatenation layer with 256 neurons, followed by two fully connected layers with 128 and 10 neurons.

We compared the baseline and proposed systems fairly by following the same training process and used 20% of the training data as validation data to choose the best system. For the OPL loss, we fixed $\gamma = 1$ and $\lambda = 1$ based on the classification accuracy obtained for the validation data.

3.5   Results and Discussion

Figure 2 shows the classification performance of the baseline and proposed systems for the main task of acoustic scene classification, along with the auxiliary tasks of recording location and device classification. The objective of the auxiliary tasks is to showcase the effectiveness of OPL loss in reducing the variation due to different recording conditions. The following inferences can be drawn from the obtained results:

1. The fusion framework obtains higher scene classification accuracy than non-fusion frameworks, i.e., Local-Baseline and Global-Baseline, which showcases the effectiveness of fusion-based acoustic scene representations. The integration of OPL in the fusion system further improves the scene classification performance.

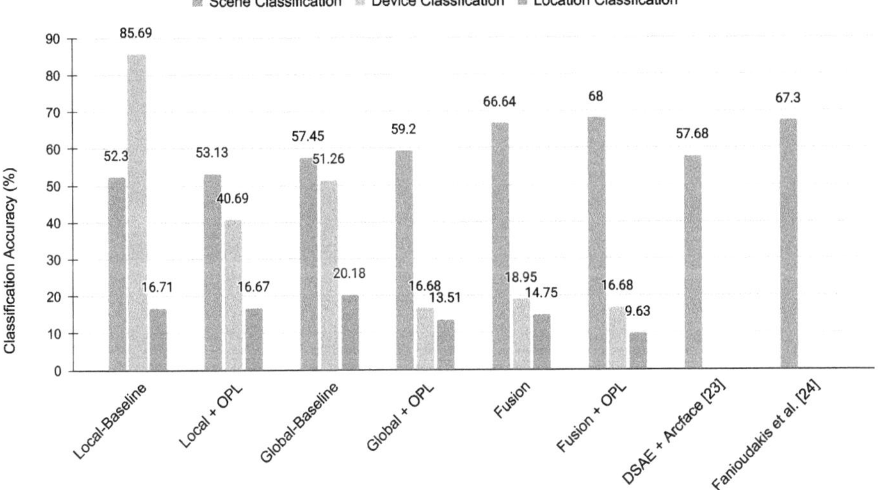

Fig. 2. Scene, Device, and Location classification accuracy for both baseline and proposed systems for the DCASE 2020 dataset as per DCASE train-test split and a comparison with other similar acoustic scene classification systems.

2. The significant decrease in the recording location and recording device classification accuracy for the systems using OPL loss (Local+OPL, Global+OPL, and Fusion+OPL) compared to systems not using OPL loss (Local-Baseline, Global-Baseline, and Fusion) demonstrates the effectiveness of OPL in reducing variation.
3. Figure 3 demonstrates the impact of OPL in decreasing inter-scene similarity and increasing intra-scene similarity. It shows a pairwise cosine similarity plot for examples from various acoustic scenes, where the blue color in the lower plot indicates a reduction in inter-scene similarity. Figure 4 highlights this observation, where the t-SNE plot with OPL (right plot) demonstrates tighter clustering of examples within the same acoustic scene and better separation between scenes compared to the plot without OPL (left plot).

Figure 2 also presents the comparison of proposed systems with other similar ASC systems; the best performing proposed system Fusion+OPL performs better than DSAE [25], which uses a margin-based loss function to handle variation and is comparable to [6] which uses Convolutional Recurrent Neural Network (CRNN)-based framework.

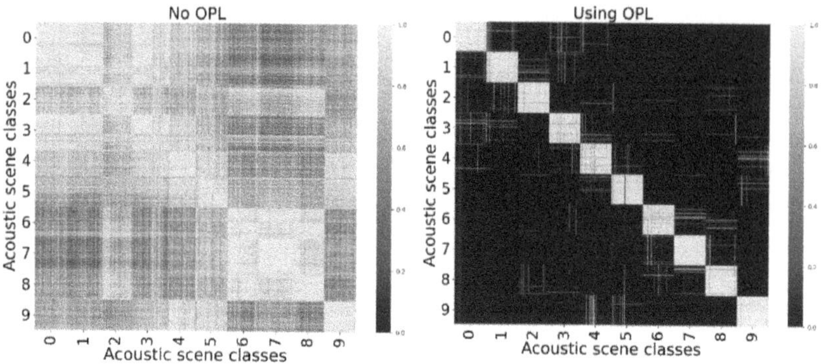

Fig. 3. The pairwise cosine similarity plot for examples from different acoustic scenes, with yellow representing high similarity and blue indicating low similarity. The left plot displays cosine similarity before OPL, while the right plot shows the results after OPL. (Color figure online)

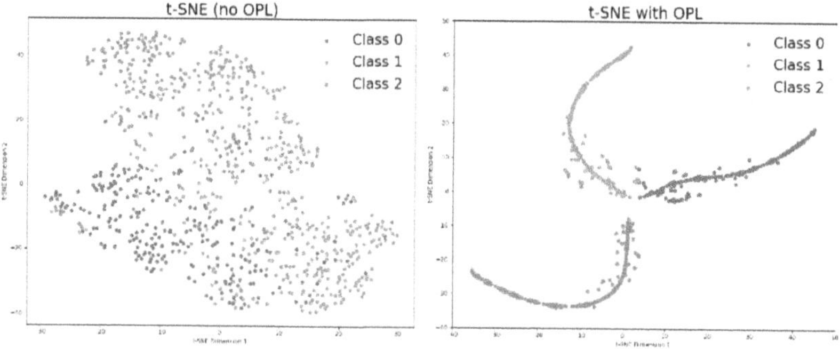

Fig. 4. A t-SNE plot for examples from three randomly selected acoustic scene classes, with and without using OPL. Different colors represent different classes; the left plot shows embeddings without OPL, while the right plot shows results with OPL.

4   Conclusion

In this paper, we showed that the fusion of CNN and transformer-based representations is more useful for classifying different acoustic scenes than using only CNN or transformer-based representations. We also validated the effectiveness of orthogonal projection loss in improving acoustic scene classification performance and decreasing variations in the acoustic scene data due to different recording locations and recording devices. All the proposed systems using OPL outperformed their respective counterparts that did not use OPL. In the future work, we would like to explore the usage of OPL on the intermediate layer representations of the OpenL3 and PaSST networks.

References

1. Aytar, Y., Vondrick, C., Torralba, A.: Soundnet: learning sound representations from unlabeled video. In: Advances in Neural Information Processing Systems (NIPS) (2016)
2. Barchiesi, D., Giannoulis, D., Stowell, D., Plumbley, M.D.: Acoustic scene classification: classifying environments from the sounds they produce. IEEE Signal Process. Mag. (2015)
3. Bear, H.L., Heittola, T., Mesaros, A., Benetos, E., Virtanen, T.: City classification from multiple real-world sound scenes. In: Workshop on Applications of Signal Processing to Audio and Acoustics (WASPAA), pp. 11–15 (2019)
4. Cramer, A.L., Wu, H.H., Salamon, J., Bello, J.P.: Look, listen, and learn more: Design choices for deep audio embeddings. In: IEEE International Conference on Acoustics, Speech and Signal Processing (ICASSP), pp. 3852–3856 (2019)
5. Deng, J., Dong, W., Socher, R., Li, L.J., Li, K., Fei-Fei, L.: Imagenet: a large-scale hierarchical image database. In: IEEE Conference on Computer Vision and Pattern Recognition, pp. 248–255 (2009)
6. Fanioudakis, E., Vafeiadis, A.: Investigating temporal and spectral sequences combining GRU-RNNs for acoustic scene classification. The Detection and Classification of Acoustic Scenes and Events (DCASE) Challenge, Technical Report (2020)
7. Gao, W., McDonnell, M., UniSA, S.: Acoustic scene classification using deep residual networks with focal loss and mild domain adaptation. In: Detection and Classification of Acoustic Scenes and Events (DCASE) Challenge, pp. 1–2 (2020)
8. Gemmeke, J.F., et al.: Audio set: an ontology and human-labeled dataset for audio events. In: IEEE International Conference on Acoustics, Speech and Signal Processing (ICASSP) (2017)
9. Gong, Y., Chung, Y.A., Glass, J.: AST: audio spectrogram transformer. In: Interspeech (2021)
10. Heise, D., Bear, H.L.: Visually exploring multi-purpose audio data. In: 23rd International Workshop on Multimedia Signal Processing (MMSP), pp. 1–6 (2021)
11. Hershey, S., et al.: CNN architectures for large-scale audio classification. In: International Conference on Acoustics, Speech and Signal Processing (ICASSP), pp. 131–135 (2017)
12. Jiang, P., Yang, Y., Zou, C., Wang, Q.: An attention-based time-frequency pyramid pooling strategy in deep convolutional networks for acoustic scene classification. IEEE Signal Process. Lett. (2024)
13. Kim, J.: Acoustic scene classification using multi-channel audio feature with convolutional neural networks and subtract filter augmentation. In: Detection and Classification of Acoustic Scenes and Events (DCASE) Challenge (2020)
14. Kosmider, M.: Spectrum correction: acoustic scene classification with mismatched recording devices. In: Interspeech (2020)
15. Koutini, K., Schlüter, J., Eghbal-Zadeh, H., Widmer, G.: Efficient training of audio transformers with patchout. In: Interspeech (2021)
16. Li, Y., Tan, J., Chen, G., Li, J., Si, Y., He, Q.: Low-complexity acoustic scene classification using parallel attention-convolution network. arXiv preprint arXiv:2406.08119 (2024)
17. Liu, Y., Zhou, X., Long, Y.: Acoustic scene classification with various deep classifiers. In: Detection and Classification of Acoustic Scenes and Events (DCASE) Challenge (2020)

18. McDonnell, M.D., Gao, W.: Acoustic scene classification using deep residual networks with late fusion of separated high and low frequency paths. In: IEEE International Conference on Acoustics, Speech and Signal Processing (ICASSP), pp. 141–145 (2020)
19. Morocutti, T., Schmid, F., Koutini, K., Widmer, G.: Device-robust acoustic scene classification via impulse response augmentation. In: 31st European Signal Processing Conference (EUSIPCO), pp. 176–180 (2023)
20. Park, D.S., et al.: Specaugment: a simple data augmentation method for automatic speech recognition. In: Interspeech (2019)
21. Ranasinghe, K., Naseer, M., Hayat, M., Khan, S., Khan, F.S.: Orthogonal projection loss. In: Proceedings of the IEEE/CVF International Conference on Computer Vision, pp. 12333–12343 (2021)
22. Seo, S., Kim, C., Kim, J.H.: Multi-channel feature using inter-class and inter-device standard deviations for acoustic scene classification. In: Detection and Classification of Acoustic Scenes and Events (DCASE) Challenge (2020)
23. Tan, Y., Ai, H., Li, S., Plumbley, M.D.: Acoustic scene classification across cities and devices via feature disentanglement. IEEE/ACM Trans. Audio Speech Lang. Process. (2024)
24. Tyagi, A., Rajan, P.: Location-invariant representations for acoustic scene classification. In: 30th European Signal Processing Conference (EUSIPCO), pp. 394–398 (2022)
25. Tyagi, A., Rajan, P.: Sparse representation frameworks for acoustic scene classification. In: International Conference on Speech and Computer, pp. 177–188 (2023)
26. Virtanen, T., Plumbley, M.D., Ellis, D.: Computational Analysis of Sound Scenes and Events, 1st edn. Springer, Cham (2017)
27. Zhang, H., Cisse, M., Dauphin, Y.N., Lopez-Paz, D.: mixup: beyond empirical risk minimization. In: International Conference on Learning Representations (2018)

A Two-Level Indexing Scheme
for Extracting Frequent Patterns
with GPUs from Symbolic Sequence Data

P. Shiridi Kumar[1(✉)], Uday Kiran Rage[2], and P. Krishna Reddy[1]

[1] International Institute of Information Technology Hyderabad, Hyderabad, India
p.shiridi@research.iiit.ac.in, pkreddy@iiit.ac.in
[2] The University of Aizu, Aizuwakamatsu, Japan
udayrage@u-aizu.ac.jp

Abstract. Symbolic sequence data is generated from applications such as human activity recognition, financial modeling, intrusion detection systems, and drug discovery tasks. Extracting frequent patterns from symbolic data is an important research problem. In the literature, several CPU-based approaches have been proposed for pattern extraction from symbolic sequence data. However, due to their sequential execution, CPU-based approaches suffer from performance issues in extracting patterns from large sequence data. In modern computing infrastructures, GPUs are playing a central role in carrying out high-performance computing tasks due to their parallel design. Consequently, researchers are investigating GPU-based data mining techniques for tasks such as pattern mining, clustering, and classification. In this paper, we propose a two-level indexing scheme that utilizes extra auxiliary memory to facilitate the efficient extraction of frequent patterns from symbolic data by exploiting the parallel processing capabilities of GPUs. Experimental results on three real-world datasets demonstrate that the proposed approach achieves substantial runtime improvements—approximately $4\times$ speedup over traditional CPU-based algorithms for smaller support values (e.g., 5), and more than $20\times$ speedup for support thresholds greater than 10—when applied to large symbolic sequence data, outperforming both CPU-based and naïve GPU-based methods.

Keywords: Symbolic Sequences · frequent pattern · parallel computing · GPU

1 Introduction

Frequent pattern mining from symbolic sequence data has several applications. A symbolic sequence represents a contiguous sequence of characters (or symbols). The symbolic data format is widely used in several scientific and computational fields. The transformation of raw data into symbolic format facilitates efficient processing and knowledge extraction in areas such as human activity recognition

R. Gupta et al. (Eds.): BDA 2025, LNCS 16041, pp. 111–126, 2026.
https://doi.org/10.1007/978-3-032-15134-6_8

[12], financial modeling [10], network security [5], temporal knowledge graph forecasting [11] and bioinformatics [3].

In bioinformatics, symbolic sequence analysis is widely used for DNA and RNA studies in disease research, drug development, and evolution [22]. Frequent pattern mining from DNA sequence involves identifying combinations of the four nucleotide bases (Adenine (A), thymine (T), Cytosine (C), and Guanine (G)) that frequently appear together in DNA sequence data. DNA sequences range from short regulatory regions to entire genomes spanning millions of base pairs [19]. Given a DNA sequence, when we say that the support of the pattern "ACTCCG" is 10, it indicates that the combination of nucleotides 'ACTCCG' appeared 10 times at different segments of a DNA sequence. Various methods have been proposed for DNA analysis, including sequence similarity measures [8], graph-based models [16], and entropy-based techniques [23].

Traditionally, frequent pattern mining [2] is a widely used data mining technique for identifying frequently occurring patterns in transactional datasets. This method has been applied across various domains such as market basket analysis [20], web usage mining [17], and social network analysis [4], enabling valuable insights into consumer behavior, website optimization, and social dynamics. Recent work in this field has focused on developing improved CPU-based approaches for frequent pattern mining from symbolic sequence data, particularly in applications such as DNA sequence analysis [3,14,18]. Although these algorithms can find patterns effectively in short-sequence datasets, they suffer from significant performance issues on large datasets, often requiring substantial computation time.

Recently, GPUs have been increasingly utilized not only as traditional graphics co-processors but also as general-purpose many-core computing devices. Initially designed for rendering tasks, modern GPUs feature a highly parallel architecture with numerous cores, enabling them to significantly outperform CPUs in compute-intensive tasks such as data mining and matrix operations. This parallel design makes GPUs particularly well-suited for large-scale, high-performance computing applications. Despite these architectural advantages, developing a GPU-based approach for mining frequent patterns in symbolic sequences is non-trivial. A naïve implementation may result in frequent kernel invocations and excessive data transfers between the CPU and GPU, ultimately negating the benefits of parallel execution. These inefficiencies pose significant challenges in fully exploiting the computational capabilities of modern GPUs.

With this motivation, we propose a novel Two-Level Indexing Scheme that leverages additional memory to optimize frequent pattern mining from symbolic sequence data. The core contribution lies in its efficient representation of sequence data, enabling parallel computation by minimizing data transfer between the GPU and CPU and reducing the number of GPU kernel invocations. By leveraging GPU capabilities, we apply this indexing scheme to perform frequent pattern mining, significantly accelerating the process through parallelism. We conduct experiments on three datasets and show that the proposed approach exploits GPU computing power and significantly improves the performance over CPU-based and naïve GPU-based approaches.

The remainder of this paper is organized as follows: Sect. 2 provides a review of related work. Section 3 presents the proposed Two-Level Indexing Scheme and the algorithm, which is designed to leverage this indexing strategy for efficient pattern mining. Section 4 presents the experimental results. Finally, Sect. 5 concludes the paper with future directions.

2   Related Work

In [2], an approach to extract frequent patterns and association rules from the given transactional database was proposed. Since then, the problem of finding these patterns in temporal databases [9], uncertain databases [1], utility databases [6], and sequential databases [15] has received a great deal of attention.

To extract frequent patterns from symbolic sequence data, a depth-first search algorithm, namely Prefix Span, is proposed in [15]. It was argued in [14] that Prefix span is impractical (or suffers from performance issues) for mining long contiguous subsequences from biological sequence datasets. Since then, several algorithms, such as macoFSpan [14], macoVSpan [14], and positional information-based approaches [3,18], have been proposed in the literature for discovering frequent patterns in very long sequence data. Notably, these algorithms, being single CPU-based sequential algorithms, suffer from performance issues while dealing with long sequences. This paper tries to solve this problem by exploiting the speedups and parallelism offered by the GPUs.

Several GPU-based algorithms have been described in the literature to find user interest-based patterns in real-world heterogeneous databases. Key developments include using GPUs to speed up important algorithms like Apriori and FP-Growth. Wang and Yuan's [21] parallel FP-Growth algorithm improved how FP-trees are built and mined, leading to faster results on test datasets. Similarly, Jian et al. [7] enhanced CU-Apriori, CU-KNN, and CU-K-means, showing great performance improvements by optimizing how GPU threads are managed and data is processed. These advancements show GPUs' cost-effectiveness and ability to handle large-scale data mining tasks. Zhang et al. [24] introduced GPApriori, a GPU-accelerated version of Apriori, which achieved up to 100x speedup over CPU versions. They used a particular memory structure that worked well with GPUs, making it very efficient for large and complex datasets. Zhou et al. [25] developed a parallel frequent pattern mining algorithm on GPUs, achieving up to 14.857x speedup with 16 threads. Their approach used a compact data structure to store the database directly on the GPU, improving both memory use and speed. These contributions highlight how GPUs make frequent pattern mining faster, more scalable, and highly useful for big data tasks in various fields.

Notably, none of the preceding approaches address the issue of developing a GPU-based algorithm to extract patterns from symbolic sequence data. Also, it can be observed that existing GPU-based pattern mining algorithms are not directly applicable to symbolic sequence databases where the objective is to identify contiguous frequent patterns.

3   Proposed Approach

In this section, we explain the problem of extracting frequent patterns from symbolic sequence data. Next, we briefly describe the CPU-based algorithm and present the proposed approach.

3.1   Problem Definition

The basic model of frequent pattern mining in symbolic sequences is as follows [3]. Let $\Sigma = \{A, B, \ldots, Z\}$ be a set of symbols (or characters)[1], and let $S = \langle s_1, s_2, \ldots, s_n \rangle$, where each $s_i \in \Sigma$, denote a sequence of length n. A pattern P is a contiguous subsequence of S, i.e., $P = \langle s_j, s_{j+1}, \ldots, s_{j+k-1} \rangle$, for $1 \leq j \leq j + k - 1 \leq n$. The number of distinct occurrences of P in S is its *support*, denoted by *support*(P). A pattern is considered **frequent** if *support*$(P) \geq min_sup$, where min_sup is a user-defined threshold. The goal is to identify all such frequent patterns in S.

Example 1. Let $\Sigma = \{A, C, G, T\}$ and $S = $ ACTGCATGCTATGCATGC. Consider the pattern $P = $ AT, which appears at positions 5, 10, and 14 in S, hence *support*$(P) = 3$. If $min_sup = 3$, then P is frequent.

3.2   Overview of CPU-Based Algorithm

We first introduce the key terminologies. Given a pattern P, the *prefix* of P, denoted as prefix(P), is defined as the substring obtained by removing its last character. Similarly, the *postfix* of P, denoted as postfix(P), is the substring obtained by removing its first character. The array Pos(P) refers to the list of starting positions in the sequence S where the pattern P occurs. Furthermore, the *shifted position array* of a pattern P, denoted as Pos$(P) - 1$, is obtained by subtracting 1 from each element in Pos(P), discarding any resulting negative values.

Example 2. Consider the sequence $S = $ "ACTGCATGCTATGCATGC" and the pattern $P = $ "ATG". Then: prefix$(P) = $ "AT", postfix$(P) = $ "TG", Pos$(P) = [5, 10, 14]$, and Pos$(P) - 1 = [4, 9, 13]$.

The approach proposed in [3] focuses on utilizing positional information to iteratively construct longer frequent patterns from a given DNA sequence S of length n. It adopts a level-wise pattern growth framework inspired by the Apriori principle, which eliminates infrequent candidates at each step, thereby improving computational efficiency. A crucial component of the algorithm is

---

[1] We use the terms 'symbols' and 'characters' interchangeably.

Property 1, which enables the computation of the support count and occurrence positions of a candidate pattern of length k by leveraging the position arrays of its $(k-1)$-length prefix and postfix patterns.

Property 1. The positions of a candidate pattern M in the sequence S are determined by the intersection of the positions of its prefix and the shifted positions of its postfix. Specifically, the positions of M are given by:

$$\mathrm{Pos}(M) = \mathrm{Pos}(\mathrm{prefix}(M)) \cap (\mathrm{Pos}(\mathrm{postfix}(M)) - 1)$$

Algorithm 1. CPU-FP()

Require: S: DNA sequence, `min_sup`: minimum support threshold
Ensure: Frequent patterns in S
 1: Initialize `position_table` as an empty hash table
 2: **for** each position i in S **do**
 3:     symbol $\leftarrow S[i]$
 4:     Append i to `position_table[symbol]`
 5: `frequent_patterns_1` $\leftarrow$ symbols with frequency $\geq$ `min_sup`
 6: $k \leftarrow 1$
 7: `position_table`$[1] \leftarrow$ `frequent_patterns_1`
 8: **while** `position_table`[k] is not empty **do**
 9:     **for** each pair (P_1, P_2) in `position_table`[k] **do**
10:         **if** $\mathrm{postfix}(P_1) = \mathrm{prefix}(P_2)$ **then**
11:             new_pattern $\leftarrow P_1 +$ last character of P_2
12:             positions $\leftarrow$ `Pos`$(P1) \cap ($`Pos`$(P2 - 1))$
13:             **if** $\mathrm{count}(\mathrm{positions}) \geq$ `min_sup` **then**
14:                 `position_table`[k+1][new_pattern] $\leftarrow$ positions
15:     $k \leftarrow k + 1$
16: Output all frequent patterns from `position_table`

CPU-Based Approach: Given a symbolic sequence S and *min_sup* threshold, the approach incrementally generates candidate patterns of length $k + 1$ by joining frequent k-length patterns based on the intersection between their postfix and prefix (Refer Algorithm 1).

1. *First iteration*: Construct a position table by scanning S. It maps each 1-length pattern to its corresponding positions in S. Patterns that satisfy the *min_sup* threshold are stored as 1-length patterns in `position_table`[1].
2. *Subsequent iterations (k $\geq$ 1)*: The candidate patterns of length $k + 1$ are generated by joining frequent k-length patterns whose postfix and prefix overlap. The positions of the new candidate pattern are computed by intersecting the positions of the prefix pattern P_1 with the shifted positions of the postfix pattern. A candidate pattern is deemed frequent and added to `position_table`[k+1] if the resulting position set has a cardinality which satisfies the *min_sup* constraint. This step is repeated by incrementing k until no more frequent patterns can be generated.

Example 3. Consider the DNA sequence $S = $ "ACTGCATGCTATGCATGCC", and let the *min_sup* threshold be 2.

1. First iteration (1-length patterns): We compute the starting positions of all 1-length patterns: $Pos(A) = [0, 5, 10, 14]$, $Pos(C) = [1, 4, 8, 13, 17, 18]$, $Pos(G) = [3, 7, 12, 16]$, and $Pos(T) = [2, 6, 9, 11, 15]$.
   All patterns have support ≥ 2 and are retained for the next level.
2. Second iteration (2-length patterns): Two-length candidate patterns are generated by joining 1-length frequent patterns with each other. Two patterns P_1 and P_2 are considered joinable if the postfix of P_1 matches the prefix of P_2 ($postfix(P_1) = prefix(P_2)$). For 1-length patterns, this condition trivially holds, as both the prefix and postfix are empty strings, thus allowing all pairs to be considered. As an example, we examine the join of A with all other 1-length patterns, and utilize Proposition 1 to compute the positions and support count of the resulting joined patterns.
   - AA: $Pos(A) - 1 = [4, 9, 13]$, $Pos(A) = [0, 5, 10, 14] \Rightarrow$ Intersection $= \emptyset \Rightarrow$ Support $= 0$ (Pruned)
   - AT: $Pos(T) - 1 = [1, 5, 8, 10, 14]$, $Pos(A) = [0, 5, 10, 14] \Rightarrow$ Intersection $= [5, 10, 14] \Rightarrow$ Support $= 3$ (Retained)
   - AG: $Pos(G) - 1 = [2, 6, 11, 15]$, Intersection with $Pos(A) = \emptyset \Rightarrow$ (Pruned)
   - AC: $Pos(C) - 1 = [0, 3, 7, 12, 16, 17]$, Intersection with $Pos(A) = [0] \Rightarrow$ Support $= 1$ (Pruned)

   Notably, only the pattern AT survives with a support of 3. A similar procedure is applied to all the join operations in the subsequent iterations.

3.3   Basic Idea

The issue with the CPU-based approach is the computation of set intersection operations, which are currently performed sequentially for each candidate pattern in Algorithm 1 (line 12). The basic idea is to optimize the computation of candidate pattern positions and their corresponding support by parallelizing the set intersection operations in batches during each iteration, using auxiliary memory for an indexed representation of the given sequence. It is observed that there is an opportunity to organize and group the positions of each pattern contiguously within a single array, which allows each thread to independently identify and process the relevant positions for the corresponding prefix and postfix patterns during set intersection, as illustrated in Fig. 2. Specifically, a *Two-Level Indexing Scheme* is employed: the pattern positions are grouped into the array $\bar{S}$, as shown in Fig. 1, while the starting indices for each pattern are maintained in the array $\bar{P}$.

A given sequence S is represented as a triplet $(L, \bar{S}, \bar{P})$, where L denotes the lexicographically ranked list of patterns, with each pattern assigned a unique index starting from 0 in increasing order based on its rank. These indices are then mapped to a second-level index $\bar{P}$, which provides pointers to the corresponding starting positions of pattern occurrences in $\bar{S}$. At iteration k, when mining frequent patterns of length $k + 1$, the pattern positions in $\bar{S}$ are maintained as an ordered sequence, grouped according to the lexicographic order specified in L.

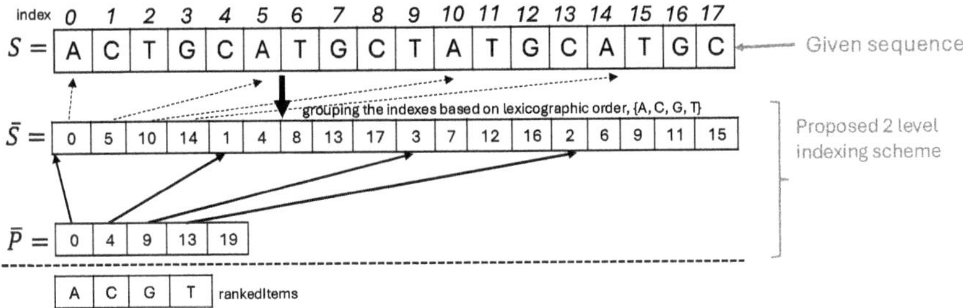

Fig. 1. Proposed Two-Level Indexing Scheme.

Example 4. Consider the sequence S = "ACTGCATGCTATGCATGC". In the proposed scheme, each pattern of length k is assigned a pattern ID based on lexicographic ordering: {A: 0, C: 1, G: 2, T: 3}. The position array $\bar{S}$, which contains the positions of the frequent 1-length patterns is grouped by pattern and ordered lexicographically, is constructed as [0, 5, 10, 14, 1, 4, 8, 13, 17, 3, 7, 12, 16, 2, 6, 9, 11, 15]. Here, all positions of pattern 'A' appear first, followed by those of 'C', 'G', and 'T'. To track the starting index in $\bar{S}$ where the positions of each pattern begin, a prefix-sum array $\bar{P}$, which acts as a second-level index, is maintained. It is defined such that $\bar{P}[i]$ gives the starting index in $\bar{S}$ for the positions corresponding to the pattern with ID i. This array $\bar{P}$ is computed as [0, 4, 9, 13, 18]. Here, $\bar{P}[0] = 0$, and for $i > 0$, $\bar{P}[i] = \bar{P}[i-1] + \text{support_counts}[i-1]$. For example, the pattern 'C' has pattern ID 1, and $\bar{P}[1] = 4$ indicates that the positions of 'C' begin from index 4 in $\bar{S}$. This structure enables constant-time access to the positional indices of each pattern in $\bar{S}$.

Space Complexity: The conventional sequence S has a space complexity of $O(n)$, where $n = |S|$ is the length of the sequence. The proposed representation incurs a space complexity of $O(n) + O(m)$, where m denotes the number of candidate patterns. The worst-case value of m at the k-th iteration, while mining $(k+1)$-length frequent patterns, is given by $(n - k + 1) \cdot |\Sigma|$, where $|\Sigma|$ denotes the number of unique symbols in the symbolic sequence data. For instance, in the case of DNA sequences, $|\Sigma| = 4$. This analysis is based on the observation that, in a sequence of length n, there can be at most $(n - k + 1)$ contiguous patterns of length k. Each of these patterns can potentially be extended by any symbol from the set Σ, resulting in a total of $(n - k + 1) \cdot |\Sigma|$ possible candidate patterns of length $(k + 1)$. The proposed scheme consumes extra space to enable efficient batch processing of k-length patterns in parallel by reducing CPU-GPU communication costs. In the proposed approach, we employ the following data structures in addition to data structure to represent the proposed triplet $(L, \bar{S}, \bar{P})$.

- *patternID*: A mapping that assigns a unique integer to each pattern, incrementing by one for each new pattern. This mapping is stored exclusively in host memory (starts with 0 at each iteration).

– *Result*: A one-dimensional array that stores the intersection results of the prefix pattern positions and the corresponding shifted postfix pattern positions for all the candidate patterns.
– *ResultOffset*: A one-dimensional array that indicates the starting index in the Result array where the intersection results of each candidate pattern should be stored.
– *GlobalPatternTable*: A hash table that stores all the resulting frequent patterns.

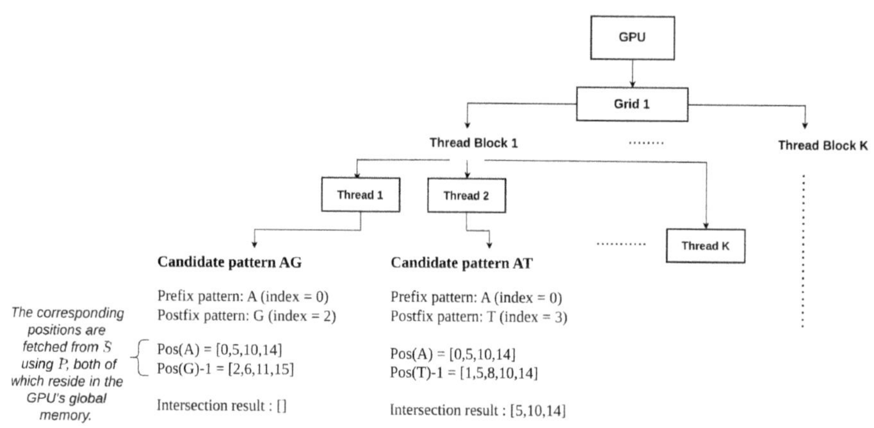

Fig. 2. Workflow of the GPU-based frequent pattern mining algorithm.

3.4   Proposed Approach: gFSP-Miner

The proposed gFSP-Miner(GPU based frequent symbolic pattern miner), similar to the Apriori algorithm, operates iteratively. At each kth iteration, it mines and computes frequent patterns of length k (The Pseudocode is given in Algorithm 2).

(i) First Iteration: In the first iteration, the input sequence S is scanned to identify frequent patterns of length 1, and patterns with support counts below the specified threshold are pruned. The positions of each frequent pattern are then stored in GPU memory in an array denoted as $\bar{S}$, organized such that the positions are grouped according to their corresponding pattern (i.e., the individual symbols). These grouped positions are ordered lexicographically based on the pattern IDs as shown in Fig. 1.

(ii) Subsequent Iterations ($k \geq 1$): For each iteration with pattern length $k \geq 1$, the algorithm generates candidate patterns of length $(k + 1)$ by extending each frequent k-length pattern with all possible symbols in lexicographic order. A candidate pattern is considered valid only if both its k-length prefix and postfix sub-patterns are frequent, in accordance with the Apriori principle. To support efficient GPU-based processing, metadata associated with each candidate pattern is computed on the CPU. This metadata includes the pattern IDs of its prefix and postfix patterns, along with result offsets derived from the minimum of their support counts. These values are organized into three arrays: `prefixPatternIDs`, `postfixPatternIDs`, and `resultOffset`, which are freshly constructed for each iteration and transferred to the GPU for parallel intersection operations.

Subsequently, a parallel *Position Intersection and Result Update* kernel, as described in Algorithm 3, performs two-pointer-based intersections between the position lists of postfix and prefix patterns. It concurrently updates the support counts and records the intersected positions of all candidate patterns. Leveraging the precomputed metadata, the kernel retrieves the pattern IDs and the starting positions of the corresponding prefix and postfix patterns from $\bar{S}$ using the prefix-sum array $\bar{P}$, as illustrated in Example 4. Following the kernel execution, infrequent patterns are discarded, and $\bar{S}$ is compressed by retaining only the positions of the frequent patterns and are stored in *GlobalPatternTable*. This compression is performed in parallel through a dedicated GPU kernel, *Compress Positions*, after which the pattern IDs are reassigned for the new $k + 1$ frequent patterns sequentially starting from zero. Finally, the prefix-sum array $\bar{P}$ is recalculated to reflect the updated support counts and pattern locations, preparing the data structures for the subsequent iteration. This iterative process continues until no new frequent patterns are discovered.

Example 5. Consider the sequence $S = $ "ACTGCATGCTATGCATGC" and a minimum support threshold 2. Each distinct 1-length pattern is assigned a pattern ID based on lexicographic order: {`A: 0`, `C: 1`, `G: 2`, `T: 3`}. In the first iteration, the position array $\bar{S}$, which groups positions of each pattern in lexicographic order, is computed as [0, 5, 10, 14, 1, 4, 8, 13, 17, 3, 7, 12, 16, 2, 6, 9, 11, 15]. The prefix-sum array $\bar{P}$, used to locate the start index of each pattern in $\bar{S}$ is computed as [0, 4, 9, 13, 18]. In the second iteration, the candidate 2-length patterns are generated by joining frequent 1-length patterns. With the same lexicographic ordering {`A: 0`, `C: 1`, `G: 2`, `T: 3`}, the candidate set is [AA, AC, AG, AT, ...]. Metadata arrays for these candidates are as follows: `prefixPatternIDs` = [0, 0, 0, 0, ...], `postfixPatternIDs` = [0, 1, 2, 3, ...] and `resultOffset` = [0, 4, 8, 12, ...]. In the `PositionIntersectUpdateResult` kernel, each thread uses the metadata to compute intersections for its assigned candidate pattern. For example, for the pattern `AC`: (i) The prefix and postfix pattern IDs are 0 and 1, respectively. (ii) The start indices for both patterns in

$\bar{S}$ are obtained from $\bar{P}$: $\bar{P}[0] = 0$ and $\bar{P}[1] = 4$ respectively. A two-pointer algorithm as depicted in (Algorithm 3) is used to compute the intersection, yielding positions [0] which are not frequent and pruned. Other threads compute intersections for the remaining candidate patterns in parallel. The updated $\bar{S}$ and $\bar{P}$ arrays after the second iteration are shown in Fig. 3.

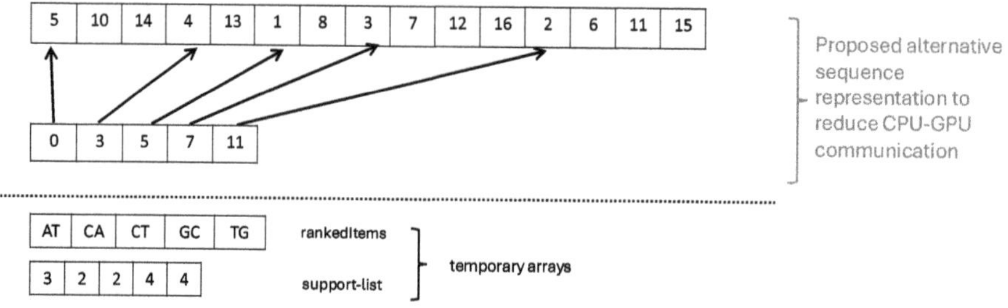

Fig. 3. Updated $\bar{S}$ and $\bar{P}$ of the proposed triplet.

Algorithm 2. gFSP-Miner()

Input: Sequence S, Minimum Support min_sup
Output: Frequent patterns
1: **Initialize:**
2: $patternId \leftarrow$ empty hash map
3: $kLength_frequent_patterns \leftarrow$ empty list
4: $\bar{S} \leftarrow$ initial positions of all symbols in S
5: $\bar{P} \leftarrow$ empty prefix sum array
6: **Compute support counts for all 1-length patterns in S:**
7: Prune infrequent patterns with support $< min_sup$
8: Update $patternId$, $\bar{S}$, and $\bar{P}$
9: **while** frequent patterns exist **do**
10:     **Generate all $(k+1)$-length candidate patterns:**
11:     **for** each frequent k-length pattern **do**
12:         Extend by appending symbols
13:         Check if postfix pattern is frequent
14:     Position Intersection and Result Update Kernel()
15:     **Update frequent patterns:**
16:     Prune infrequent patterns
17:     Update $\bar{P}$ and compact $\bar{S}$ by removing unused spaces in the result computed in Position Intersection and Result Update Kernel, and also removing the infrequent pattern positions
18:     Set $k = k + 1$
19: **Return** all frequent patterns

Algorithm 3. Position Intersection and Result Update Kernel

Input: Prefix and Postfix Pattern IDs, Position Arrays $\bar{S}$, $\bar{P}$, Offset Array, min_sup
Output: Updated Support Counts and Result Array
 1: **Parallel Execution for Each Candidate Pattern** i:
 2: Retrieve start and end indices for prefix and postfix patterns from $\bar{P}$
 3: Initialize $support \leftarrow 0$
 4: **while** prefix positions and shifted postfix positions are not exhausted **do**
 5:     **if** Prefix position matches shifted Postfix position **then**
 6:         Record position in result array
 7:         Increment $support$
 8:         Move both prefix and postfix pointers forward
 9:     **else if** Prefix position is greater **then**
10:         Move postfix pointer forward
11:     **else**
12:         Move prefix pointer forward
13: Update $support_counts[i]$ with $support$

---

4 Performance Evaluation

This section presents the performance evaluation of the proposed algorithm, conducted using three real-world genomic datasets from the NCBI Nucleotide database [13]. These datasets represent distinct bacterial strains, each from a different taxonomic group, with varying genome sizes. The first dataset (**D1**) pertains to *Domibacillus* sp. DTU_2020_1001157, a Gram-positive bacterium with a sequence length of 1,711,846 base pairs. This genus holds significant potential in biotechnology, particularly in areas related to metabolic processes and stress response mechanisms. The second dataset (**D2**) is derived from *Vibrio cholerae* strain VCKR3, with a sequence length of 737,849 base pairs. This Gram-negative bacterium, associated with cholera outbreaks, provides crucial genomic insights into virulence factors, genetic diversity, and antimicrobial resistance. The third dataset (**D3**) represents a genomic sequence from *Escherichia coli* strain EC4439, with a sequence length of 201,548 base pairs. *E. coli*, a highly adaptable species, serves as an important model for studying genetic adaptation, metabolic pathways, and antibiotic resistance mechanisms. Collectively, these datasets are invaluable resources for comparative genomics, offering a broad understanding of bacterial genome diversity and the underlying mechanisms of bacterial adaptation.

Table 1. Details of the Datasets

Dataset	Species	Sequence Length (base pairs)
D1	*Domibacillus* sp. DTU_2020_10011576	17,11,846
D2	*Vibrio cholerae* strain VCKR3	7,37,849
D3	*Escherichia coli* strain EC4439	2,01,548

All algorithms were implemented in C++ using GCC 7.5.0, targeting the `x86_64-linux-gnu` architecture. Experiments were conducted on a CentOS 7 server equipped with an Intel Xeon E5-2640 v4 CPU (2.40 GHz), 32 GB RAM, and an NVIDIA GeForce RTX 2080 Ti GPU (4,352 CUDA cores, 11 GB GDDR6, 616 GB/s bandwidth). The source code is publicly available[2].

4.1 Approaches Implemented

We evaluated our proposed GPU-based approach against two CPU-based algorithms: Tanvee et al. [18] (CPU-FP1) and Deng et al. [3] (CPU-FP2). Both algorithms utilize positional information for support computation. Notably, Tanvee et al.'s method has been shown to outperform other position-based and non-positional techniques, such as macoVSpan [14], and thus serves as our primary point of comparison.

A naïve GPU-based baseline was also implemented for comparison. (i) In the first iteration, the naïve GPU algorithm scans the input sequence in parallel using a CUDA kernel to identify frequent 1-length patterns (individual symbols) that satisfy the minimum support threshold. Each frequent symbol is assigned a unique ID based on lexicographic order, and the sequence is transformed into an array of identifiers, referred to as `PatternInfo`. (ii) In subsequent iterations ($k \geq 1$), candidate patterns of length $k+1$ are generated by extending frequent k-length patterns, with validation based on the Apriori principle—only candidates with both frequent prefix and postfix sub-patterns are considered valid. The occurrence of a candidate pattern at position i is identified by checking whether `PatternInfo[`i`]` and `PatternInfo[`$i + 1$`]` correspond to the IDs of the prefix and postfix patterns, respectively. This process is performed in parallel across the sequence by multiple GPU threads. The above process continues until no new frequent patterns are found.

Example 6. Consider the sequence S = "ACTGCATGC" with `minSup` = 2. The frequent symbols {A, C, G, T} are assigned IDs {A:1, C:2, G:3, T:4} based on lexicographic order. The transformed sequence becomes [1, 2, 4, 3, 2, 1, 4, 3, 2]. The pattern "AT" (represented by prefix pattern ID 1 and postfix pattern ID 4) occurs at position 5 in the sequence.

The primary limitation of the naïve approach lies in its requirement to invoke a separate GPU kernel for each individual candidate pattern, leading to excessive kernel launches and significant communication overhead between the GPU and CPU.

4.2 Experimental Results

Runtime vs. Minimum Support: Figures 4(a) to 4(c) illustrate the runtime performance across datasets D1, D2, and D3, where the dataset sizes follow the

---

[2] https://github.com/pshiridikumar/Symbolic_sequence_mining.

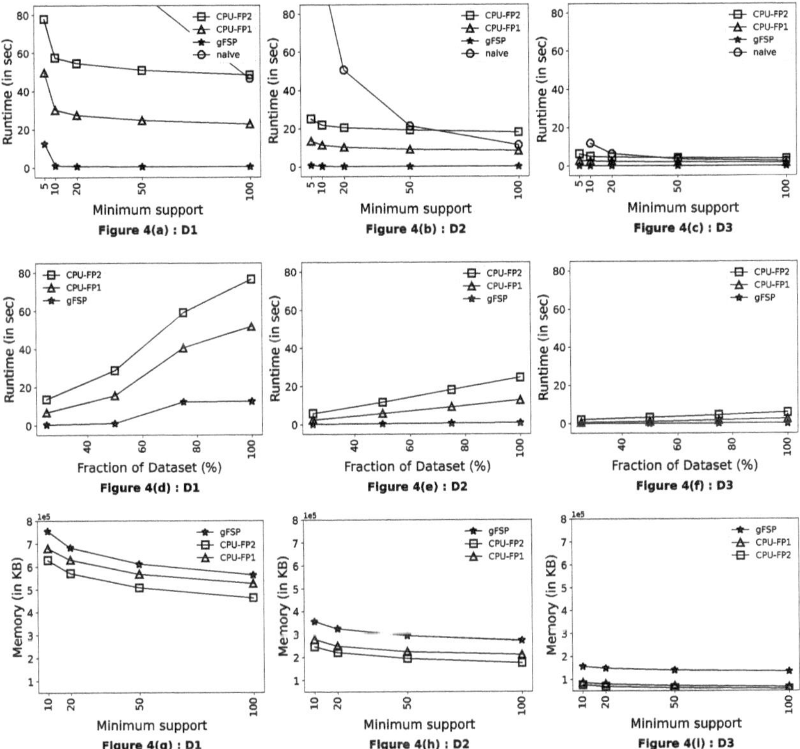

Fig. 4. Performance comparison of gFSP, naïve, and existing algorithms across Runtime, Scalability, and Memory utilization experiments.

order D1 > D2 > D3 (refer to Table 1 for details). These results correspond to minimum support thresholds ranging from 5 to 100. (**i**) As expected, gFSP consistently achieves the lowest runtime due to its optimized GPU utilization and batch processing of candidate pattern support count computations. Notably, the performance of existing CPU-based methods is significantly worse on dataset D1 (Fig. 4(a)) due to sequential execution. While performance improves slightly for D2 and D3 (Figs. 4(b) and 4(c)), which are smaller in size compared to D1, it still falls short of the performance achieved by gFSP. The naïve CUDA baseline exhibits an exponential increase in runtime as the support threshold decreases across all three datasets. This is primarily due to the frequent invocation of GPU kernels when handling a large number of candidate patterns, highlighting the inefficiency of this approach in processing large candidate sets. (**ii**) At higher support thresholds, the naïve CUDA method outperforms existing CPU-based methodologies, indicating its relative effectiveness and better GPU utilization under relaxed mining conditions.

Scalability with Dataset Size: Figures 4(d) to 4(f) present runtime scalability on datasets of increasing sizes (25%, 50%, 75%, 100%) at a fixed minimum support threshold of 5 on Dataset D1. **(i)** *gFSP* maintains near-linear scalability across all datasets, showcasing its robustness against data volume growth. **(ii)** The runtime of existing methodologies increases sharply with dataset size, in contrast to the stable performance of *gFSP*. **(iii)** These results affirm the suitability of *gFSP* for large-scale pattern mining, offering

Memory Consumption Analysis: Figures 4(g) to 4(i) depict the USS (Unique Set Size) memory consumption of *gFSP* and existing CPU-based approaches across datasets D1, D2, and D3 at varying minimum support thresholds. *gFSP* exhibits higher memory usage due to its two-level indexed data representation, auxiliary structures, and storage requirements for the final set of frequent patterns. While this results in increased memory overhead, the significant performance gains observed in Figs. 4(a) to 4(c) justify this trade-off.

The detailed results of these experiments are presented in Fig. 4.

Table 2. Number of GPU Kernel Calls for the naïve GPU Baseline and the Proposed gFSP Approach across Different Datasets and Minimum Support Thresholds

Min Sup	D1		D2		D3	
	naïve	gFSP	naïve	gFSP	naïve	gFSP
20	214118	22	92372	18	24977	16
50	84620	20	37232	16	10290	14
100	42264	18	18543	14	4887	12
200	21212	18	9248	14	2644	12

Table 2 presents the number of GPU kernel calls made by the naïve and proposed gFSP algorithms across different datasets (D1, D2, and D3) and varying minimum support thresholds. As shown, the naïve method requires a significantly larger number of kernel invocations due to its lack of optimization in managing candidate pattern positions. In contrast, gFSP consistently reduces the number of GPU kernel calls across all datasets and support levels, highlighting its efficiency in minimizing CPU-GPU communication overhead through batch processing and optimized position maintenance.

5   Conclusions and Future Work

This paper presented an efficient GPU-based algorithm for mining frequent patterns in large symbolic sequence databases using a novel Two-Level Indexing Scheme. By utilizing extra auxiliary memory and GPU parallelism, the approach achieves significant runtime improvements—up to 4× speedup for low support values and more than 20× for higher thresholds. Future work includes adapting this scheme to distributed paradigms such as MapReduce and MPI, optimizing memory usage.

Acknowledgment. This research is supported by iHub-Data, IIIT Hyderabad, Telangana, India.

References

1. Aggarwal, C.C., Li, Y., Wang, J., Wang, J.: Frequent pattern mining with uncertain data. In: Proceedings of the 15th ACM SIGKDD International Conference on Knowledge Discovery and Data Mining, pp. 29–38 (2009)
2. Agrawal, R., Srikant, R.: Fast algorithms for mining association rules. In: Proceedings of the 20th International Conference on Very Large Data Bases (VLDB), pp. 487–499. Morgan Kaufmann Publishers Inc. (1994)
3. Deng, N., Chen, X., Li, D., Xiong, C.: Frequent patterns mining in DNA sequence. IEEE Access **7**, 108400–108410 (2019)
4. Erlandsson, F., Bródka, P., Borg, A., Johnson, H.: Finding influential users in social media using association rule learning. Entropy **18**(5), 164 (2016)
5. Eskin, E., Lee, W., Stolfo, S.: Anomaly detection using self-similarity estimation. In: Proceedings of the 7th International Conference on Data Mining and Knowledge Discovery (KDD), pp. 17–22. ACM (2000)
6. Gan, W., Lin, J.C.W., Fournier-Viger, P., Chao, H.C., Tseng, V.S., Philip, S.Y.: A survey of utility-oriented pattern mining. IEEE Trans. Knowl. Data Eng. **33**(4), 1306–1327 (2019)
7. Jian, L., Wang, C., Liu, Y., Liang, S., Yi, W., Shi, Y.: Parallel data mining techniques on graphics processing unit with compute unified device architecture (CUDA). J. Supercomput. **64**, 942–967 (2013)
8. Jin, X., et al.: Similarity/dissimilarity calculation methods of DNA sequences: a survey. J. Mol. Graph. Model. **76**, 342–355 (2017)
9. Kiran, R.U., Watanobe, Y., Chaudhury, B., Zettsu, K., Toyoda, M., Kitsuregawa, M.: Discovering maximal periodic-frequent patterns in very large temporal databases. In: 2020 IEEE 7th International Conference on Data Science and Advanced Analytics (DSAA), pp. 11–20. IEEE (2020)
10. Lin, J., Keogh, E., Lonardi, S., Chiu, B.: A symbolic representation of time series, with implications for streaming algorithms. In: Proceedings of the 8th ACM SIGMOD Workshop on Research Issues in Data Mining and Knowledge Discovery (DMKD), pp. 2–11. ACM (2003)
11. Liu, Y., Ma, Y., Hildebrandt, M., Joblin, M., Tresp, V.: Tlogic: temporal logical rules for explainable link forecasting on temporal knowledge graphs. In: Proceedings of the AAAI Conference on Artificial Intelligence, vol. 36, pp. 4120–4127 (2022)
12. Mejía, M., Tran, A., Cohen, P.: Human activity recognition using symbolic sequences. ARPN J. Eng. Appl. Sci. **11**(21), 12571–12581 (2016)
13. National Center for Biotechnology Information (NCBI): NCBI - National Center for Biotechnology Information (2024). https://www.ncbi.nlm.nih.gov/. Accessed 22 Nov 2024
14. Pan, J., Wang, P., Wang, W., Shi, B., Yang, G.: Efficient algorithms for mining maximal frequent concatenate sequences in biological datasets. In: Proceedings of the 5th International Conference on Computer and Information Technology, pp. 98–104. IEEE (2005)
15. Pei, J., et al.: Prefixspan: mining sequential patterns efficiently by prefix-projected pattern growth. In: Proceedings of the 17th International Conference on Data Engineering, pp. 215–224. IEEE (2001)

16. Qi, X., Wu, Q., Zhang, Y., Fuller, E., Zhang, C.Q.: A novel model for DNA sequence similarity analysis based on graph theory. Evol. Bioinforma. **7**, 149–158 (2011)
17. Srivastava, J., Cooley, R., Deshpande, M., Tan, P.N.: Web usage mining: discovery and applications of usage patterns from web data. ACM SIGKDD Explor. Newsl. **1**(2), 12–23 (2000)
18. Tanvee, M.M., Kabeer, S.J., Chowdhury, T.M., Sarja, A.A., Shuvo, M.T.: Mining maximal adjacent frequent patterns from DNA sequences using location information. Int. J. Comput. Appl. **76**(15) (2013)
19. Touchon, M., Rocha, E.P.C.: Causes of insertion sequences abundance in prokaryotic genomes. Mol. Biol. Evol. **24**(4), 969–981 (2007)
20. Ünvan, Y.A.: Market basket analysis with association rules. Commun. Stat.-Theory Methods **50**(7), 1615–1628 (2021)
21. Wang, F., Yuan, B.: Parallel frequent pattern mining without candidate generation on GPUs. In: 2014 IEEE International Conference on Data Mining Workshop, pp. 1046–1052. IEEE (2014)
22. Watson, J.D., Baker, T.A., Bell, S.P., Gann, A., Levine, M., Losick, R.: Molecular Biology of the Gene, 7th edn. Pearson (2013)
23. Xie, X., Guan, J., Zhou, S.: Similarity evaluation of DNA sequences based on frequent patterns and entropy. BMC Genom. **16**, 1–10 (2015)
24. Zhang, F., Zhang, Y., Bakos, J.: Gpapriori: GPU-accelerated frequent itemset mining. In: Proceedings of the 2011 IEEE International Conference on Cluster Computing, pp. 590–594. IEEE (2011)
25. Zhou, J., Yu, K.M., Wu, B.C.: Parallel frequent patterns mining algorithm on GPU. In: Proceedings of the 2010 IEEE International Conference on Systems, Man, and Cybernetics, pp. 435–440. IEEE (2010)

Robust Spoken Language Identification for Indian Languages Using Phonetic Representations

Shubham Sharma, Shilpa Chandra, and Padmanabhan Rajan[✉]

School of Computing and Electrical Engineering, IIT Mandi, Mandi, India
{s23044,s22004}@students.iitmandi.ac.in, padman@iitmandi.ac.in

Abstract. Automatic language identification (LID) systems typically perform well when dealing with high-resource languages, but their performance declines with low-resource languages. Moreover, when training data consists mainly of clean, read speech and test data includes real-world situations such as spontaneous speech from noisy or uncontrolled environments, the system's performance suffers due to domain shift. This paper addresses the challenge of domain shift, focusing on 12 low-resource Indian languages, many of which belong to the same language family. We propose two frameworks that integrate phonotactic features with advanced deep learning frameworks to enhance LID performance under domain shift conditions. By leveraging self-supervised learning techniques, we extract phoneme posterior probabilities, improving the representation of acoustic features beyond low-level features. Our experiments demonstrate the effectiveness of this approach.

1 Introduction

In recent years, advances in deep learning have resulted in advances in various aspects of speech technology. The use of deep learning pipelines such as convolutional neural networks (CNNs), self-attention modules, and more recently, self-supervised learning (SSL) have resulted in reduced error rates in various applications including automatic speech recognition (ASR), speaker and language identification and keyword spotting.

Agents such as voice assistants typically show excellent performance when native speakers speak in high-resource languages (such as American English), due to the abundance of large labeled datasets. In multilingual societies, speech interfaces need to use an automatic spoken language identification (LID) stage to effectively process input speech. Building LID systems in low-resource settings is challenging due to limited availability of data. Moreover, many languages may be derived from the same family and may be closely related, increasing the chances of misclassifications. An additional difficulty is to develop acoustic models that generalize well to unseen conditions. For example, if the training data is predominately read speech in clean conditions (such as news broadcasts) and the test data is spontaneous speech in uncontrolled conditions (such as

R. Gupta et al. (Eds.): BDA 2025, LNCS 16041, pp. 127–141, 2026.
https://doi.org/10.1007/978-3-032-15134-6_9

from social media or from noisy environments), it is likely that the LID system will perform poorly, which in turn results in poor performance in downstream applications. This situation is an example of *domain shift*, and can be attributed to the train-test mismatch of various factors including variations in speaking style, acoustic environment, speakers and dialects.

Many recent LID systems rely on feature representations derived form hand-crafted features, such as Mel frequency cepstral coefficients (MFCC) or Mel filterbank energies. These basic features in-turn are passed though learning frameworks such as attention modules and time-delay neural networks resulting in effective acoustic representations. The fundamental hypothesis we make in this paper is that low-level features such as Mel-based representations are prone to domain shift. Thus, the objective is to come up with representations which are more robust to such changes. In particular, for LID, we focus on *phonotactics*, which are the rules governing the sequence of admissible sound units (or phonemes) of a language. To decompose a speech utterance into its constituent phonemes, we require a phonemes recognizer. For this, we utilize a recently proposed model based on self-supervised learning.

India has the fourth-highest number of languages in the world, and has 22 official languages [1]. Two major divisions of Indian languages are the Indo-Aryan family and the Dravidian family. As a result of geographic and cultural proximity, languages from the same family have similar characteristics, both from the linguistic and perceptual points-of-view. Moreover, the set of phonemes is common for majority of Indian languages. Our experiments on langauge identification using phonotactics combined with powerful learning frameworks reveal their effectiveness in classifying 12 Indian languages, especially in conditions with domain shift.

Contributions: The contributions of this paper are summarized below:

- We use features derived from phonotactics combined with advanced architectures such as conformers and knowledge distillation, replacing low-level features such as MFCCs.
- We conduct extensive experiments on 12 Indian languages in both in-domain and out-of-domain conditions to evaluate the proposed models. We also compare performance with various baseline systems.

2   Related Work

In recent years, several works have addressed automatic language identification (LID). These include using time-delay neural networks (TDNN) that apply statistics pooling, resulting in the x-vector representations and their variants [2]. Several works have also explored integrating LID into automatic speech recognition (ASR) systems. Lightweight LID modules embedded in streaming ASR pipelines have achieved high prediction accuracy with minimal latency [3], achieving high accuracy across diverse language groups.

Other approaches have used bidirectional long short-term memory (BLSTM) networks to extract bottleneck features [4], which are then classified using support vector machines (SVMs) [5]. In speech processing, frame-level analysis divides an audio signal into short frames, usually 20–30 milliseconds long. Features are extracted for each frame, capturing fine-grained changes in the signal. Speech remains nearly stationary over such short periods, making frame-level features a meaningful representation. The MERLIon CCS Challenge baseline [6] used a statistics pooling layer to compress frame-level outputs into a fixed-size vector for a bilingual task: distinguishing English from Mandarin. The model was inspired from [7] in which a conformer based LID system was used.

In [8], the authors tackle the problem of language identification (LID) for low-resource languages using a convolutional and long short-term Memory (CLSTM) neural network. This architecture improves feature extraction and captures long-range temporal patterns. They introduce a two-dimensional attention mechanism that applies attention across both time and frequency, helping the model focus on more informative parts of the input. The CLSTM model outperforms traditional x-vector systems, particularly in identifying low-resource languages. Yet another work introduces uses novel loss functions [9] to improve performance in domain shift scenarios. On similar lines, [10] used a transformer-based architecture with adversarial training and hierarchical attention to handle domain shift.

Liu et al. [11] trains an x-vector model on mel-filterbank features with self-attention. Liu et al. [12] enhances this with a dual-mode approach, using knowledge distillation. The full mode (teacher) guides the short mode (student) through a combined loss of cross-entropy and a distillation loss, enabling the student to capture the rich patterns learned by the teacher.

Another popular approach to learn speech representations is to utilise self-supervised learning (SSL.) These are predominately of two types. The first uses contrastive objectives, aligning embeddings of the same utterance with different augmentations while separating embeddings from different utterances. Wav2vec 2 [13] and its variants use this approach. The second employs predictive objectives, masking parts of the input and training the model to reconstruct discrete or continuous targets. Examples include w2v-BERT [14], HuBERT [15], and BEST-RQ [16]. SSL-based approaches give rise to sophisticated representations which can be used in various tasks, including LID.

More recently, wav2vec2-based architectures have been fine-tuned for phonological feature learning in low-resource settings [17]. They utilize a combined architecture that leverages both acoustic and phonotactic features to process short segments of speech, which is critical given the rapid switching between languages in conversations.

Finally, off-the-shelf pretrained models like the VoxLingua107 ECAPA-TDNN [18, 19] and OpenAI's Whisper-base [20] offer powerful foundations for language identification tasks, leveraging their training on extensive and diverse datasets. VoxLingua107 ECAPA-TDNN, trained on the VoxLingua107 dataset spanning 107 languages, excels in identifying the dominant language in speech recordings. Similarly, Whisper-base, designed for automatic speech recognition (ASR), provides robust transcription capabilities that can be adapted for language detection by analyzing the language patterns within its outputs. Both models eliminate the need for large-scale training from scratch, enabling researchers to fine-tune these systems for specific multilingual or code-switching scenarios. By leveraging their pretrained architectures and vast linguistic knowledge, these models offer scalable and efficient solutions for language identification in diverse real-world applications.

3   Proposed Approach

Phonotactics regulate the sequence of phonemes in a speech utterance. We utilise the pre-trained *wav2vec2phone* model [21] to convert a speech utterance into a sequence of phoneme posterior probabilities. The sequence of phoneme posteriors encode the temporal dynamics between the various phonemes in a language. To use this information to effectively discriminate between languages, we use two pipelines:

1. Multiple conformer layers driven by a cross-entropy objective derived from [6] (henceforth termed ConfPhoneme)
2. x-vectors with self-attention, with combined knowledge distillation and cross entropy objectives derived from [12] (henceforth termed XvecPhoneme)

3.1   Phoneme Posterior Feature

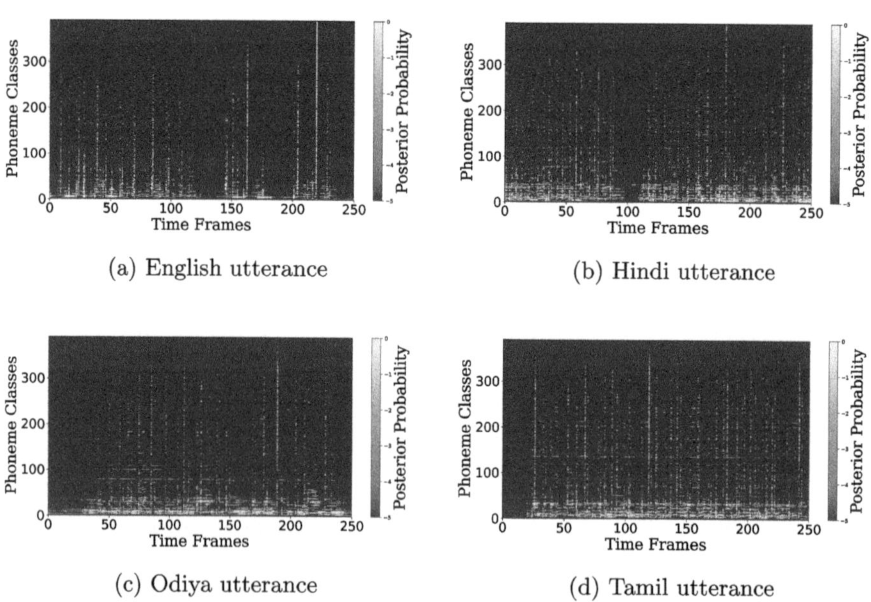

(a) English utterance   (b) Hindi utterance

(c) Odiya utterance   (d) Tamil utterance

Fig. 1. Phoneme posterior probabilities over time ($392 \times T$) for utterances from four languages: English, Hindi, Odiya, and Tamil. T depends on the duration of the utterance. For better visualization, logarithmic compression is used in these figures.

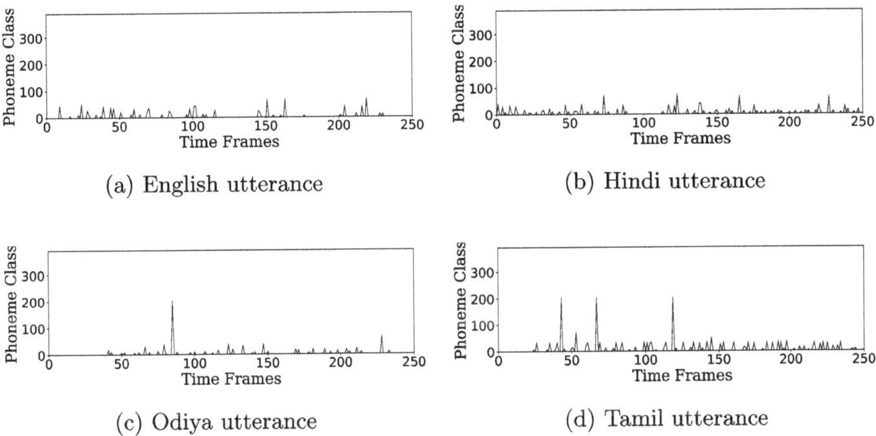

(a) English utterance

(b) Hindi utterance

(c) Odiya utterance

(d) Tamil utterance

Fig. 2. Maximum phoneme probabilities over time for four languages: English, Hindi, Odiya, Tamil.

We use a self-supervised learning (SSL) model, based on wav2vec 2.0, as the foundation for phoneme extraction. Wav2vec 2.0 is pre-trained on raw 16 kHz audio using contrastive learning to learn meaningful speech representations without transcripts. It consists of a convolutional feature encoder that downsamples the input audio, followed by a transformer that models long-range dependencies. The model is trained on multilingual data, making it suitable for cross-lingual speech processing. Using the wav2vec 2.0 model, the phoneme extractor wav2vec2phoneme is proposed in [22], which produces 392-dimensional phoneme-level outputs. It is fine-tuned using *eSpeak* [22] symbols[1], which allows it to handle languages it hasn't seen before. The model uses articulatory features—such as place and manner of articulation—to align phonemes across different languages, in addition to a suitable decoder and language model.

Given an input speech utterance, wav2vec2phone outputs a $392 \times T$ matrix, where each column gives frame-level posterior probabilities of the 392 phoneme classes. The matrix captures detailed phonetic information that helps in capturing phonotactics of the language and are used as frame-level features.

Figure 1 shows the phoneme posterior features for individual utterances from four languages. The relationship between neighboring phonemes are captured using these features and they contain discriminative information across languages. Figure 2 gives the maximum phone (via argmax) for each of the T columns for the utterances above. It can be seen that certain phonemes are more prevalent in some languages, which can help in discriminating them.

---

[1] eSpeak is a compact open source software speech synthesizer for English and other languages.

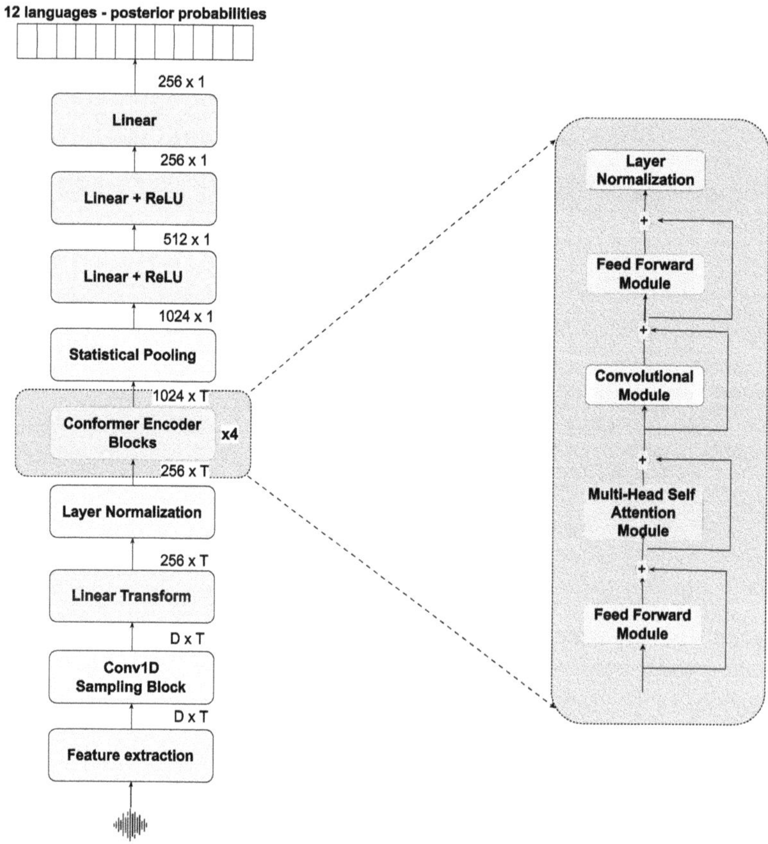

Fig. 3. Flow diagram of conformer model architecture, with the proposed Conf-Phoneme model utilising phoneme posterior features (yellow box). (Color figure online)

3.2   ConfPhoneme Model Architecture

The proposed ConfPhoneme architecture is based on the conformer model [6], which was previously used for LID. This model combines convolutional layers and transformer self-attention. The convolutional layers capture local patterns, while the self-attention layers model global context across the audio. To generate an utterance-level output (a single vector) from frame-level outputs, pooling is used. We apply a statistics pooling layer to compress the frame-level outputs into a fixed-size vector. The model is trained using cross-entropy loss to predict language labels.

Figure 3 shows the architecture of the conformer-based model. D represents the feature dimension of the input, with $D = 392$ for the proposed ConfPhoneme model.

3.3 XvecPhoneme Architecture

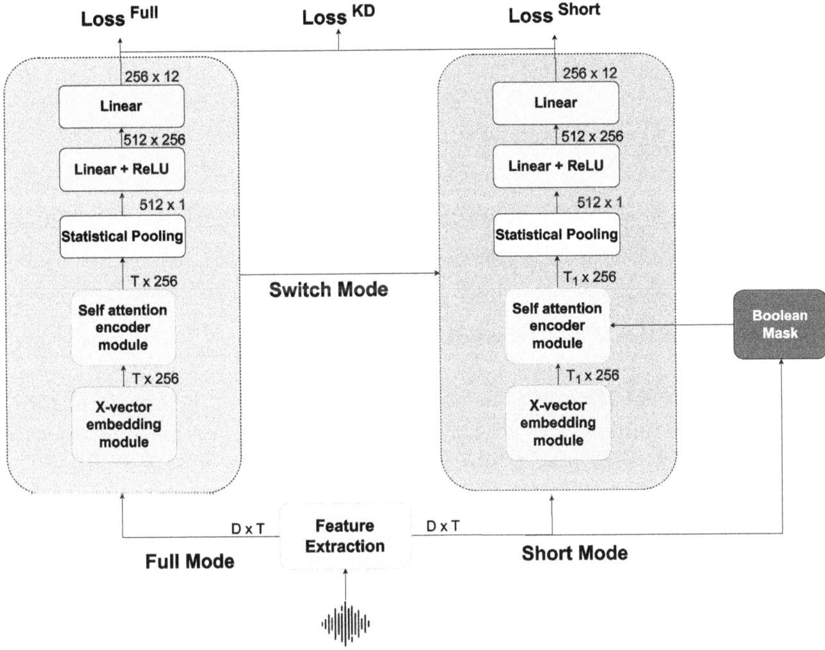

Fig. 4. XvecPhoneme architecture, which uses phoneme posterior features (yellow box). (Color figure online)

The XvecPhoneme architecture, illustrated in Fig. 4, builds upon [12], which targets LID for both long and short utterances. This is achieved through a dual-mode framework that integrates cross-entropy and knowledge-distillation (KD) objectives. The system operates in two modes—*full mode* and *short mode*— with both pathways sharing the same model weights. In full mode, the model processes the entire utterance, allowing it to capture broad language patterns such as prosody, phonotactics, and syntactic structures. In contrast, in short mode, the model processes random 1–2-second segments selected using a boolean mask, focusing learning on more limited and variable input spans.

In XvecPhoneme, posterior probabilities are used as input features, extracted from wav2vec2phoneme. As shown in Fig. 4, these features pass through an x-vector embedding module and a self-attention encoder module, which capture frame-level phonetic details and temporal dependencies, respectively. A statistical pooling layer aggregates the sequence into a fixed-length vector, which is then passed through two linear layers (with ReLU activations) for language classification into 12 categories.

During training, three losses are computed: the cross-entropy loss on the full-mode output L^{Full}, the cross-entropy loss on the short-mode output L^{Short}, and the KD loss

$$L^{\text{KD}} = \text{KL}\big(\log P(X),\, P(X_s)\big),$$

where X denotes the full utterance input and X_s its randomly masked short-segment input, and $P(X)$ and $P(X_s)$ are the softmaxed outputs of the full and short branches, respectively. These are combined into the overall objective

$$L^{\text{Dual}} = \alpha\, L^{\text{Full}} + \beta\, L^{\text{Short}} + \big(1 - \alpha - \beta\big) L^{\text{KD}},$$

where α and β are hyperparameters that weigh the contributions of the full-mode and short-mode cross-entropy losses, respectively.

4   Experiments and Results

This section presents the experimental setup and results. It first describes the dataset used for training and evaluation, followed by the experimental setups. The model architecture and training settings are then detailed. Finally, the results are presented and analyzed.

4.1   Indian Languages Data

We worked with 12 Indian languages: Assamese, Bengali, English, Gujarati, Hindi, Kannada, Malayalam, Marathi, Odia, Punjabi, Tamil, and Telugu. These languages come from different regions but often share similar sounds [1]. This made them a strong choice for testing how well our model can separate closely related languages, especially when training data is limited.

We used two Indian language datasets: the Open-Speech EkStep dataset (also commonly referred to as Open-Speech EkStep-DS is split into Ekstep-DS-1 and Ekstep-DS-2) and the IIT-Mandi dataset (also commonly referred to as IIT-Mandi-DS is split into IIT-Mandi-1 and IIT-Mandi-2). For Ekstep-DS-1, Ekstep-DS-2, and IIT-Mandi-1, 80% of each data was used for training, and the remaining 20% of each was used for testing. The training splits from each of these three datasets were pooled together to form the combined training set. Similarly, the 20% test splits from each of these datasets were pooled to create the *seen test* set, meaning the test data comes from the same domain as the training data. In contrast, the IIT-Mandi-2 dataset, which was not used at all during training, served as the *unseen test* set. Figure 5 gives an overview.

Ekstep-DS, the Ekstep Foundation released this open-source dataset to support the Vakyansh ASR toolkit [23]. It includes audio clips collected from various online sources [24]. For our experiments, two subsets were used: Ekstep-Multi-Domain (Ekstep-DS-1) and Ekstep-TVnews (Ekstep-DS-2), designed to test model performance under different training conditions. Ekstep-DS-1 contains audio from TV shows, and topics like sports, technology, religion, and

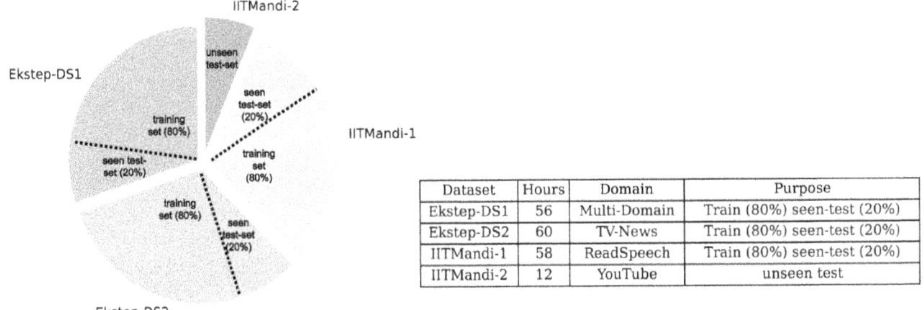

Dataset	Hours	Domain	Purpose
Ekstep-DS1	56	Multi-Domain	Train (80%) seen-test (20%)
Ekstep-DS2	60	TV-News	Train (80%) seen-test (20%)
IITMandi-1	58	ReadSpeech	Train (80%) seen-test (20%)
IITMandi-2	12	YouTube	unseen test

Fig. 5. Indian Language Dataset Composition.

education. Ekstep-DS-2 focuses on TV news and represents a limited-resource setting to test models with less training data.

IIT-Mandi DS, The IIT-Mandi-1 dataset (Readspeech-DS) [9] comprises approximately five hours of studio-recorded speech per language, sourced from All India Radio (AIR) broadcasts and featuring at least 15 speakers. The IIT-Mandi-2 dataset (YouTube-DS) [24] includes around one hour of speech per language, collected from YouTube videos across various domains (e.g., teaching, interviews, vlogs) with diverse recording devices and real-world background conditions. It is used as the *unseen test* set.

4.2   Experiments

We perform two sets of experiments. In the first set of experiments, we train three baseline systems on the seen data. These three baselines are referred to as Bxvec, BmfccConf, and Bw2v2Conf in this paper. They are described below.

Baseline 1: It is an x-vector-based LID system [10] that uses time-delay neural network (TDNN) layers to extract temporal features, followed by a statistics pooling layer that computes the mean and standard deviation. These are concatenated into an x-vector, which is classified to predict the language label. It uses 80-dimensional bottleneck features (BNFs) from the BUT/Phonexia Extractor [4], which are computed from 25 ms speech frames with a 10 ms shift. Each BNF covers 31 frames (approximately 325 ms) of speech, forming a sequence representing the entire utterance. This baseline will be referred to as Bxvec throughout the rest of the paper.

Baseline 2: This is a conformer based baseline as shown in Fig. 3. In this baseline, raw MFCCs combined with their first- and second-order derivatives are used for feature extraction, resulting in 39-dimensional features that serve as input D in Fig. 3. This baseline will be referred to as BmfccConf throughout the rest of the paper.

Baseline 3: Similarly, this is also a conformer based baseline as shown in Fig. 3. In this baseline, 768-dimensional embeddings are generated by wav2vec

2.0 model, serving as the input D in Fig. 3. This baseline will be referred to as Bw2v2Conf throughout the rest of the paper.

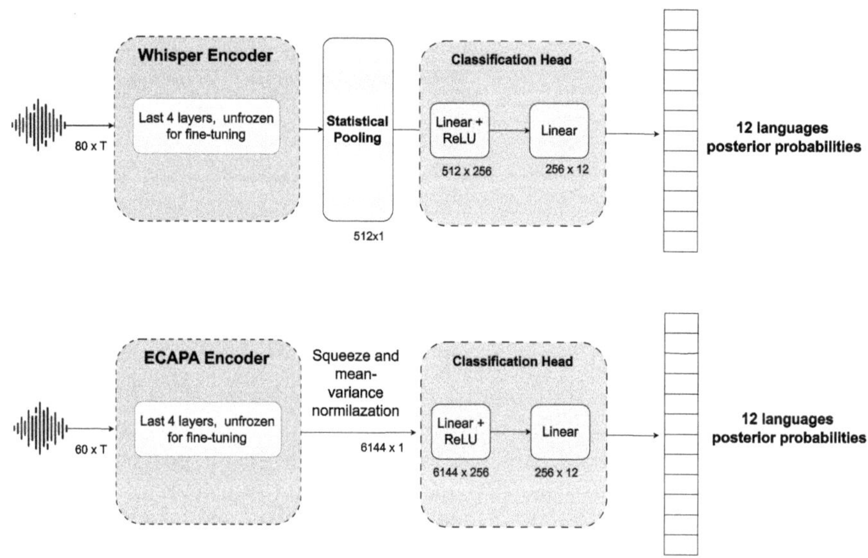

Fig. 6. Flow diagram of Whisper-base (top) and VoxLingua107 ECAPA-TDNN (bottom).

The second set of experiments uses the off-the-shelf models of OpenAI's Whisper-base & VoxLingua107 ECAPA-TDNN. Fine-tuning was performed since the Whisper-base was not trained with all 12 languages under consideration Assamese, Malayalam, and Odia are missing from Whisper-base's practical support (not present in benchmarks or with negligible training data). The 12 languages are present in VoxLingua107 ECAPA-TDNN but, except for English and Hindi, generally have high error rates due to limited training data. Hence, both models were fine-tuned for our dataset (Fig. 6).

4.3   Implementation Details

In our first set of experiments, we trained the TDNN-based x-vector model (baseline 1) for 30 epochs using the Adam optimizer with a fixed learning rate of 1×10^{-4}, weight decay of 5×10^{-5}, $\beta_1 = 0.9$, $\beta_2 = 0.98$, and $\epsilon = 1 \times 10^{-9}$. Training used a batch size of 32, while evaluation used a batch size of 1. For Baselines BmfccConf and Bw2v2Conf, as well as the proposed ConfPhoneme model, we trained for five epochs with a batch size of 16 and applied gradient accumulation. The XvecPhone model, trained for 20 epochs with a batch size of 32, used knowledge distillation with KL divergence and a learning rate warm-up from zero to 1×10^{-4} over 2,400 steps, followed by a cosine decay schedule to ensure stable learning and reduce overfitting.

In the second set of experiments, we fine-tuned off-the-shelf models for language identification tasks. For the Whisper model, we froze all but the last four encoder layers, pooled the outputs over time, and passed them through two fully connected layers with 256 and 12 units, respectively. Fixed-length padding of 3,000 frames was applied. Training ran for five epochs with a batch size of 16, using the AdamW optimizer with a learning rate of 10^{-5}. For the ECAPA-TDNN–based model, we used separate learning rates of 10^{-5} for the encoder—since we are fine-tuning its weights—and 10^{-4} for the projection and classifier layers, as those weights are being trained from scratch. The model was trained for 20 epochs with a batch size of 16, using cross-entropy loss for the 12-class classification task.

All experiments were conducted on a single NVIDIA RTX A5000 GPU (24 GB RAM).

4.4   Results and Discussion

Table 1. Experiment set 1: Comparison of baseline Systems on Seen and Unseen Test Sets, accuracy and EER in %

LID System	Seen Test Set		Unseen Test Set	
	Accuracy (%) ↑	EER (%) ↓	Accuracy (%) ↑	EER (%) ↓
Bxvec [10]	80.66	8.6	51.23	19.45
BmfccConf [6]	92.7	3.4	17.7	40.6
Bw2v2Conf	92.4	3.9	29.9	33.1
ConfPhoneme (ours)	82.4	9.1	**72.4**	**11.5**
XvecPhoneme (ours)	84.88	5.6	**72.6**	**9.8**

In LID tasks, the standard metrics are equal error rate (EER) and accuracy, and our results are also presented using these measures. In our first set of experiments, we compared three baseline systems on both seen test data and unseen test data (see Table 1). The BmfccConf performed best on seen data (92.7% accuracy, 3.4% EER) but performed poorly on unseen data (17.7% accuracy, 40.6% EER), showing its vulnerability to domain shifts and overfitting. The Bw2v2Conf matched its seen performance (92.4% accuracy, 3.9% EER) and handled unseen data better (29.9% accuracy, 33.1% EER) due to its contextual embeddings. The Bxvec, utilizing BNF features, achieved lower accuracy on seen data (80.66% accuracy, 8.6% EER) but showed robustness to domain shift, with 51.23% accuracy and 19.45% EER on unseen data.

Our proposed systems—ConfPhoneme and XvecPhoneme—demonstrated better performance under unseen conditions, highlighting the robustness of phoneme-based features. ConfPhoneme achieved 82.4% accuracy (9.1% EER) on seen data and 72.4% accuracy (11.5% EER) on unseen data. XvecPhoneme, which integrates audio and phoneme cues, further improved results, with 84.88% accuracy (5.6% EER) on seen data and 72.6% accuracy (9.8% EER) on unseen data. These findings emphasize the advantages of phoneme features in addressing domain shifts and enhancing model generalization, though with a small loss performance drop in the seen data as compared with two baselines.

Table 2. Comparison of Off-the-shelf systems on seen and unseen test sets, accuracy and EER in %

LID System	Seen Test Set		Unseen Test Set	
	Accuracy (%) ↑	EER (%) ↓	Accuracy (%) ↑	EER (%) ↓
Whisper-base*	97.95	2.4	88.1	4.5
VoxLingua107 ECAPA-TDNN*	44.4	24.7	15.45	41.3

* These off-the shelf model are fine-tuned according to our need.

In the second set of experiments, we evaluated off-the-shelf models pre-trained on large datasets and fine-tuned with our data. These models showed diverse results. The Whisper-base model performed exceptionally well, achieving 97.95% accuracy (2.4% EER) on seen data and 88.1% accuracy (4.5% EER) on unseen data, indicating strong generalization across domains. Although the Whisper-base model outperforms our proposed systems, it has 12 million trainable parameters—significantly more than ConfPhoneme with 2 million and XvecPhoneme with 5 million. This corresponds to approximately 83% and 58% reductions in model size, respectively, leading to substantially lower training costs while still achieving competitive performance. In contrast, the ECAPA-TDNN model struggled, with low performance on both seen (44.4% accuracy, 24.7% EER) and unseen data (15.45% accuracy, 41.3% EER) (Table 2).

In Fig. 7, the t-SNE plots illustrate that the ConfPhoneme model forms clearer and more distinct clusters between languages, making it better at distinguishing closely related languages. On the other hand, the Bw2v2Conf model shows more overlapping clusters, which could make it difficult to classify languages.

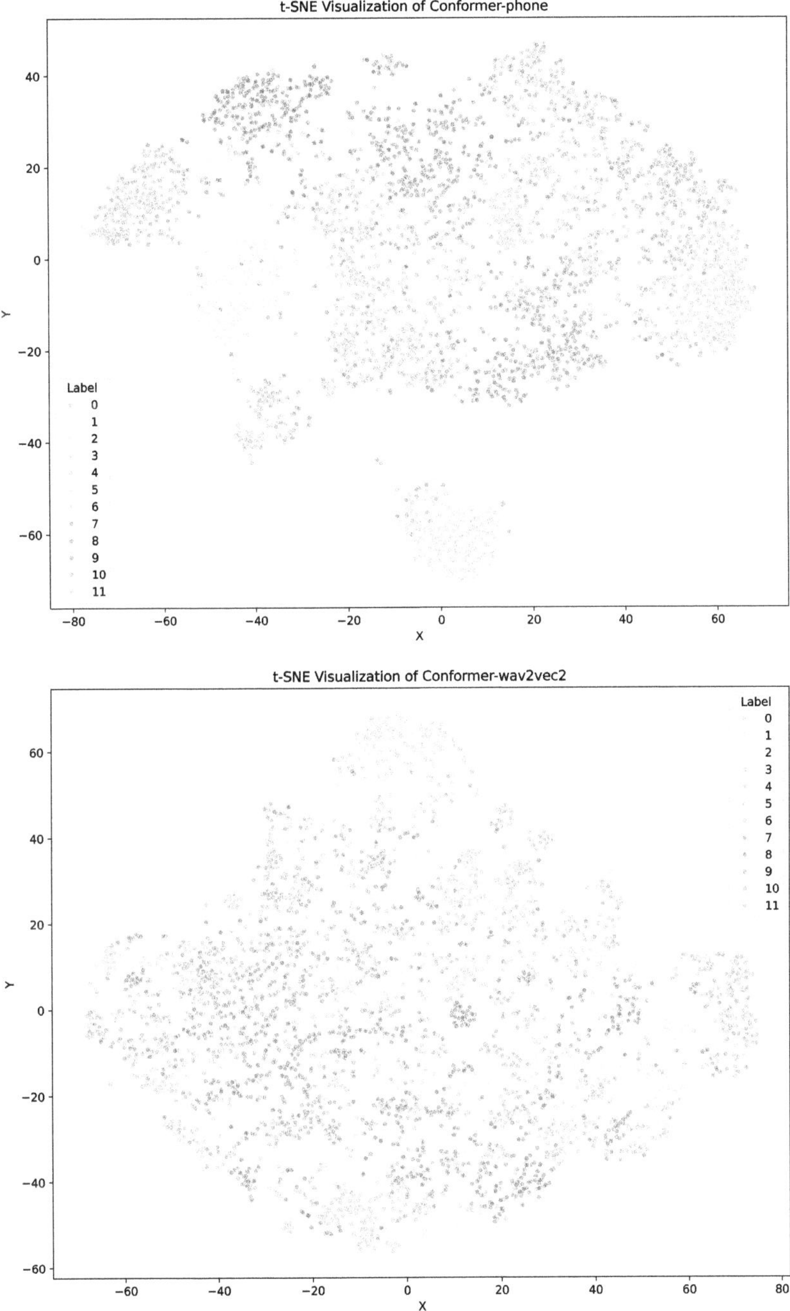

Fig. 7. t-SNE plots of embeddings from the Conformer model trained with phoneme-posterior features (top) and wav2vec2 features (bottom). The embeddings are extracted from unseen test data. The labels represent languages as follows: Assamese (0), Bengali (1), English (2), Gujarati (3), Hindi (4), Kannada (5), Malayalam (6), Marathi (7), Odiya (8), Punjabi (9), Tamil (10), and Telugu (11).

5 Conclusion

In this paper, we investigated robust techniques for spoken language identification, with a focus on closely related Indian languages. Incorporating phoneme posterior features with advanced deep learning architectures, such as ConfPhoneme and XvecPhoneme, resulted in improved classification performance. Traditional low-level features, such as MFCCs, were found to be prone to domain shifts and lacked in distinguishing similar languages. In contrast, the proposed phoneme posterior approach maintained performance under domain-mismatched conditions.

References

1. Dey, S., Sahidullah, M., Saha, G.: An overview of Indian spoken language recognition from machine learning perspective. ACM Trans. Asian Low-Resour. Lang. Inf. Process. **21**(6), 1–45 (2022)
2. Snyder, D., Garcia-Romero, D., Sell, G., Povey, D., Khudanpur, S.: X-vectors: robust DNN embeddings for speaker recognition. In: IEEE International Conference on Acoustics, Speech and Signal Processing, ICASSP 2018, pp. 5329–5333. IEEE (2018)
3. Zhang, C., et al.: Streaming end-to-end multilingual speech recognition with joint language identification, 28 September 2023, uS Patent App. 18/188,632
4. Silnova, A., et al.: BUT/Phonexia bottleneck feature extractor. Odyssey **2018**, 283–287 (2018)
5. Muralikrishna, H., Sapra, P., Jain, A., Dinesh, D.A.: Spoken language identification using bidirectional LSTM based LID sequential senones. In: IEEE Automatic Speech Recognition and Understanding Workshop, ASRU 2019, pp. 320–326. IEEE (2019)
6. Chua, V.Y.H., et al.: MERLIon CCS challenge: a English-mandarin code-switching child-directed speech corpus for language identification and diarization. In: Interspeech 2023, pp. 4109–4113 (2023)
7. Wang, Q., Yu, Y., Pelecanos, J., Huang, Y., Moreno, I.L.: Attentive temporal pooling for conformer-based streaming language identification in long-form speech. In: The Speaker and Language Recognition Workshop (Odyssey 2022), pp. 255–262 (2022)
8. Miao, X., McLoughlin, I., Yan, Y.: A new time-frequency attention mechanism for TDNN and CNN-LSTM-TDNN, with application to language identification. In: Interspeech 2019, pp. 4080–4084 (2019)
9. Muralikrishna, H., Kapoor, S., Dinesh, D.A., Rajan, P.: Spoken language identification in unseen target domain using within-sample similarity loss. In: IEEE International Conference on Acoustics, Speech and Signal Processing, ICASSP 2021, pp. 7223–7227. IEEE (2021)
10. Goswami, U., Muralikrishna, H., Dinesh, A., Thenkanidiyoor, V.: Adversarially trained hierarchical attention network for domain-invariant spoken language identification. In: International Conference on Speech and Computer, SPECOM 2023, pp. 475–489. Springer, Cham (2023)
11. Liu, H., et al.: End-to-end language diarization for bilingual code-switching speech. In: Interspeech 2021, pp. 1489–1493 (2021)

12. Liu, H., Perera, L.P.G., Khong, A.W., Dauwels, J., Styles, S.J., Khudanpur, S.: Enhancing language identification using dual-mode model with knowledge distillation. In: Odyssey 2022 – The Speaker and Language Recognition Workshop, pp. 248–254 (2022)
13. Baevski, A., Zhou, Y., Mohamed, A., Auli, M.: wav2vec 2.0: a framework for self-supervised learning of speech representations. In: Advances in Neural Information Processing Systems, vol. 33, pp. 12 449–12 460 (2020)
14. Chung, Y.-A., et al.: W2v-BERT: combining contrastive learning and masked language modeling for self-supervised speech pre-training. In: IEEE Automatic Speech Recognition and Understanding Workshop, ASRU 2021, pp. 244–250. IEEE (2021)
15. Zanon Boito, M., Iyer, V., Lagos, N., Besacier, L., Calapodescu, I.: mHuBERT-147: a compact multilingual HuBERT model. In: Interspeech 2024, pp. 3939–3943 (2024)
16. Vashishth, S., et al.: Label aware speech representation learning for language identification. In: Interspeech 2023, pp. 5351–5355 (2023)
17. Shahin, M., Nan, Z., Sethu, V., Ahmed, B.: Improving wav2vec2-based spoken language identification by learning phonological features. In: Interspeech 2023, pp. 4119–4123 (2023)
18. Ravanelli, M., et al.: SpeechBrain: a general-purpose speech toolkit (2021). arXiv:2106.04624
19. Valk, J., Alumäe, T.: VoxLingua107: a dataset for spoken language recognition. In: IEEE SLT Workshop 2021 (2021)
20. Radford, A., Kim, J.W., Xu, T., Brockman, G., McLeavey, C., Sutskever, I.: Robust speech recognition via large-scale weak supervision. In: International Conference on Machine Learning, ICML 2023, pp. 28 492–28 518. PMLR (2023)
21. Xu, Q., Baevski, A., Auli, M.: Simple and effective zero-shot cross-lingual phoneme recognition. In: Interspeech 2022, pp. 2113–2117 (2022)
22. Phoneme features and the international phonetic alphabet. https://github.com/espeak-ng/espeak-ng/blob/master/docs/phonemes.md. Accessed Mar 2025
23. Open-speech-ekstep dataset. https://github.com/Open-Speech-EkStep. Accessed Mar 2025
24. Chadha, H.S., et al.: Vakyansh: ASR toolkit for low resource Indic languages. arXiv preprint arXiv:2203.16512 (2022)

ML Frameworks and System-Level Intelligence

End-To-End Modular Intelligence Through Machine Learning

Ravikant[✉], Abhishek Sengupta, Ravi Tanwar, Vijay Pandey, Jayit Saha, and C. Ashwini

Walmart Global Tech, Bangalore, Karnataka, India
`ravikant.ravikant@walmart.com`

Abstract. The strategic arrangement and availability of products on store shelves, known as Store Modulars, play a critical role in driving Walmart sales. Insights derived from snapshots of a shelf at any given time can be used to drive actions in the form of evaluating their standings in a store/region, address missed sales opportunities, prioritize store activities, etc. to help optimize the performance of the assortment. Modular Intelligence provides an end-to-end scalable solution, starting from shelf image capture at stores to generating insights and recommending strategies of action; helping Walmart Suppliers, Merchants and Store Managers get a holistic perspective of their physical shelves. Modular Intelligence will be used for auto-scheduling of tasks for store associates or field representatives to facilitate efficient corrective actions pertaining to shelf management, enhancing overall operational effectiveness.

Keywords: Modular Intelligence · Assortment · Computer Vision · End-to-End

1   Introduction

Walmart has always been known for its enormous volume of data that is captured through the huge magnitude of transactions happening throughout the year. Information on sales, inventory, customer footfall, etc. Powers a diverse number of solutions like demand forecast, inventory management, supply chain management, etc. Enabling Walmart to stay on top of the chain.

In addition to the transaction data, Walmart also captures the actual Images of the physical shelves in the stores, which are the source of information for on-shelf availability, erroneous item placements, shelf state, price tag availability etc. The shelf image capture data enables Machine Learning modules to provide a whole new type of insights, alerts and recommendations on important sales and inventory KPIs to suppliers.

Modular Intelligence is a pioneer in building this capability at scale by providing a complete solution, starting with image capture of Walmart shelves at a regular cadence (Sect. 3.1), followed by an image annotation module to identify the shelf products using Computer Vision models (Sect. 3.2).

R. Gupta et al. (Eds.): BDA 2025, LNCS 16041, pp. 145–155, 2026.
https://doi.org/10.1007/978-3-032-15134-6_10

146   Ravikant et al.

Subsequently, a novel Shelf Health Score (SHS) is derived which provides a single metric gauging the overall health of a shelf (Sect. 3.3), highlighting areas of concern, root causes and potential impacts. Along with SHS, Modular Intelligence provides additional key insights and recommendations (Sect. 3.4) by overlaying shelf information with Walmart Data (sales, inventory, planogram, etc.) using Data Science Models. One such insight is quantification of sales impact due to modular changes and optimal modular recommendation. This is achieved through a novel Planogram Similarity Matching (PSM) algorithm which is a cornerstone innovation highlighted in this paper (Sect. 3.5).

The solution is developed using a robust MLOps Architecture (Sect. 4) to handle scale and enable real-time generation of insights from the time of image capture. It is also built on a plug-and-play framework fostering cross-collaboration opportunities across multiple teams for image capture, image annotation and other capabilities.

The solution is surfaced to the users through a real-time dashboard (Sect. 5), providing a data-driven snapshot of their shelves the moment a new shelf image comes in, enabling them to swiftly act on critical issues.

2   Terminologies

Following terms are used through out the paper (see Fig. 1)

Shelf: A horizontal surface arranged within fixtures to display goods for sale.

Section: A collection of shelves stacked vertically. Sections are usually 4 feet, 6 feet and 8 feet long.

Modular: A collection of Sections arranged horizontally. A Modular will contain items of a single category and can have sections ranging from 1 to 30+.

Planogram: Detailed blueprint outlining the intended layout and arrangement of products within a modular including product placement, quantities and facing directions.

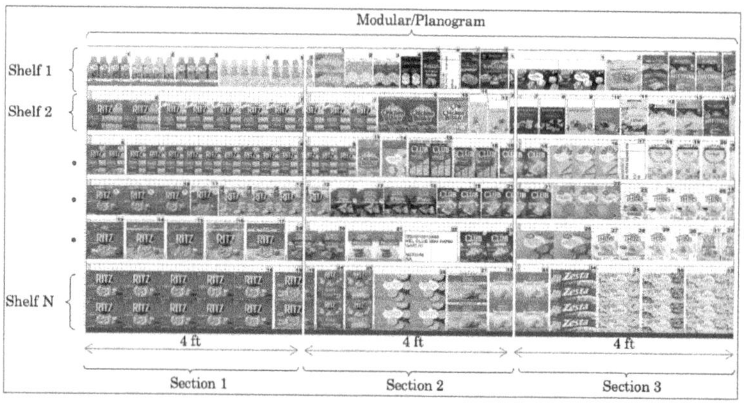

Fig. 1. Shelf, Section and Modular Illustration.

3   Solution

As mentioned above, Modular Intelligence encompasses the below key modules:

- Shelf Image Capture
- Shelf Annotation
- Shelf Health scoring and impact
- Shelf insights and recommendations

3.1   Shelf Image Capture

For shelf image capture, we have currently partnered with a 3P Vendor, who manages field representatives who can visit Walmart stores and perform store activities (item restocking, etc.). As part of the contract made with 3P, these field representative visit stores at a bi-weekly (twice a week) cadence and capture images of each modular.

3.2   Shelf Annotation

The image annotation module (see Fig. 2) consists of 2 steps:

- Bounding Box Regression
- Product Identification

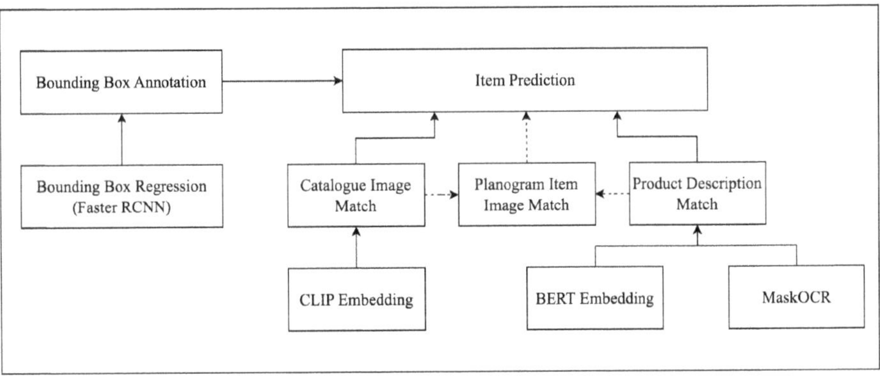

Fig. 2. Image Annotation Architecture.

For Bounding Box Regression, which involves detection of each individual product in the shelf, a Faster RCNN [6] algorithm is used.

For product identification in each bounding box, a three-tier matching algorithm is used:

- Catalogue image matching: For each bounding box, multimodal embeddings of the product (inside the bounding box) is generated using CLIP [4]. The same is matched with Walmart Catalogue images for the category in question (having generated CLIP embedding for the catalogue images as well) and the top K items with the highest cosine similarity scores(Sim_Cat_i) are retrieved.
- Product Description matching: The product description is extracted from each product through MaskOCR [3] and matched with the item descriptions (using BERT [5] embeddings and cosine similarity) to retrieve the top K products with the highest cosine similarity scores (Sim_Pr_i).
- Price Tag Matching: The price tags below each product are also extracted from the images and is used to break ties between different sizes of the same products (higher price indicates larger size).

The final UPC (item) prediction UPC_{pred} is an ensemble of Catalogue Image Prediction and Product Description prediction wherein:

$$UPC_{pred} = argmax_i(0.7 * Sim_Cat_i + 0.7 * Sim_Pr_i) \tag{1}$$

Annotation Results. For evaluation, a semi-automated Ground Truth annotation framework is created. A test image dataset is annotated using this framework against which the predictions are evaluated. The image annotation module has a precision of 80% with recall of 70% and performs comparably similar across different type of products in size and orientations.

3.3   Shelf Health Score

Based on the annotated images, a novel metric called Shelf Health Score (SHS) is derived as a single metric denoting the overall shelf/section health. SHS is constituted of the below features: Shelf Health Score

- osa_{zero} : Zero on-shelf availability (an item is completely out of stock from the modular)

$$osa_{zero} = \frac{|\{p_k, \forall p_k \notin A\}|}{|p|} \tag{2}$$

- $osa_{partial}$: Partial on-shelf availability (the item is not fully out-of-stock, has partial facings on shelf)

$$osa_{partial} = \frac{\Sigma_{i=1}^{k} I(\frac{I(p_i \in P)}{I(a_i \in A)} > 0.3)}{|p|} \tag{3}$$

- $wanderer$: Items from the same modular have been placed in the shelf but in a different location

$$wanderer = \frac{|\{s_{ij} \neq l_{ij}, \forall a_k \in P\}|}{|A|} \tag{4}$$

- *invader* : Items from a different modular have been placed in the shelf

$$invader = \frac{|\{a_k, \forall a_k \notin P\}|}{|A|} \tag{5}$$

- *ptm* : Price tag Missing (Items without a price tag in the modular)

$$ptm = 1 - \frac{|\{t_i, \forall a_i \in A\}|}{|A|} \tag{6}$$

To obtain the above features, annotated images are overlaid with corresponding planogram, and SHS is derived as below:

$$SHS = 1 - \frac{\Sigma_i w_i m_i}{\Sigma_i w_i} \tag{7}$$

where
$m_i \in \{osa_{zero}, osa_{partial}, wanderer, invader, ptm\}$
$P = \{p_i | p_i \in Planogram\ Items\}$
$S = \{s_{ij} | s_{ij} \in Item\ Positions\ in\ Planogram\}$
$A = \{a_i | a_i \in Annotated\ Items\}$
$L - \{l_{ij} | l_{ij} \in Item\ Positions\ in\ annotated\ image\}$
$T = \{t_i | t_i \in Item\ Price\ tags\}$

Shelf Health Score $\in [0, 1]$, with a higher SHS indicating well-stocked, well-arranged, and modular compliant shelves. This metric has been demoed to and vetted by multiple marquee suppliers.

For Modulars with low shelf health, UPC level actions are suggested (see Fig. 3) for Low SHS in the form of which items to "Reposition" to a different section of the modular and which items need a "Restock".

UPC #	UPC Name	Shel Cap Qty	Actual Position	Planned Position	Status	Action
012345678901	Hershey's Milk Chocolate Giant Candy, Bar 7.56 oz Strike Zone	32	S01-14	S01-12, S05-15	Incorrect Position	Reposition
012345678901	Reese's Milk Chocolate Giant Candy, Bar 7.56 oz Strike Zone	44	-	S01-13	Missing	Restock
012345678901	Hershey's White Chocolate Giant Candy, Bar 7.56 oz Strike Zone	20	S01-14	S01-14	Ok	No Action

Fig. 3. UPC Level Actions for Low SHS.

Additionally, potential sales impact of low SHS is estimated using an XGBoost Regressor [1] to regress each component of SHS with baseline expected sales by aggregating demand forecast over a period.

Figure 4 shows SHS and its breakdown, in a period of one week for one supplier and one category.

Health Component	Health Score	Change from last Image capture	8 week trend
Overall	85%	↓	
Breakdown			
• No Shelf Availability (Zero OSA)	12%	↑	
• Partial OSA	14%	↑	
• Mod Non-Compliance	48%	↓	
• Wanderers	48%	↓	
• Invaders	0.01%	↓	
• PTM (Price Tag Missing)	9%	↑	

Fig. 4. SHS Breakdown.

3.4 Insights and Recommendations

Overlaying shelf information with other Walmart data sources enables Modular Intelligence to provide additional key insights, which can be grouped under the below two buckets:

- Near Real Time(NRT) Inventory and Online Pickup and Delivery(OPD) Insights
- Planogram Insights

NRT insights involve invoking the NRT Inventory API to retrieve real-time on-shelf, back-room and in-transit inventory (at the time of image capture); while OPD insights involve retrieving the scheduled OPD orders for the end of day. These are then merged with the shelf/modular data to provide insights and alerts on the below (see Fig. 5):

- Mismatch in actual on-shelf and recorded on-shelf inventory
- Items to be restocked from backroom
- Items without enough inventory to meet OPD requirements
- Impact on Nil-Picks due to item unavailability

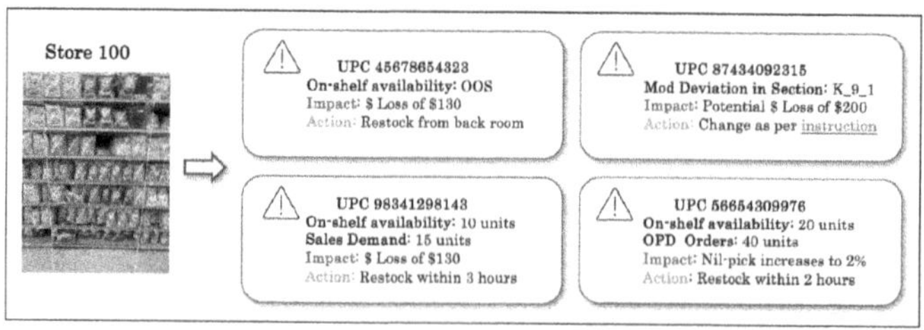

Fig. 5. NRT Insights.

Note: Impact on Nil-Picks is computed in a similar way as impact on sales is computed in Sect. 3.3. For brevity, details of the same are excluded from this paper.

Planogram insights, on the other hand, focus on quantifying the efficacy of an implemented planogram and the potential impact of shifting to a new planogram, which can help the user (supplier) to make informed decisions when negotiating modular arrangements with Walmart. This is achieved through a novel Planogram Similarity Matching (PSM) algorithm (Sect. 3.5), which is another key innovation proposed in this paper.

3.5 Planogram Similarity Matching

Planogram Similarity Matching (PSM) is a novel algorithm introduced in this paper which aims to compare similarities between planograms and cluster them, with the goal of quantifying the sales impact due to a planogram change.

For each supplier, the algorithm first converts the planogram data into an HSV image (Hue, Saturation, Value) taking into consideration the item arrangement, shelf capacity and supplier share. Each unique UPC is represented as a unique base color (Hue) and the saturation denotes how densely they are packed in each shelf location.

Figure 6 Planogram to Image Conversion shows the pictorial representation of two planograms, where non grayed-out cells represent items belonging to the supplier.

As can be seen, the top planogram represents a scenario where the supplier items are not placed together (sparsely packed) and items from another supplier are placed in between. The lower planogram on the other hand represents a densely packed planogram for the supplier along with a higher shelf capacity (darker saturation).

This is the main novelty of the algorithm, which, by converting a tabular planogram data into an image, automatically encodes the positional information without having to compute and store pairwise distances for each item-item combination (a time and space intensive operation). Once converted into an image, standard image clustering algorithms like Spectral Clustering [2] can be leveraged to cluster similar images (and thus similar planograms) together.

Each cluster is then profiled based on the type of planogram it represents:

- Densely/sparsely packed
- High/med/low shelf capacity
- Top/bottom/middle heavy shelves
- Other cluster types (Work in Progress)

The sales impact of each planogram cluster is computed (Similar to impact computation in Sect. 3.3) and the difference highlights the impact due to inter-cluster planogram shifts (intra-cluster planogram shifts are considered to have negligible impact in sales).

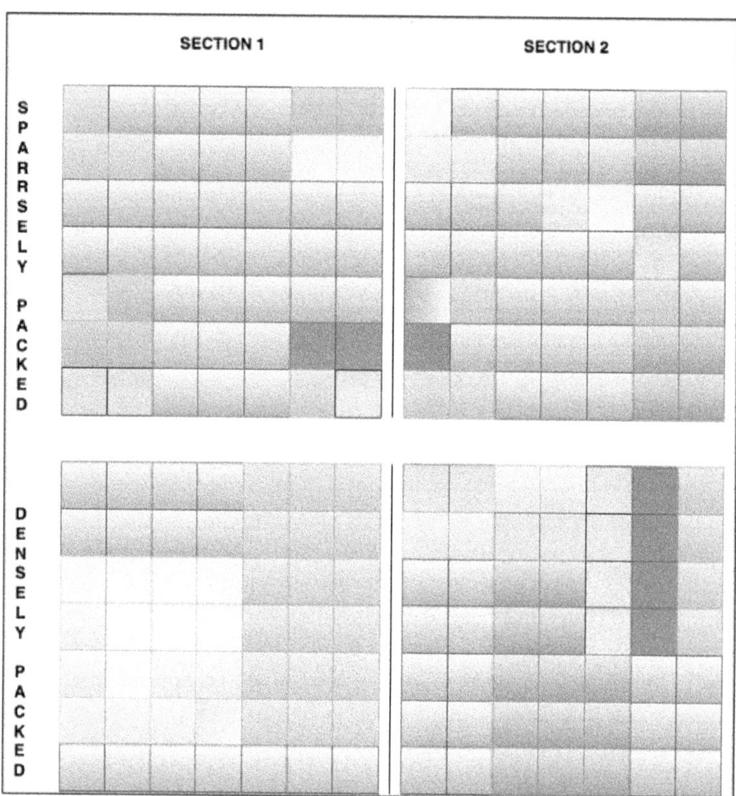

Fig. 6. Planogram to Image Conversion.

4   Scale and Architecture

Once the shelf images are captured and annotated (Image Capture Service), the annotated data is pushed to Kafka, which is then consumed by a spark streaming application for the generation of insights in real-time(see Fig. 7).

Other input datasets like planogram, sales and inventory data are pulled from a data consumption layer which is refreshed at a daily cadence. As the images for a modular plan across different stores come in a staggered fashion, the output is persisted in Cosmos and pulled back for aggregation.

Planogram insights are generated as a batch process through Spark and are refreshed only when a new planogram is implemented in a store.

The aggregated insights are finally pushed to the back-end database and UI consequently from Kafka.

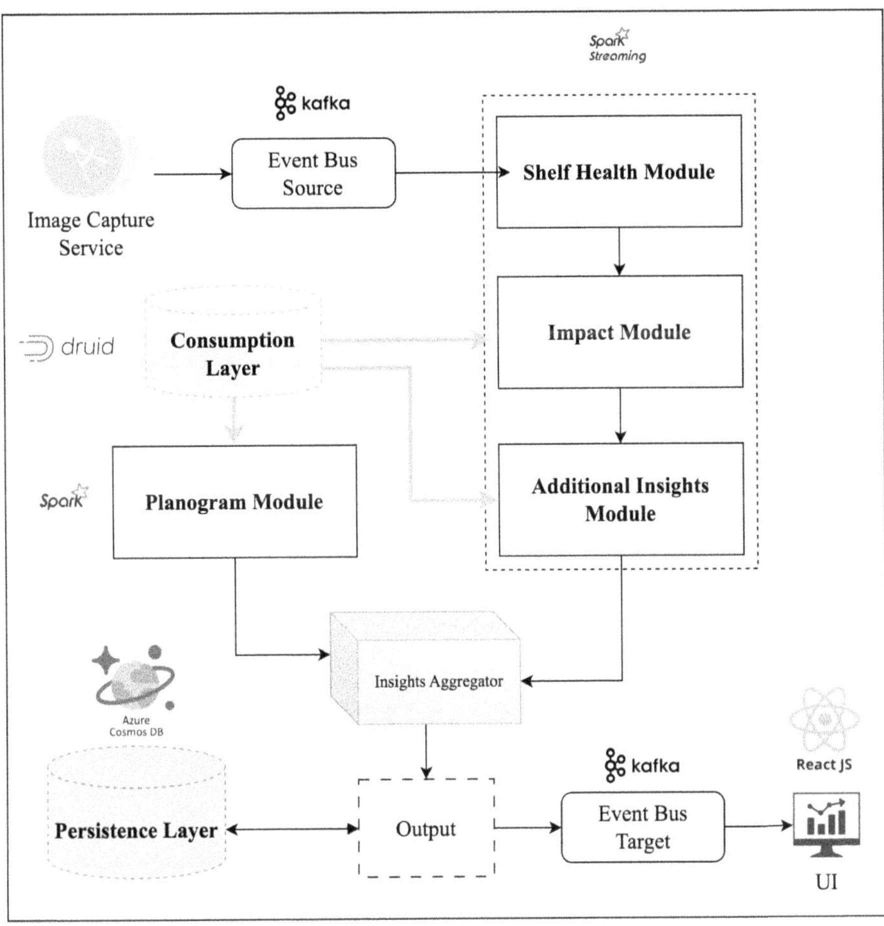

Fig. 7. Modular Intelligence Architecture.

5   Dashboard

The insights are displayed to the user via a Real- Time React based Dashboard which enables the user (supplier) to select their categories of interest and get insights on the category health, root cause analysis on stores with low health, alongside other key insights and recommendations. The Dashboard (see Fig. 8) provides flexible drill down options starting from the overall category level, down to the lowest granularity of a store section UPC level.

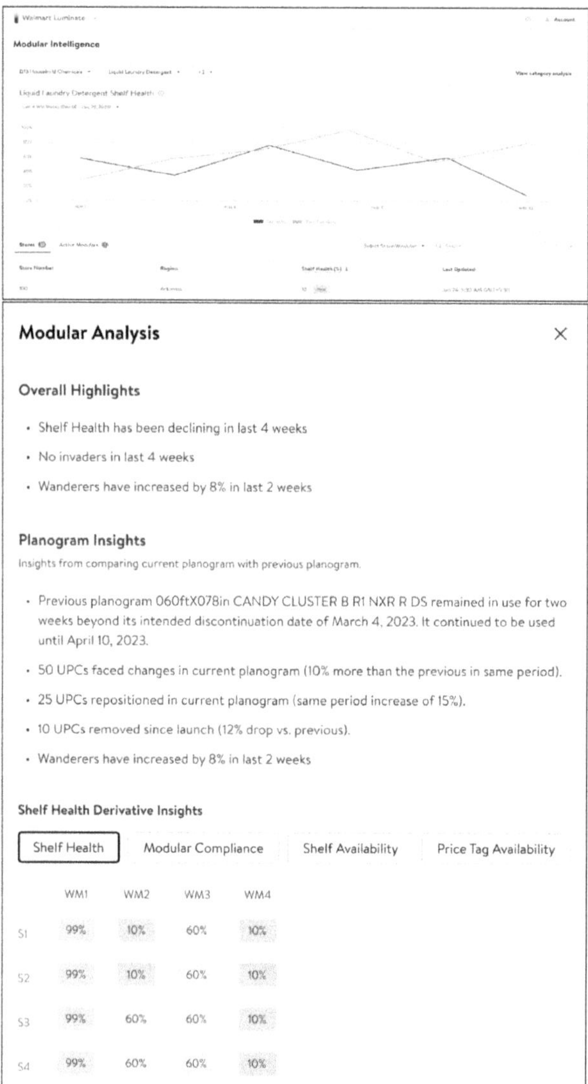

	WM1	WM2	WM3	WM4
S1	99%	10%	60%	10%
S2	99%	10%	60%	10%
S3	99%	60%	60%	10%
S4	99%	60%	60%	10%

Fig. 8. Modular Intelligence Dashboard.

6 Conclusion

Modular Intelligence is the first scalable offering from Walmart providing modular insights, alerts and prescriptive actions; opening a new realm of insights for the Walmart landscape. The plug-and-play architecture of Modular Intelligence enables it to switch to more advanced image capture mechanisms (automated high-resolution captures) leading to more frequent insights, as well as more advanced computer vision models to enable better detection of small or partially occluded items.

Given the unprecedented nature of this type of data capture, all insights have been derived based on the data capture in a period. Future work includes tuning the models to capture even more complex patterns such as seasonality, cyclic patterns, etc. And provide more nuanced insights as more and more data continues to get generated.

References

1. Chen, T., Guestrin, C.: Xgboost: A scalable tree boosting system. In: Proceedings of the 22nd ACM SIGKDD International Conference on Knowledge Discovery and Data Mining, pp. 785–794. KDD '16, ACM (2016). https://doi.org/10.1145/2939672.2939785
2. von Luxburg, U.: A tutorial on spectral clustering (2007). https://arxiv.org/abs/0711.0189
3. Lyu, P., et al.: Maskocr: Text recognition with masked encoder-decoder pretraining (2023). https://arxiv.org/abs/2206.00311
4. Radford, A., et al.: Learning transferable visual models from natural language supervision. CoRR **abs/2103.00020** (2021). https://arxiv.org/abs/2103.00020
5. Reimers, N., Gurevych, I.: Sentence-bert: Sentence embeddings using Siamese Bert-networks (2019). https://arxiv.org/abs/1908.10084
6. Ren, S., He, K., Girshick, R., Sun, J.: Faster R-CNN: towards real-time object detection with region proposal networks. Adv. Neural Inf. Process. Syst. **28** (2015)

CycleAware: A Machine Learning Framework for Menstrual Phase Prediction and Athlete Performance Tracking

Shreya Srikant[✉], Shreya Naveen, Aditya Verulkar, V. Shrujan,
and N. Chaitra

Department of Computer Science Engineering, PES University, Bangalore, India
`srikantshreya@gmail.com`, `chaitran@pes.edu`

Abstract. Despite growing recognition of the menstrual cycle's influence on female athletic performance, personalized, data-driven solutions remain scarce in mainstream sports technology. This paper introduces CycleAware, a novel machine learning-powered web platform that bridges the gap between menstrual health and athletic optimization. By augmenting a real-world dataset to ensure balanced representation across all menstrual phases, we enable accurate modeling of nuanced physiological trends. A deep-tabular hybrid architecture, TabNet for interpretable feature learning combined with Random Forest for robust classification, was developed and benchmarked against a classical SVC with PCA pipeline. The hybrid model's ability to autonomously prioritize key features while capturing complex interdependencies results in significantly improved predictive accuracy and generalization. Integrated into a user-centric Django web application, CycleAware empowers athletes and coaches to log cycle and training data, visualize historical trends, and receive real-time, phase-aware performance insights. This work pioneers a scalable, AI-integrated approach to menstrual cycle prediction, promoting inclusivity and data-informed training in competitive sports - a space where personalization and female physiology have long been overlooked.

Keywords: Hybrid Machine Learning · Smart Wellness Monitoring · Sports Analytics · Intelligent Athlete Support Systems · Physiological Data Analysis · Tabular Data Classification

1 Introduction

Menstrual health is a significant yet an overlooked factor in maximizing the athletic performance of women. In contrast to male athletes, females undergo periodic fluctuations in hormones, primarily in estrogen and progesterone, with

© The Author(s), under exclusive license to Springer Nature Switzerland AG 2026
R. Gupta et al. (Eds.): BDA 2025, LNCS 16041, pp. 156–170, 2026.
https://doi.org/10.1007/978-3-032-15134-6_11

profound influences on energy metabolism, muscle repair, endurance, neuromuscular control, and training adaptation. In spite of increasing recognition, menstrual cycle phases are seldom considered when designing training programs, and thus many athletes are left without specific guidance.

The menstrual cycle is divided into four different phases: menstruation, follicular, ovulation, and luteal, each with corresponding distinct endocrinological changes. During the follicular phase, there is high estrogen, which enhances muscle repair, carbohydrate metabolism, and endurance, and hence this phase is most appropriate for high-intensity training. Estrogen also allows protein synthesis and glycogen use, leading to less perceived exertion and faster recovery [9].

Conversely, the luteal phase is progesterone-dominated, increasing body temperature and perceived exertion and diverting metabolism to fat metabolism, usually at the expense of endurance performance [3,5,8]. Augmenting coordination and explosive power at ovulation, but also ligament laxity, hence increasing the risk of injury, especially for ACL tears.

Periodization of training according to cycle phases may enhance performance and reduce the risk of injury. For example, intense effort is appropriate during the follicular phase, and the luteal phase may be complemented with recovery-based, low-impact training [6].

Overreaching and energy underconsumption can cause Relative Energy Deficiency in Sport (RED-S) and hypothalamic amenorrhea, a state of persistent low energy availability. RED-S deranges hormonal homeostasis, leading to menstrual dysfunction and adversely impacting bone health, metabolism, cardiovascular function, and immune response [1]. Recognition of these hormonal impacts is key to preventing overtraining and maintaining realistic expectations. Athletes with irregular cycles may find it hard to identify their phase, making it challenging to adjust their training strategies effectively. For example, underperformance in the luteal phase may be a function of physiological as opposed to motivational factors—whereas underperformance in the follicular phase may be an indication of needing to try harder.

Hormonal fluctuations have a significant impact on psychological health, affecting emotions, motivation, and self-confidence. By monitoring their menstrual cycles, athletes can modify their training schedules, establish realistic targets, and enhance physical and mental endurance.

To enable this, we created CycleAware—a machine learning-based platform predicting menstrual phases based on training data. Implemented as part of a Django web application, it provides real-time predictions and phase-targeted training recommendations for more intelligent, cycle-aware athletic planning.

2   Related Works

Oliveira et al. [10] developed a state-space modeling approach to predict menstrual cycle lengths in athletes, addressing the challenges posed by cycle variability. Utilizing data from 22 athletes over an average of 15 cycles each, the model incorporated individual-specific parameters to account for personal menstrual patterns. The findings demonstrated that the state-space model provided more accurate predictions compared to traditional methods, highlighting its potential for personalized monitoring and management of menstrual health in athletic populations. Urteaga et al. [12] introduced a novel approach to personalized modeling of the female hormonal cycle by integrating mechanistic models with Gaussian process regression. The study aimed to predict menstrual cycle phases by accounting for individual variations in hormonal patterns. Using a validated mechanistic model, the authors generated realistic hormonal data and applied Gaussian processes to estimate hormone levels over time. They evaluated the feasibility of predicting menstrual cycle phases under various conditions, including different training cycle counts, hormonal measurement noise levels, sampling rates, and informed versus agnostic sampling strategies. The findings suggest that Gaussian processes are effective in modeling the female menstrual cycle, highlighting the potential for personalized healthcare applications in reproductive health. Calcaterra et al. [2] investigated the prevalence and impact of menstrual dysfunction in adolescent female athletes, emphasizing the role of energy availability in menstrual health. The study highlights that intense training combined with inadequate nutritional intake can lead to conditions such as delayed menarche, oligomenorrhea, and amenorrhea. These dysfunctions are linked to adverse health effects, including decreased bone mineral density and an increased risk of fractures.

Statham et al. [11] conducted a comprehensive review of the effects of the menstrual cycle on exercise performance in eumenorrheic women. The study analyzed hormonal fluctuations across the menstrual cycle and their potential impacts on various performance metrics, including strength, endurance, and thermoregulation. The authors found that while some phases of the menstrual cycle may influence specific aspects of performance, the overall effects are highly individualistic.

Meignié et al. (2021) [7] conducted a systematic review examining the effects of menstrual cycle phases on elite athletes' performance. The study highlighted variations in physiological and psychological factors across different phases, influencing endurance, strength, power output, neuromuscular function, and cognitive performance. While some studies reported declines in performance during the luteal phase due to increased fatigue and hormonal fluctuations, others found minimal or no impact. The review emphasized the need for individualized approaches to training, considering hormonal variability and athlete-specific responses.

3   Methodology

3.1   Dataset Description

The dataset utilized in this study was obtained from a publicly available research on menstrual phase prediction in athletes [4]. It comprises of physiological and performance-related attributes critical for phase classification.

- **Phase**: Menstrual phase category (target variable).
- **Durata**: Duration of the session (in minutes).
- **DistTOT**: Total distance covered (in meters).
- **HSR**: High-speed running distance (in meters).
- **ACC**: Number of accelerations (count).
- **DEC**: Number of decelerations (count).
- **RPE**: Rate of perceived exertion (self-reported scale).
- **sRPE**: Session rating of perceived exertion (computed metric).

Feature Engineering. To enhance model performance and ensure robust classification, the following feature transformations were applied:

- **Feature Scaling:** Standardization using *Z-score normalization* (Standard-Scaler) was performed on physiological and performance-related metrics to maintain consistency and improve model convergence.
- **Categorical Encoding:** The target variable, representing menstrual cycle phases, was *label-encoded* into numerical values (1, 2, 3, and 4) representing menstrual, follicular, ovulation and luteal respectively to facilitate classification.
- **Feature Selection:** *Recursive Feature Elimination (RFE)* was utilized to identify the most predictive attributes, optimizing model interpretability and reducing dimensionality while preserving critical information.

This preprocessing strategy enhances model generalization, minimizes bias due to scale variations, and ensures numerical stability for downstream machine learning algorithms.

3.2 Data Augmentation

To overcome data imbalance and improve the stability of the model, cyclic augmentation was utilized. The data set was increased from 64 to 4064 samples by applying repeated augmentation cycles. 32 original samples were reserved for validation and 4032 for training and testing to ensure unbiased training. Two augmentation techniques—Synthetic Minority Over-sampling Technique (SMOTE) and Generative Adversarial Networks (GANs)—were tested to create synthetic samples for minority menstrual phases. They were compared against Kolmogorov-Smirnov (KS) statistics and Wasserstein distance to test statistical similarity to the original data.

SMOTE: SMOTE generated synthetic samples through interpolation, preserving natural variation and avoiding overfitting. It attained an average KS statistic of 0.0949 and an 81.33% similarity score, which is high fidelity to the original data.

The distributions of the important features, such as DistTOT, HSR, ACC, and DEC, showed lower KS statistics and Wasserstein distances, indicating a minimal divergence from true real-world patterns. The deterministic nature of the approach ensured that the newly generated data preserved statistical and physiological consistency. In addition, the computational efficiency and interpretability of this approach further validate its suitability for augmenting structured physiological datasets.(See Table 1)

Table 1. Distribution Similarity Metrics for SMOTE

Metric	KS Stat	p-val	Wasserstein Dist
Durata	0.107	0.438	3.61
DistTOT	0.060	0.966	183.05
HSR	0.101	0.503	102.96
ACC	0.083	0.746	8.71
DEC	0.091	0.634	10.32
RPE	0.123	0.275	0.53
sRPE	0.099	0.534	50.45
Overall Similarity Score	0.8133 (81.33%)		

GAN: GANs, although useful for image data, did not work well with this structured data. They produced unrealistic outliers and had strong statistical divergence: The metrics of evaluation indicated that data generated with GAN possessed a greater KS statistic (0.3608) and a lesser similarity measure (65.71%), which meant synthetic distributions were significantly different from the original data.

Feature comparison indicated that Durata, sRPE, and DistTOT exhibited the largest discrepancies, with KS statistics over 0.5 and large Wasserstein distances. The samples generated included outliers and implausible values, indicating that the GAN could not model the intricate physiological patterns in menstrual phase data. GANs also needed considerable tuning of hyperparameters and computational resources, making them less convenient for this research. (See Table 2)

Why SMOTE Was Preferred Over GAN for Data Augmentation?

Superior Feature Distribution Retention: SMOTE effectively preserved the distributions of key physiological metrics such as DistTOT, HSR, ACC, and DEC. In contrast, GAN-generated data exhibited major deviations, particularly in Durata and sRPE, with KS stats of 0.582 and 0.533, and Wasserstein distances exceeding 600 in some cases. These discrepancies indicate that GAN outputs were less physiologically plausible and diverged significantly from real-world training patterns.

Table 2. Distribution Similarity Metrics for GAN

Metric	KS Statistic	KS p-value	Wasserstein Distance
Durata	0.5820	1.35×10^{-18}	20.42
DistTOT	0.2676	4.62×10^{-4}	626.81
HSR	0.2910	9.85×10^{-5}	373.43
ACC	0.1543	0.1227	13.37
DEC	0.3184	1.36×10^{-5}	19.80
RPE	0.3789	8.52×10^{-8}	1.91
sRPE	0.5332	1.81×10^{-15}	379.29
Overall Similarity Score			**0.6571 (65.71%)**

SMOTE's Consistency in Data Augmentation: Unlike GANs, SMOTE uses deterministic interpolation between minority class samples, ensuring the augmented data remains within the natural range of the original distribution. This makes it especially effective for structured numerical data, such as menstrual cycle training metrics. In contrast, GANs introduced high-dimensional distortions and outliers, as reflected by significantly lower KS p-values (e.g., 1.35e-18 for Durata, 8.52e-08 for RPE), indicating poor alignment with real-world distributions and reduced reliability for predictive modeling.

Stability and Interpretability of SMOTE: SMOTE provides more stability and interpretability in tabular data than GANs, which involve extensive hyperparameter tuning and mode collapse, a type of failure prevalent in GAN's training process in which the generator generates low-diversity samples.

Since GANs struggle to model complex dependencies in small, structured data sets, they could produce unrealistic values.

Computational Efficiency and Practicality: From a computational viewpoint, SMOTE is significantly less costly than GANs. Unlike the process of training GANs with huge computational loads and repeated iteration of generator and discriminator networks, SMOTE accomplishes its job through a mathematical interpolation, generating new data points in a single processing step.

Inference from the Feature Distribution Graphs: The SMOTE-synthesized feature data distributions closely resemble the original distributions. (Refer to Fig. 1) The orange (augmented) and blue (original) density plots overlap significantly, which means that SMOTE preserves the inherent variance of the dataset. In contrast, the GAN-generated data show more visible deviations from the original distribution (see Fig. 2). The density curves of some features have significantly altered, indicating that GAN has trouble maintaining the exact statistical properties of the original data.

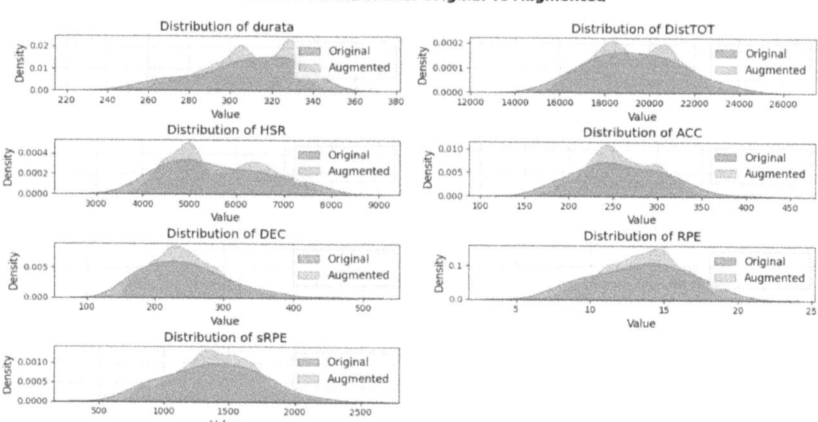

Fig. 1. Feature distributions of original vs augmented data using SMOTE.

Since CycleAware prioritizes reliability, interpretability, and efficiency, SMOTE was the preferred data augmentation technique over GANs.

3.3 Model Framework

We developed and evaluted two distinct machine learning and deep learning pipelines and chose the most suitable approach for menstrual phase prediction. The approaches are as follows:

TabNet with Random Forest. TabNet applies sequential attention to select the most relevant features dynamically, minimizing the requirement for manual feature engineering. During training, it employed a momentum factor of 0.98, allowing stable updates with the capacity to handle changes in feature significance between training iterations. The three-stage decision procedure within the model allowed TabNet to gradually narrow feature selections to the most relevant aspects of the input data where necessary. Additionally, a feature sparsity regularization term ($\lambda = 0.001$) prevented over-exposure to redundant features, allowing interpretability by imposing selective feature utilization without compromising critical information.

Following training, the learned feature embeddings were extracted from the last layer of TabNet and fed as inputs to a Random Forest classifier, chosen for its ensemble-based stability. The Random Forest classifier was configured to 50 decision trees, each with a maximum depth of 8, striking a balance between model complexity and generalization. By limiting tree depth, the model avoids overfitting while still learning informative patterns from the extracted TabNet embeddings. The Gini impurity measure was employed to estimate feature splits, preferring partitions maximizing class purity at each node. This setup allows

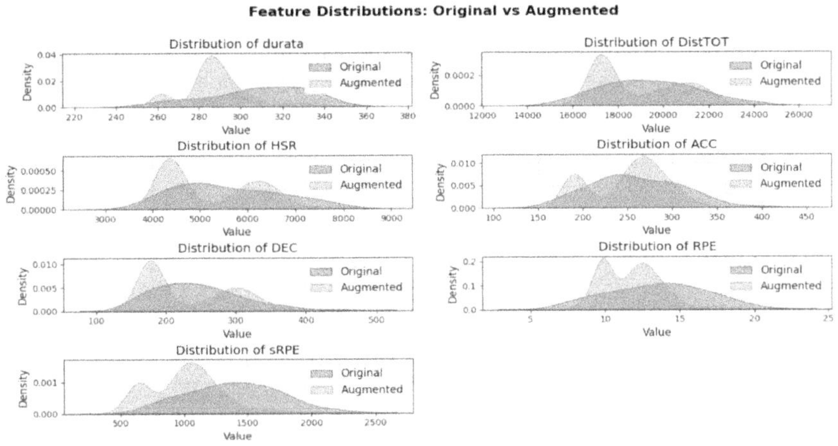

Fig. 2. Feature distributions of original vs augmented data using GAN.

the ensemble to leverage diverse decision paths while offering robustness and interpretability.

This hybrid approach combined the strength of deep learning in detecting nuanced feature interactions with the strength of Random Forest for non-linear classification to avoid overfitting and improve generalization.

Support Vector Classifier with PCA. The Radial Basis Function (RBF) Support Vector Classifier (SVC) is well-suited for classification problems involving high-dimensional data. In this work, Principal Component Analysis (PCA) was applied before training to reduce dimensionality while retaining 95% of the original variance. This transformation preserved the dominant variance components, reduced noise, and improved computational efficiency. Whitening was also applied to normalize the transformed features and reduce scale-related bias.

The SVC model was trained using an RBF kernel with γ set to 'scale' for automatic variance-based tuning. A high regularization parameter ($C = 10.0$) was employed to prioritize accurate classification, even at the cost of model simplicity. Five-fold cross-validation ensured robustness and generalization across different data partitions.

While PCA expedited training and mitigated redundancy, it also potentially masked nuanced interdependencies between features that are critical in modeling menstrual phase transitions. This occurs because PCA identifies orthogonal principal components based on global variance, not task-specific discriminative power. Features that contribute minimally to overall variance but significantly to phase classification (e.g., slight changes in high-speed running during ovulation) may be de-emphasized or omitted in the reduced space. Consequently, PCA may suppress low-variance, high-importance signals essential for accurate classification in physiological contexts. Although the RBF kernel enables SVC to capture non-linear relationships, and advanced kernels such as polynomial or sigmoid are capable of modeling even more complex interactions, kernel-based SVCs still lack the capability for context-specific feature prioritization. Even with improved classification performance from alternative kernels, SVCs continue to treat all features uniformly and lack an internal mechanism to adaptively highlight the most informative features for each decision instance.

3.4   Performance Evaluation Metrics

To comprehensively assess and compare the performance of both models, multiple evaluation metrics were considered:

- **Accuracy (Acc):** Measures the proportion of correctly classified instances. While useful for general evaluation, it may be misleading on imbalanced datasets.
- **Precision** (P) : Indicates the number of predicted positives that are actual positives, important in reducing false positives.

$$P = \frac{TP}{TP + FP}$$

where TP represents True Positives and FP represents False Positives.
- **Recall** (R): Reflects the model's ability to identify all actual positives, critical when missing a positive case is costly.

$$R = \frac{TP}{TP + FN}$$

where TP represents True Positives and FN represents False Negatives.
- **The F1-Score** (F_1): Harmonic mean of precision and recall, ideal for imbalanced datasets where both false positives and false negatives matter.

$$F_1 = 2 \times \frac{P \times R}{P + R}$$

where: - P is Precision - R is Recall
- **Cross-Validation Accuracy (CV_Acc):** Reflects model stability by averaging over multiple data splits, assisting in detecting overfitting and ensuring generalization.

3.5   Observations and Results

The higher accuracy of TabNet + Random Forest as shown in the table 3 indicates improved generalization to novel data and improved pattern recognition. With a higher cross-validation accuracy , The TabNet + RF hybrid model exhibits higher stability with different data partitions with less sensitivity to variability in training data.

TabNet+RF's F1 score performs better showing a better precision-recall balance in classification across menstrual phases, thereby reducing both false positives and negatives.

Table 3. Key findings

Model	TabNet + RF	SVC with PCA
Test Accuracy	91.25%	88.90%
Cross Validation Accuracy	90.2%	86.17 %
F1 score	0.91	0.89

Confusion Matrix Analysis: TabNet + Random Forest model(refer Fig. 3) demonstrates high diagonal dominance, indicating high accuracy across menstrual phases. Low False Negatives suggest low misclassification of actual phases, and low False Positives (FP) ensure well-balanced predictions whereas SVC with PCA model depicts higher rate of misclassification especially with adjacent phases like ovulation and late follicular. The model overestimates the transitions in some cases and, therefore, produces more False Positives as well as somewhat fewer True Positives.

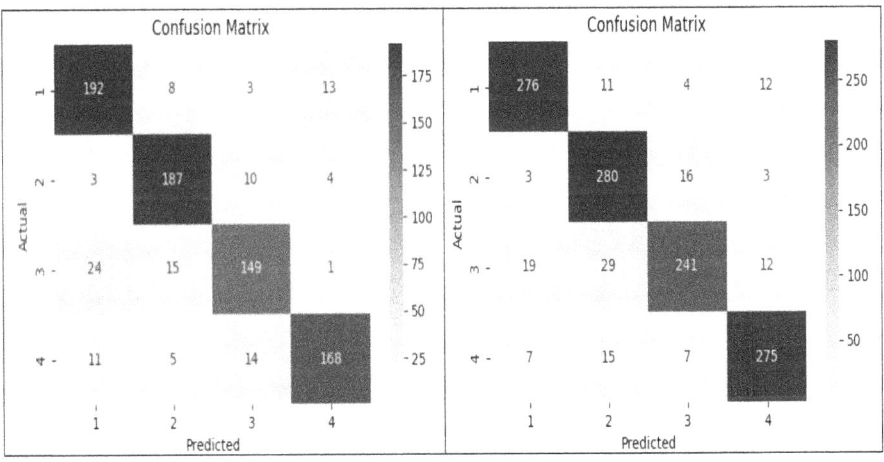

Fig. 3. a. Confusion matrix of the SVC with PCA model b. Confusion matrix of Tabnet + Random Forest model.

The superior performance of the TabNet + Random Forest hybrid model can be attributed to:

Dynamic Feature Learning: The superior performance of the TabNet + Random Forest hybrid can be attributed to its ability to prioritize and interpret feature relevance dynamically. Unlike SVC, which processes input features uniformly, TabNet's sequential attention mechanism and feature sparsity regularization allows the model to focus on the most discriminative inputs during different phase classifications. This is especially valuable for physiological data, where certain metrics (e.g., high-speed running or RPE) may vary in importance across menstrual phases.

Ensemble-Based Robust Classification: The Random Forest classifier, with its use of many decision trees, effectively reduces variance and eases overfitting, thus encouraging more stable predictions across different subsets of data. dataset.

Although SVC with PCA is still a top contender because of its computational ease and stability, its application of linear transformations may compromise its capacity to identify subtle non-linear relationships within the data. Furthermore, PCA-based dimensionality reduction, although useful in the interest of computational practicability, may cause masking of subtle but significant variations in the prediction of the menstrual phase. In conclusion, the TabNet with Random Forest model is widely recognized as the better approach, employing deep learning for feature representation and ensemble learning to obtain effective classification, hence being the best choice for menstrual phase prediction.

4   Web Framework

For improved athlete and coach use, the CycleAware system was deployed as a Django web application with an SQLite database for structured storage.

4.1   Monitoring New Entries

This module allows individuals with regular cycles to record training and symptom information on a standardized form to enable consistent tracking over time. Important inputs are menstrual cycle length, last menstrual period, cramps severity, mood changes, training status and performance ratings.

4.2   Historical Overview

This module offers a full historical review of all previously documented menstrual and training data. The main aim is to determine recurring patterns and correlations between the menstrual cycle phases and performance.

4.3   Track Athlete Performance

For irregular menstrual cycle athletes, estimation of the current menstrual phase must be done based on training parameters. This module considers several indirect physiological indicators such as:

- **Duration (minutes) and total distance (m):** Measures endurance capacity.
- **High-speed running (m), accelerations, and decelerations:** Evaluates neuromuscular demand and sprint efforts.
- **Rate of Perceived Exertion (RPE) and session RPE:** Captures subjective training load perception.

These are then used by a trained DL model to predict the menstrual phase (Menstruation, Follicular, Ovulation, Luteal) and offer a phase-specific training recommendation. For example, if the athlete enters:

- **Duration:** 340 min
- **Total Distance:** 19,056 m
- **High-Speed Running:** 5,012.4 m
- **Accelerations:** 212
- **Decelerations:** 298
- **RPE:** 13, **Session RPE:** 1125

The system predicts the corresponding **menstrual phase** and produces an adaptive training message, informing athletes of their physiological status. This aspect removes psychological frustration by putting performance fluctuations within the menstrual cycle, suggesting data-driven training adaptations (Fig. 4).

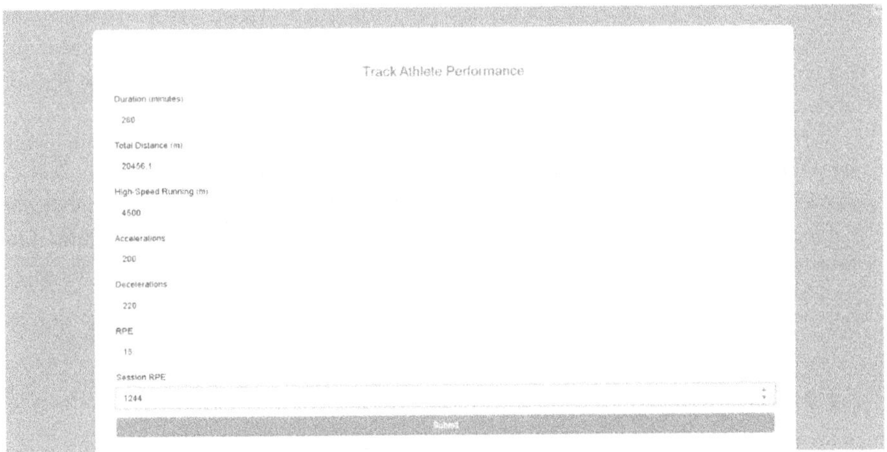

Fig. 4. Tracking an athlete's performance by providing necessary parameters.

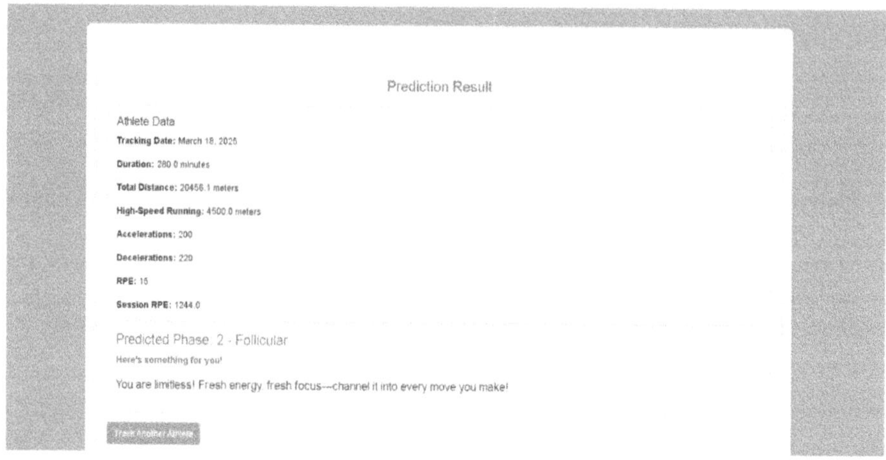

Fig. 5. Prediction result and phase wise message generation.

5 Limitations

- The dataset used for training is relatively small and may not adequately reflect the physiological variability present across a larger, more diverse population. Although augmentation techniques such as SMOTE were applied to balance the dataset and improve stability, synthetic data cannot fully substitute for real-world physiological diversity. This may introduce bias and reduce the model's generalizability when deployed at scale.
- Several key input features, such as Rate of Perceived Exertion (RPE) and mood, are self-reported and inherently subjective. These metrics can be influenced by external factors including psychological stress, sleep quality, and environmental conditions. Such variability can lead to noise in the data and may affect the reliability of model predictions, particularly when these inputs dominate phase classification.
- The model currently does not utilize direct biomarkers such as hormone levels, heart rate variability (HRV), or basal body temperature. These parameters are critical for accurately tracking physiological changes across the menstrual cycle. Their absence limits the model's ability to capture more granular and biologically grounded phase transitions, especially in athletes with irregular or atypical cycles.

6 Future Scope

Integration of Wearable Device Data: Future versions of CycleAware will aim to reduce reliance on self-reported metrics by integrating data from wearable devices. Physiological signals such as heart rate variability, sleep quality, and stress levels can serve as objective proxies for exertion and recovery. This will enhance data reliability, reduce user burden, and improve the accuracy of phase-aware predictions (Fig. 5).

Adaptive Training Recommendations: AI-based models can be expanded from phase prediction to dynamically modify training recommendations based on menstrual cycle fluctuations.

Physiological Signal Integration and Hormonal Monitoring: To improve prediction accuracy and physiological relevance, future iterations of CycleAware will explore the integration of direct biomarkers such as hormone levels, heart rate variability (HRV), and basal body temperature. This can be achieved through partnerships with wearable device platforms and at-home hormone testing kits. Incorporating such data will allow the model to more precisely detect subtle phase transitions, particularly in users with irregular or complex menstrual patterns.

Scalability and Data Privacy Considerations: To support broader user adoption, future versions of CycleAware will migrate from a local SQLite based backend to a scalable, cloud-native infrastructure. This will enable support for large user bases and real time model updates. To ensure data privacy and athlete confidentiality, we plan to implement robust security measures, including end-to-end encryption, role-based access controls, and compliance with data protection regulations such as GDPR. Users will retain control over their data, with transparent consent and opt-out mechanisms.

7 Conclusion

This study introduces a machine learning-based model for the prediction of menstrual cycle phase using important physiological and statistical parameters for enhanced classification performance. By taking into account variables such as cycle history, rated perceived exertion, and physical activity patterns, the introduced model offers a data-driven approach for describing menstrual phase variation. The results showcase the potential of predictive analytics to menstrual monitoring, providing informative insights for individuals, athletes, and clinicians.

The use of artificial intelligence to predict menstrual phases is a milestone in personalized health care, in which people can now make informed decisions regarding training, recovery, and overall health based on physiological variations. Through the use of machine learning methods, the present study fills a core research gap in menstrual health, in which menstrual cycles have previously been poorly represented in the research on performance and health.

This research points to the growing relevance of artificial intelligence-based technologies in menstrual health and demands continued research in this field. By bridging menstrual health research with data-driven technologies, this research helps form a more holistic approach to health observation, ultimately helping people derive personalized insights for enhanced cycle management and well-being.

References

1. Cabre, H.E., Moore, S.R., Smith-Ryan, A.E., Hackney, A.C.: Relative energy deficiency in sport (red-s): scientific, clinical, and practical implications for the female athlete. Deutsche Zeitschrift für Sportmedizin **73**(7), 225–234 (2022). https://doi.org/10.5960/dzsm.2022.546
2. Calcaterra, V., et al.: Menstrual dysfunction in adolescent female athletes. Sports **12**(9), 245 (2024). https://doi.org/10.3390/sports12090245
3. Carmichael, M.A., Thomson, R.L., Moran, L.J., Wycherley, T.P.: The impact of menstrual cycle phase on athletes' performance: a narrative review. Int. J. Environ. Res. Public Health **18**(4), 1667 (2021). https://doi.org/10.3390/ijerph18041667
4. Chiara, G.: Dataset: menstrual cycle phase and elite female soccer during training. Mendeley Data (2024)
5. Delp, M., et al.: Higher rating of perceived exertion and lower perceived recovery following a graded exercise test during menses compared to non-bleeding days in untrained females. Front. Physiol. **14** (2024). https://doi.org/10.3389/fphys.2023.1297242
6. Herzberg, S.D., Motu'apuaka, M.L., Guise, J.M.: The effect of menstrual cycle and contraceptives on acl injuries and laxity: a systematic review and meta-analysis. Orthopaedic J. Sports Med. **5**(7) (2017). https://doi.org/10.1177/2325967117718781
7. Meignié, A., et al.: The effects of menstrual cycle phase on elite athlete performance: a critical and systematic review. Front. Physiol. **12**, 654585 (2021). https://doi.org/10.3389/fphys.2021.654585
8. Niering, M., Wolf-Belala, N., Seifert, J., Tovar, O., Coldewey, J., Kuranda, J., Muehlbauer, T.: The influence of menstrual cycle phases on maximal strength performance in healthy female adults: a systematic review with meta-analysis. Sports **12**(1), 31 (2024). https://doi.org/10.3390/sports12010031
9. Oosthuyse, T., Bosch, A.N.: The effect of the menstrual cycle on exercise metabolism: implications for exercise performance in eumenorrhoeic women. Sports Med. **40**(3), 207–227 (2010). https://doi.org/10.2165/11317090-000000000-00000
10. de P. Oliveira, T., Bruinvels, G., Pedlar, C.R., Moore, B., Newell, J.: Modelling menstrual cycle length in athletes using state-space models. Sci. Rep. **11**, 16972 (2021). https://doi.org/10.1038/s41598-021-95960-1
11. Statham, G.: Understanding the effects of the menstrual cycle on training and performance in elite athletes: a preliminary study. In: Progress in Brain Research, vol. 253. Elsevier (2020). https://doi.org/10.1016/bs.pbr.2020.05.028
12. Urteaga, I., Albers, D.J., Vlajić, M., Elhadad, N.: Towards personalized modeling of the female hormonal cycle: experiments with mechanistic models and gaussian processes. arXiv preprint arXiv:1712.00117 (2017). https://doi.org/10.48550/arXiv.1712.00117

SiteSense: Optimizing Restaurant Site Selection Using a Comprehensive Data-Driven Framework

Anvitha Swaroop[iD], Kritika Prasad[iD], and Sonia Khetarpaul[✉][iD]

Department of Computer Science and Engineering, Shiv Nadar Institution of Eminence, Delhi NCR, India
{as918,kp836,sonia.khetarpaul}@snu.edu.in

Abstract. Restaurant site selection is a critical decision-making process that impacts the success and longevity of dining establishments. In Bangalore, where diverse culinary preferences and economic disparities coexist, selecting optimal locations for new restaurants poses a significant challenge. Traditional methods of location selection rely on population density and foot traffic, often resulting in an oversaturation of restaurants in certain areas and a lack of options in underrepresented neighbourhoods. In this, an approach has been proposed that combines sentiment analysis of customer reviews with geographic and socio-economic data to inform restaurant site selection. Here sentiment analysis is leveraged to identify areas with high customer satisfaction and frequent visits while incorporating socio-economic data to align restaurant placement with local demand and purchasing power. By utilising datasets from platforms like Zomato and government sources, a comprehensive framework has been developed for optimising restaurant locations across Bangalore. This approach has led to a 39% improvement in identifying underserved neighborhoods and a 57% increase in restaurant success rate by matching location with customer preferences and economic conditions.

Keywords: Location Based Social Networks · Sentiment Analysis · Clustering Analysis · Recommendation Model

1 Introduction

This paper addresses the challenge of optimizing restaurant site selection in Bangalore by integrating sentiment analysis and geographic data. Traditional methods, which primarily rely on population density and foot traffic, often overlook customer preferences and local economic conditions, leading to inefficient location choices. In contrast, our approach combines sentiment analysis with location-based data to develop a more data-driven and comprehensive system for restaurant site recommendations. Sentiment analysis, a key technique in natural language processing (NLP), is employed to analyze customer reviews from platforms like Zomato, extracting sentiment to gauge customer satisfaction and

R. Gupta et al. (Eds.): BDA 2025, LNCS 16041, pp. 171–187, 2026.
https://doi.org/10.1007/978-3-032-15134-6_12

identify popular dining areas. Additionally, geographic data is used to map the distribution of restaurants across Bangalore, identifying underserved areas with potential for new establishments.

To enhance the accuracy of our recommendations, datasets from the Karnataka government and Zomato are integrated, providing insights into population demographics and economic conditions. This integration aligns restaurant placements with local demand and economic viability. The proposed method will employ evaluation metrics to assess the effectiveness of the recommendations and improve decision-making in restaurant site selection. Building on existing research in sentiment analysis and location-based decision-making, our work aims to provide a comprehensive framework that assists restaurant owners in making informed, data-driven site selection decisions in Bangalore's dynamic culinary landscape.

1.1 Objective

This research aims to develop a data-driven framework for optimising restaurant site selection in Bengaluru. By integrating sentiment analysis of customer reviews with geographic and socio-economic data, the paper aims to:

1. Identify neighborhoods with high density of restaurants and their cuisines as well as customer satisfaction scores.
2. Develop a recommendation model which suggests the ideal site for a restaurant of a particular cuisine or the ideal cuisines for a given restaurant location, for the most profitable buisness.
3. Suggest the ideal average cost for two for the restaurant recommendations in order to build the most profitable business.

1.2 Motivation and Contributions

Bangalore's rapid urban growth and its status as India's IT hub have transformed it into a culturally diverse city with a highly competitive restaurant market. Traditional site selection methods—focused on foot traffic and population density—fall short in such a dynamic environment, overlooking critical factors like customer preferences, economic conditions, and market saturation. This often leads to suboptimal placement and financial inefficiencies.

This research addresses these challenges by introducing a recommendation model that integrates sentiment analysis, geographic data, and socio-economic factors to comprehensively tackle restaurant site selection. The model provides targeted insights, such as identifying the top locations for specific cuisines or recommending the best cuisines for given locations. It also considers factors like price range (affordable vs. fine dining) and economic viability, ensuring contextually relevant and profitable decisions. By leveraging advanced analytics and actionable insights, this research offers a data-driven solution to meet the evolving demands of Bangalore's diverse and dynamic culinary landscape. The following are the key contributions of this paper:

1. Dynamic Recommendation Capabilities: Identifies the top three ideal locations for a given cuisine in Bangalore, leveraging sentiment analysis and geographic clustering techniques. As well as suggests the top three cuisines best suited for a specific restaurant location, considering local customer preferences and market demand.
2. Price Range Optimization: Determines the ideal price range (affordable or fine dining) for each recommendation, ensuring alignment with the target demographic's spending capacity and maximizing profitability.
3. Integration of Multimodal Data Sources: Combines sentiment data from customer reviews with geographic and socio-economic datasets, providing a holistic view of the market landscape.
4. Clustering Techniques: Employs algorithms such as k-Means, DBSCAN, and BIRCH to uncover patterns in customer preferences, restaurant distribution, and pricing dynamics, offering precision in identifying underserved areas and optimizing site selection.
5. Interactive Visualization Tool: Features an interactive map that visually represents recommendations, enabling stakeholders to explore restaurant density, sentiment hotspots, and ideal business opportunities in real time.

2   Related Work

Restaurant site selection has drawn substantial research interest due to its critical role in business success. Studies have explored aspects like geospatial analytics, sentiment analysis, and clustering techniques, but often in isolation, failing to provide integrated frameworks for comprehensive decision-making.

Zheng et al. [7] highlighted LBSNs' potential in creating recommendation systems through customer behavior prediction but did not apply these insights to specific industries like restaurants. Similarly, Fayyaz et al. [8] examined recommendation algorithms and business opportunities but overlooked integrating geographic and sentiment data for restaurant-specific applications. Wang et al. [2] used sentiment analysis to identify popular dining areas but focused on aggregated feedback, neglecting specific cuisines or economic influences. Kumar et al. [6] applied sentiment analysis to assess individual restaurant performance but missed geographical and socio-economic dimensions essential for site selection. Khetarpaul et al. [4] utilized K-Means clustering to identify suitable restaurant locations but relied solely on geographic data, excluding sentiment and socio-economic factors. Ghosh et al. [5] explored geospatial and sentiment data integration for urban planning but did not extend these insights to restaurants. Zhao et al. [3] analyzed sentiment data from restaurant reviews but failed to include geospatial or economic elements, reducing their findings' practical utility. Cheng [9] proposed a sentiment-based recommendation architecture but omitted geospatial and economic integration. Similarly, Cao et al. [17] used LBSNs for point-of-interest recommendations but did not incorporate sentiment or socio-economic layers, focusing on user mobility patterns.

Existing research on restaurant site selection has critical gaps, including the lack of integration between geospatial data and sentiment analysis, and the neglect of socio-economic factors like income levels and population density. Many studies rely on static data for clustering and fail to incorporate real-time sentiment or advanced algorithms. Most approaches offer generic solutions, limiting their relevance to specific industries like restaurant site selection. These gaps highlight the need for a comprehensive, multi-dimensional framework.

SiteSense addresses these gaps by combining geospatial data, sentiment analysis, and socio-economic indicators. Using advanced clustering algorithms (K-Means, DBSCAN, BIRCH), it identifies underserved regions and evaluates restaurant density. Incorporating customer feedback and sentiment trends, Site-Sense offers precise recommendations for cuisines, price ranges, and site selection.

3   Methodology

The workflow of the proposed *SiteSense* system is shown in Fig. 1. The subsections go through each step in detail:

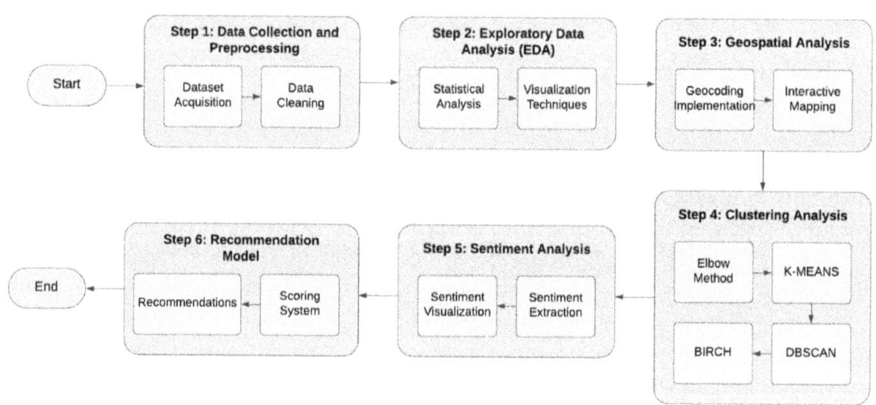

Fig. 1. SiteSense Workflow.

3.1   Data Collection and Pre-Processing

Data Collection and Pre-Processing phase focuses on acquiring, cleaning, and structuring the Zomato Bengaluru restaurants. The initial dataset contained restaurant details, including location, cuisines, ratings, costs, and reviews.

Zomato Restaurant Dataset [12]: It contains detailed information on Bangalore-based restaurants, including location, cuisines, customer reviews, ratings, cost, and votes. It is used to extract customer sentiment data through review analysis, understand the cuisines in all the areas of Bangalore and identify restaurant density by area.

Karnataka Government Dataset [16]: This dataset provides demographic and socio-economic data, including income levels, population density, and economic activity by locality. Used to map out economic characteristics of different localities, which helps determine suitable restaurant locations based on local demand and spending capacity.

The data cleaning process involved removing irrelevant columns, handling missing values in critical fields like ratings, costs, locations, and cuisines, and standardizing data formats. Ratings were cleaned by converting "NEW" and "-" to NaN, extracting numeric values, and converting them to floats. Cost data was cleaned by removing commas and converting to numeric format, while missing votes were filled with 0 and converted to integers for consistency.

3.2   Exploratory Data Analysis (EDA)

Statistical analysis explored restaurant distribution, cuisine popularity, and average costs across locations. Visualizations were created to showcase regional trends. These insights help understand local preferences and dining patterns. Bar plots were created to visualize the top 10 locations by restaurant count and the locations with the highest average costs. Appropriate color schemes like viridis and coolwarm were applied to enhance visual clarity, while proper labeling and formatting ensured the visualizations were clear and informative.

3.3   Geospatial Analysis

Geospatial analysis enables the visualisation and understanding of restaurant distribution patterns across Bangalore, providing crucial insights into market coverage and concentration areas.

Geocoding Implementation: The Nominate geocoder was utilized for address-to-coordinate conversion, with a dedicated function implemented to streamline the process. Error handling was incorporated to ensure reliability and accuracy during the geocoding process.

Interactive Map: Interactive maps were developed using Folium, featuring marker clustering for enhanced visualization and cuisine-specific filtering capabilities. Popup information was also included for each restaurant location to provide detailed insights.

3.4   Clustering Analysis

Clustering analysis helps identify natural groupings of restaurants based on geographical location, enabling a better understanding of restaurant distribution patterns and market segmentation.

K-Means Clustering. k-Means [13], a widely used clustering algorithm, partitions data into groups (clusters) by minimizing the squared distances between data points and their cluster centroids. This analyzes restaurant data based on location, cuisines, ratings, and costs. Using the elbow method, the within-cluster sum of squares(WCSS) was plotted against the number of clusters and the point on the graph where adding more clusters results in a diminishing decrease in WCSS four clusters were identified, each representing distinct restaurant features. Cluster centers, or centroids, reveal the average feature values of each group, offering insights into their defining characteristics.

DBSCAN. Density-Based Spatial Clustering of Applications with Noise (DBSCAN) [14] identifies clusters based on point density, differing from centroid-based algorithms like K-Means by not assuming spherical clusters or a predefined number. It groups closely packed points (high-density regions) and labels low-density points as noise or outliers. For location analysis, DBSCAN pinpointed dense restaurant zones and under-served regions. It also identified clusters based on specific cuisines or high ratings, aiding site selection decisions. Additionally, noise filtering highlighted anomalies, such as outliers in cost or rating, offering valuable insights.

BIRCH Clustering: Balanced Iterative Reducing and Clustering using Hierarchies (BIRCH) [15] is a hierarchical clustering algorithm optimized for large datasets. It constructs a memory-efficient Clustering Feature (CF) tree to represent data and is particularly adept at handling varying densities and hierarchical structures. BIRCH grouped restaurants by geographical and culinary proximity, revealing regional cuisine preferences and underrepresented areas. It also uncovered cost patterns among similarly rated restaurants, identifying cost-effective zones and premium dining hubs. Its hierarchical approach enabled both macro and micro-level analyses, providing detailed insights for restaurant site selection.

3.5   Sentiment Analysis

Sentiment analysis [1] is key to this approach, extracting and evaluating customer feedback to assess satisfaction with various restaurants. By analyzing reviews, overall positive, neutral, or negative sentiments associated with restaurants, cuisines, and locations are determined. This aids in understanding public opinion and informing business decisions.

Sentiment Scoring: Customer reviews are analyzed to determine sentiment-positive, neutral, or negative—using sentiment scores that reflect satisfaction across regions. Each review is assigned a sentiment score that represents customer satisfaction.

- **Positive Sentiment:** Scores closer to $+1$
- **Neutral Sentiment:** Scores around 0
- **Negative Sentiment:** Scores closer to -1

3.6 Recommendation Model

The recommendation model delivers personalized suggestions for selecting restaurant locations, cuisines, and price ranges by analyzing factors critical to customers and businesses.

Scoring System: A weighted scoring system ranks options based on four factors

- Ratings (40%): Reflects customer evaluations [10] of restaurants.
- Review Sentiments (40%): Shows how positively or negatively people feel about a restaurant [11].
- Average Income (10%): Matches economic profiles of locations.
- Population (10%): Reflects how many potential customers are in an area.

Recommending Locations: To find the ideal places to open a restaurant of a specific cuisine, the model filters restaurants offering a specific cuisine, groups them by location, calculates a weighted score, and ranks the locations to suggest the top areas where the cuisine is most likely to succeed.

Recommending Cuisines: If someone has a location in mind and wants to know what cuisines would work best there, the model filters data to include only restaurants in that area, groups them by cuisine, calculates scores using a weighted formula, and suggests the top cuisines that are both popular and highly rated.

Recommending Price Ranges: The model recommends price ranges by analyzing the average and median costs of filtered restaurants. It categorizes recommendations into affordable dining (50 – 70% of the median), fine dining (150 – 250% of the median), and general dining (80 – 120% of the median), with practical minimum thresholds for each category. By balancing factors like ratings, reviews, income, and population, the model offers personalized, data-backed suggestions. It helps customers and businesses make informed decisions, from selecting dining options to identifying optimal restaurant sites.

This model balances ratings, sentiments, income, and population for actionable, realistic recommendations. It empowers customers and businesses to make informed decisions, whether selecting dining options or identifying ideal restaurant sites.

4 Observations and Results

The following observations were made based on the results generated by SiteSense.

4.1 Exploratory Data Analysis Observations

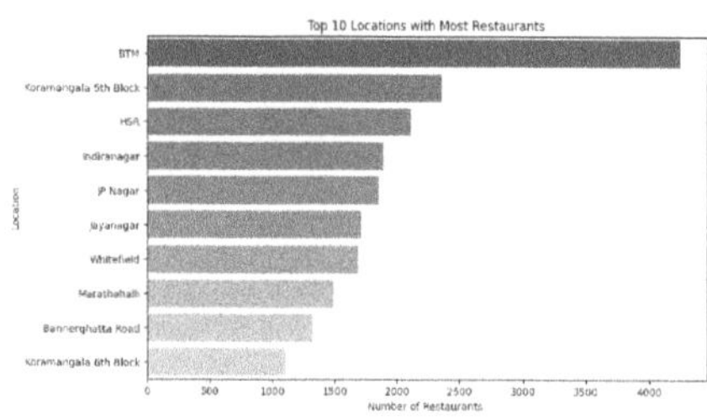

(a) Locations with Highest Restaurants

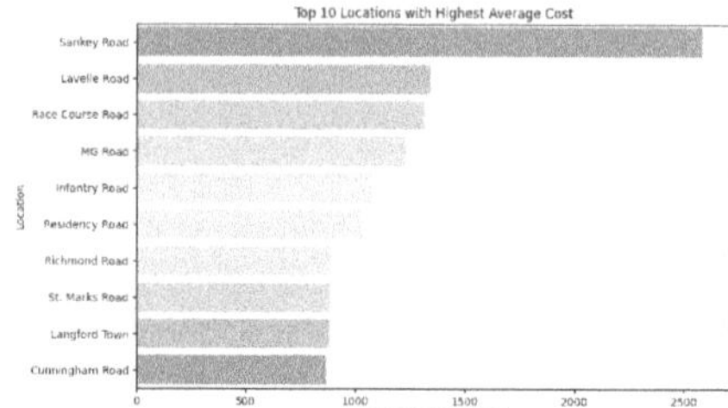

(b) Locations with highest average cost

Fig. 2. Comparison of Locations.

*Figure*2(a) identifies BTM as the area with the highest density of restaurants in Bangalore, exceeding 4,000 establishments, followed by Koramangala 5th Block and HSR Layout. These high-density zones reflect substantial demand and significant foot traffic, positioning them as favorable locations for restaurant ventures.

*Figure*2(b) highlights Sankey Road as the area with the highest average cost for two people, exceeding ₹2,500. Other premium locations such as Lavelle Road, Race Course Road, and MG Road exhibit costs ranging from ₹1,500 to ₹2,000. These areas cater to affluent individuals, presenting opportunities to target high-income clientele.

4.2 Sentiment Analysis Observations

The scatter plot in *Fig.*3(a) demonstrates that areas with consistent rating-sentiment correlations are more predictable, whereas those with numerous outliers may signify customer indecision. The diamond-shaped distribution exhibits the widest variation within mid-range ratings (3.0 – 4.0), including notable outliers such as high ratings (4.0+) coupled with negative sentiment and low ratings (2.0 – 3.0) paired with positive sentiment. *Fig.*3(b) displays a bimodal distribution, with sentiments predominantly between 0.0 and 0.5, and a positive skew, suggesting overall customer satisfaction. These observations emphasize the importance of focusing on areas with strong correlations and managing feedback expectations realistically.

Figure 4(a) indicates that dessert-focused establishments achieve the highest sentiment scores, with most restaurants clustering between 0.2 and 0.4. Food courts exhibit the highest levels of satisfaction, suggesting potential in multi-vendor setups, while quick-service restaurants may face challenges. High-performing categories, such as desserts, imply reduced market saturation, while casual dining points to segmentation opportunities. Hybrid concepts that integrate top-performing categories with favorable sentiment could enhance success rates. *Fig.*4(b) underscores the need to prioritize high-satisfaction areas for

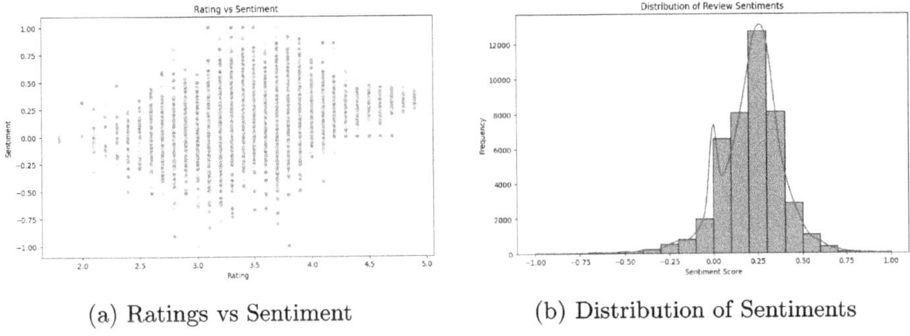

(a) Ratings vs Sentiment   (b) Distribution of Sentiments

Fig. 3. Comparison of Locations.

improved operations, whereas low-scoring regions may reflect market saturation, demanding customers, and elevated operational costs.

4.3   Geospatial Analysis Observations

The interactive map presented in *Fig.* 5 illustrates the spatial distribution of Chinese restaurants in Bangalore. The markers on the map denote restaurant density, with warmer colors (ranging from yellow to red) signifying higher concentrations, while cooler colors (green) indicate lower densities. To enhance readability, closely situated restaurants are aggregated into single markers. This tool serves as a valuable resource for restaurant proprietors to evaluate competitive landscapes and identify under-served areas. Additionally, it enables consumers to explore local dining options and supports urban planners in analyzing dining patterns and trends across various neighborhoods.

4.4   Clustering Analysis Observations

K-Means Clustering: K-Means clustering was applied to the location, cuisine, rating, and cost, with results visualized in 3D (*Fig.* 7(a)) and 2D (Location-Cuisine and Rating-Cost) (*Fig.* 6). Each cluster, distinguished by unique colors and centroids represented by black "X" markers, groups restaurants based on similarities in geographic, culinary, rating, and cost attributes. This analysis identifies dominant cuisines and highlights the relationship between customer satisfaction and pricing. The overlap observed in *Fig.* 6(a) suggests shared cuisine preferences or geographic proximity, while *Fig.* 6(b) reveals variations in ratings for similar costs, reflecting different perceptions of value.

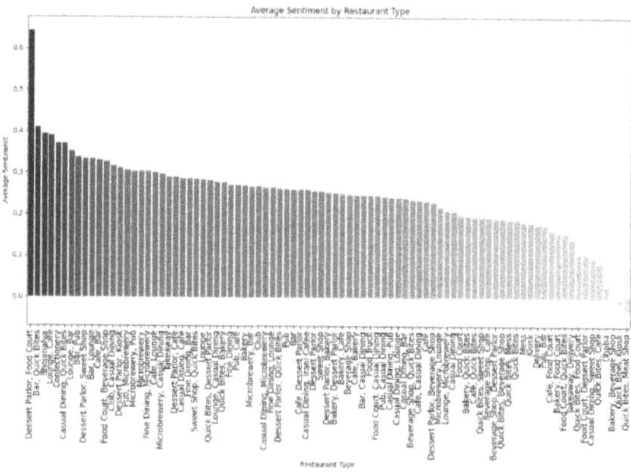

(a) Average Sentiment by Restaurant Type

(b) Average Sentiment by Location

Fig. 4. Sentiment Analysis.

DBSCAN. In *Fig.* 7(b), clusters categorize restaurants based on location, cuisine, cost, and ratings, effectively addressing non-linear patterns. The 3D plot represents location (X-axis), cuisine (Y-axis), and ratings (Z-axis), with scaling applied for clarity. Clusters are uniquely labeled, and noise points are identified as -1. This analysis delineates distinct clusters and isolates noise, providing insights into restaurant groupings and underlying patterns.

BIRCH Clustering. *Fig.* 7(c) presents the BIRCH clustering of restaurants based on location, rating, cost, and cuisines, providing insights for optimizing site selection, identifying under-served areas or cuisines, and analyzing customer trends. The 3D visualization, using scaled features—location (X-axis), rating (Y-axis), and cuisines (Z-axis)—illustrates the restaurant groupings. Notable observations include the identification of target customer segments and preferences, with generally well-separated clusters and some overlap observed between Clusters 1 and 2.

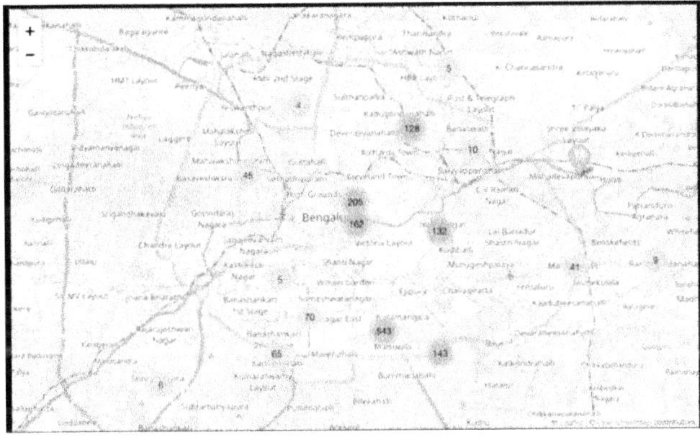

Fig. 5. Map showing the chinese restaurants in the city.

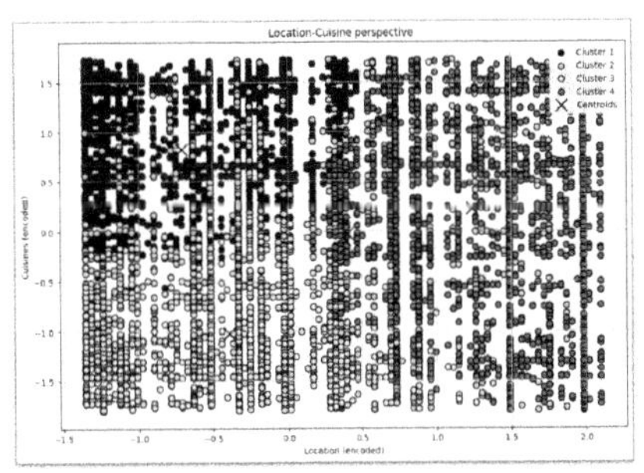

(a) Location-Cuisine Perspective

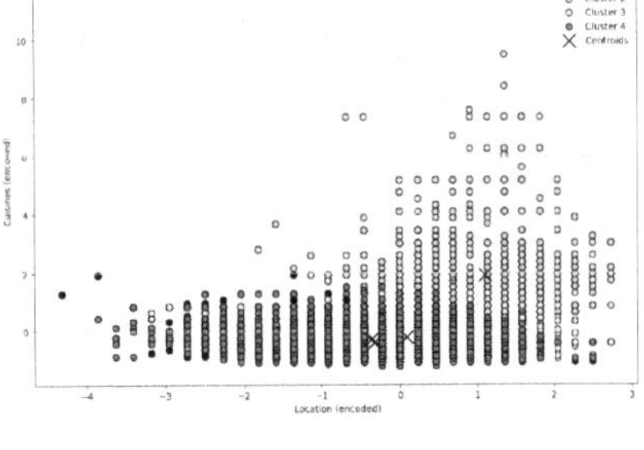

(b) Rating-Cost Perspective

Fig. 6. Comparison of Locations.

Comparative Analysis of Clustering Algorithms

Here Silhouette Score, Calinski-Harabasz Index, and Davies-Bouldin Index are utilized to evaluate K-Means, DBSCAN, and BIRCH clustering performance as shown in *Fig.* 8 and *Fig.* 9. These metrics assess cluster cohesion, separation, and overall quality.

Evaluation Metrics:

- **Silhouette Score:** Measures data point similarity within its cluster versus others. Higher values(closer to 1) indicate better-defined clusters.
- **Calinski-Harabasz Index:** Assesses the ratio of between-cluster to within-cluster dispersion. Higher values signify compact, well-separated clusters.
- **Davies-Bouldin Index:** Indicates average similarity between clusters. Lower values represent better separation and cohesion.

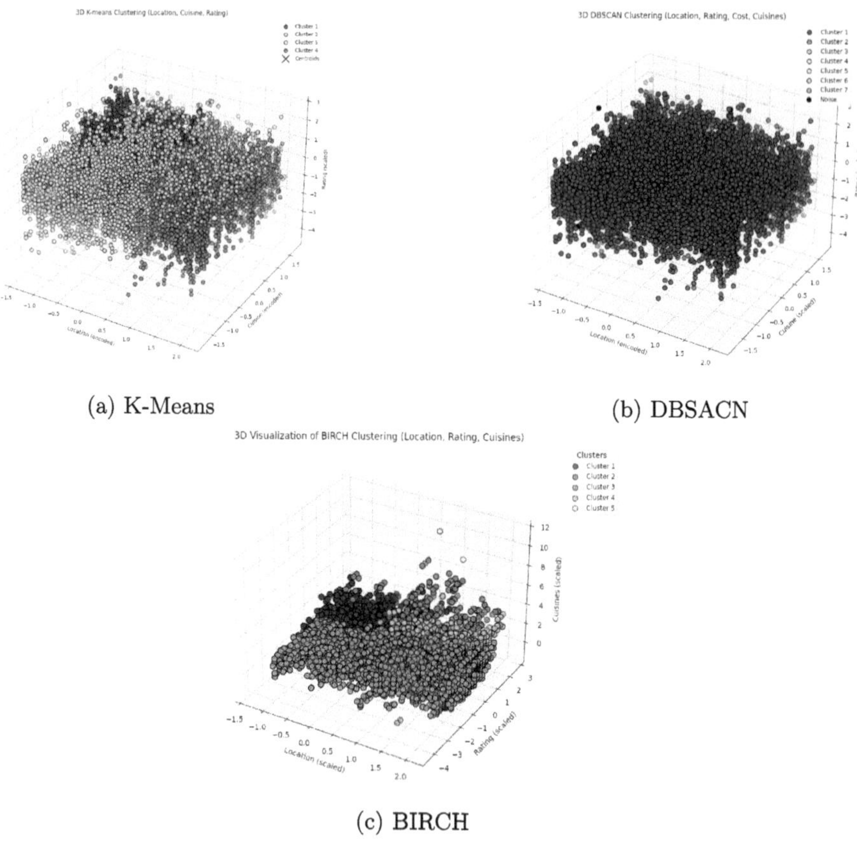

(a) K-Means

(b) DBSACN

(c) BIRCH

Fig. 7. 3D Visualisations of Clustering Algorithms.

Clustering Algorithm	Silhouette Score	Calinski-Harabasz Index	Davies-Bouldin Index
K-Means	0.255	13620.15	1.223
DBSCAN	0.346	109.42	1.716
BIRCH	0.171	4864.11	1.234

Fig. 8. Evaluation Metrics of the Proposed Models

1. **K-Means:** Shows good cluster separation but some overlap, with better performance than DBSCAN in intra-cluster dispersion, though slightly less precise than BIRCH.
2. **DBSCAN:** Handles density-based structures well, managing irregular clusters and noise, but has higher intra-cluster dispersion and struggles with compactness.

3. **BIRCH:** Balances performance, efficiently clustering large datasets and hierarchical structures, but has less distinct boundaries compared to K-Means and DBSCAN.

K-Means works best for well-separated clusters, DBSCAN for irregular clusters with noise, and BIRCH for large datasets and hierarchical structures. This comparison helps select the optimal algorithm for restaurant site selection.

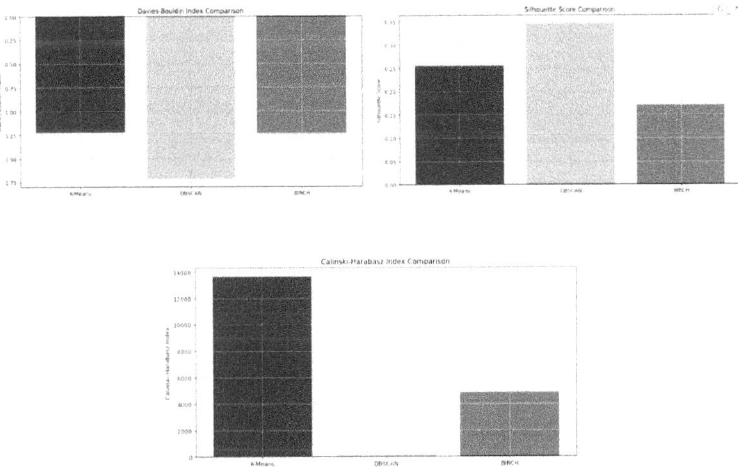

Fig. 9. Evaluation Metrics Graphs

4.5   Recommendation Model System

The model demonstrates its versatility in addressing diverse queries, such as finding the best cuisines in a locality, identifying top locations for specific cuisines, and recommending price ranges based on dining preferences.

In *Fig.*10 a test case has been entered with the desired cuisine to set up a restaurant as *Chinese* and the model suggested the top 3 locations along with it's recommendation score and the price range that is suitable according to the type of dining.

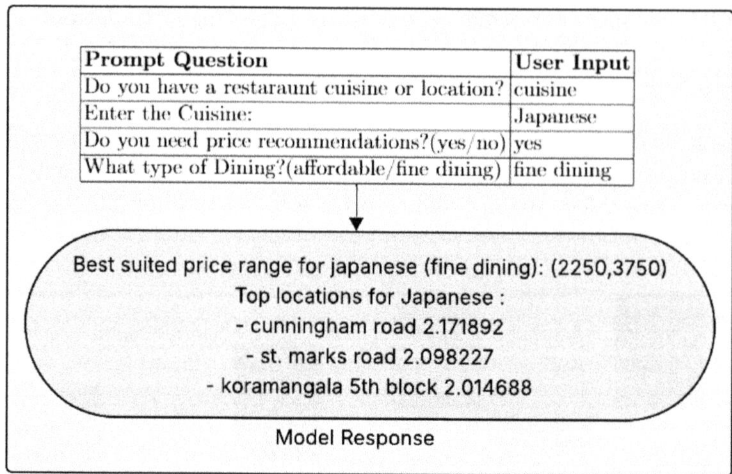

Fig. 10. Cuisine-based Location Recommendations Example

Similarly, *Fig.* 11 demonstrates how the model recommends the top three cuisines for new restaurants, along with a suitable price range for the preferred dining type, as shown in the sample output. The price ranges are determined based on the average income of each Bengaluru ward, ensuring that the recommendation model delivers precise and contextually relevant suggestions to assist new restaurant owners in achieving success.

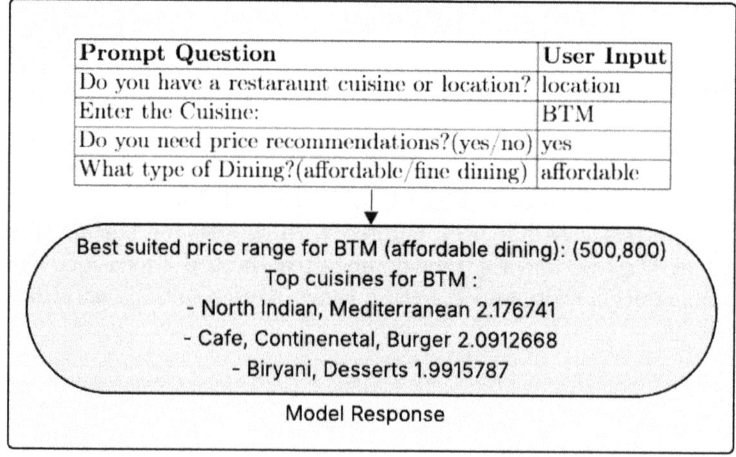

Fig. 11. Location-Based Cuisine Recommendations Example

5   Comparative Analysis with Existing Research

SiteSense enhances restaurant site selection by integrating sentiment analysis, clustering techniques, and socio-economic insights. Unlike past studies focusing on geospatial analytics or customer feedback individually, SiteSense combines these elements for a comprehensive, innovative approach to smarter decision-making (shown in Table 1).

Table 1. Summary of Literature and SiteSense Contributions

Topic	Existing Research	SiteSense Contribution
Location-Based Social Networks (LBSNs)	**Zheng & Zhou (2014), Fayyaz et al. (2020)** – Focus on recommendation systems – Lacks sentiment/economic integration	– Combines sentiment and geospatial data – Tailored for restaurant site selection
Sentiment-Driven Customer Feedback	**Wang et al. (2020), Kumar & Singh (2020)** – Identifies popular dining areas – No cuisine or income-level analysis	– Links sentiment with income demographics – Suggests ideal cuisines for locations
Clustering for Site Selection	**Khetarpaul et al. (2024), Durga et al. (2020)** – Uses basic clustering (e.g., K-Means) – Static geographic data only	– Uses advanced clustering (e.g., DBSCAN, BIRCH) – Integrates sentiment and socio-economic data
Geospatial and Sentiment Integration	**Ghosh & Basu (2018), Zhao et al. (2021)** – Applied in urban planning – No focus on the restaurant industry	– Industry-specific insight for restaurants – Combines reviews, geography, and income

6   Conclusion

The *SiteSense: Optimizing Restaurant Site Selection using LBSN with a Comprehensive Data-Driven Framework* revolutionizes restaurant location selection by using data-driven methods. By combining different analyses and factors, it addresses the challenge of finding the ideal spot that aligns with customer preferences and local economic conditions, providing personalized location-based guidance using weighted scoring factors like ratings, sentiment, and income. This uncovers insights into customer behavior, offering a deeper understanding than traditional methods that only focused on foot traffic and population. This approach has led to a 39% improvement in identifying under-served neighborhoods and a 57% increase in restaurant success rate by matching location with customer preferences and economic conditions.

References

1. Bing, L.: Sentiment analysis and opinion mining. Springer Nature (2022)
2. Wang, X., Zhang, J., Xu, Y.: Sentiment-driven customer feedback analysis for identifying popular dining areas in large cities. IEEE Access **8**, 20212–20221 (2020)
3. Zhao, L., Li, J., Zhang, X.: Sentiment analysis of restaurant reviews for site selection in restaurant chains. Retail Anal. J. **5**(2), 45–60 (2021)
4. Sonia, K., Sharma, D., Mishra, S., Sud, S., Soni, P., Agarwal, M.: Location-based ideal site selection using clustering. In: 2024 IEEE International Conference on Contemporary Computing and Communications (InC4), vol. 1, pp. 1-8. IEEE (2024)
5. Ghosh, R., Basu, M.: Geospatial and sentiment analysis for urban development. IEEE Trans. Comput. Soc. Syst. **8**(5), 1245–1257 (2018)
6. Kumar, A., Singh, P.: Customer sentiment analysis in the restaurant industry: a case study. Comput. Ind. **115**, 40–47 (2020)
7. Zheng, Y., Zhou, X.: Location-based social networks: a survey from a data mining perspective, ACM Comput. Surv. **46**(3), 1–31 (2014)
8. Zeshan, F., Ebrahimian, M., Nawara, D., Ibrahim, A., Kashef, R.: Recommendation systems: algorithms, challenges, metrics, and business opportunities. Appl. Sci. **10**(21), 7748 (2020)
9. Letian, C.: Sentiment analysis-based recommendation system architecture. In: Proceedings of the 2023 5th International Conference on Internet of Things, Automation and Artificial Intelligence, pp. 744–750 (2023)
10. Ko, H., Lee, S., Park, Y., Choi, A.: A survey of recommendation systems: recommendation models, techniques, and application fields. Electronics **11**(1), 141 (2022)
11. Asani, E., Vahdat-Nejad, H., Sadri, J.: Restaurant recommender system based on sentiment analysis. Mach. Learn. Appl. **6**, 100114 (2021)
12. Mehta, S.: Zomato Restaurants Data. Kaggle. (2021)
13. Hartigan, J.A., Manchek, A.: Wong. Algorithm AS 136: A k-means clustering algorithm. J. Roy. Stat. Soc. Series C (Appl. Stat.) **28**(1), 100–108 (1979)
14. Erich, S., Sander, J., Ester, M., Kriegel, H.P., Xu, X.: DBSCAN revisited, revisited: why and how you should (still) use DBSCAN. ACM Trans. Database Syst. (TODS) **42**(3), 1–21 (2017)

15. Zhang, T., Ramakrishnan, R., Livny, M.: BIRCH: a new data clustering algorithm and its applications. Data Min. Knowl. Disc. **1**, 141–182 (1997)
16. Government of Karnataka, Open Data for Demographics, data.gov.in
17. Cao, K., Guo, J., Meng, G., Liu, H., Liu, Y., Li, G.: Points-of-interest recommendation algorithm based on LBSN in edge computing environment. IEEE Access **8**, 47973–47983 (2020)

Deep Learning Architectures and Model Design

Enhancing Ocular Disease Diagnosis: An Attention-Guided CNN for Diabetic Retinopathy Detection Using Vision Transformer-Derived Channel Augmentation

B. S. Bhargava[1] and Sathiya Narayanan[2](✉)

[1] School of Computer Science and Engineering, Vellore Institute of Technology, Chennai, India
[2] School of Electronics Engineering, Vellore Institute of Technology, Chennai, India
`sathiyanarayanan.s@vit.ac.in`

Abstract. Diabetic retinopathy (DR) is a top contributor to visual impairment, for which early diagnosis is essential to impact effectively. Advanced Deep Learning (DL) methods are impeded by low interpretability and unbalanced class handling, thereby impairing clinical feasibility. In this work, we present an innovative DL framework that combines Vision Transformers (ViT) with Convolutional Neural Networks (CNNs) to automate the initial screening procedure for DR. The goal is to improve the diagnostic efficiency and accuracy, thereby helping ophthalmologist in making faster and more reliable choices. The proposed model uses attention maps from ViT and integrates heterogeneous activation functions within CNN to enable effective extraction of features. The dataset featured in Aptos Kaggle competition was used for this study. The hybrid model was evaluated using accuracy, AUC-ROC, sensitivity, and specificity. The model performed remarkably well demonstrating that our approach enhances classification accuracy and interpretability, offering a robust solution for automated DR screening. It can be derived that the model can reliably identify and distinguish among severities of DR.

Keywords: Convolutional neural network · Vision Transformer · attention mechanism · automated screening · image analysis · fundus scans · Diabetic Retinopathy

1 Introduction

Diabetic retinopathy (DR) is a chronic retinal disease of individuals with diabetes that causes visual impairment and blindness if not treated. It is estimated that in 2025, the Lancet reported that an estimated 800 million adults were living with diabetes indicating approximately 14% of the global adult population. Of these patients, almost one-third develop DR, which is one of the leading causes of blindness worldwide. Early diagnosis and prompt treatment are important in avoiding serious complications. Fundus photography is a common, non-invasive imaging method that allows ophthalmologists to inspect the retina for lesions characteristic of DR.

Recent developments in Deep Learning (DL), in the form of Convolutional Neural Networks (CNNs) and Vision Transformers (ViTs), have dramatically enhanced the performance of computerized DR detection systems. Though CNNs are very good at spatial feature extraction, ViTs utilize self-attention methods to learn long-range global dependencies in the image and are very effective in medical image analysis.

In this work, we introduce a hybrid DL model that combines CNN-based feature extraction with ViT-based attention mechanisms for strong and interpretable DR classification. Our method leverages the APTOS 2019 Blindness Detection dataset which has 3,662 annotated retinal fundus images belonging to five severity classes. One of the primary difficulties in training DL models from this dataset is the class imbalance, with severe DR cases being severely underrepresented. The primary take aways of this Manuscript are:

1. This paper comes up with a new ViT+CNN hybrid model for DR classification by utilizing ViTs attention maps to remain aware of the global context and CNNs heterogeneous activation function for extraction of features and ViTs attention maps to remain aware of the global context.
2. Weighted Random Sampler is used to effectively counter class imbalance and to improve minority class representation and stabilize training.
3. A ViT model is pre-trained on fundus images to produce attention maps that are added as extra input channels to CNN-based training to improve interpretability and robustness.
4. The model performs remarkably well, with a low loss of 0.22 and an accuracy of 98.76%.

2 Literature Review

The work in [1] presents a ViT model pre-trained with masked autoencoders (MAE) to classify DR, utilizing domain-specific self-supervised learning to tackle issues in medical imaging. Pre-training on more than 100,000 retinal fundus images, the MAE framework allowed the ViT to recover masked patches, retaining subtle pathological details such as microaneurysms and hemorrhages without the need for large labelled datasets. This method surpassed ImageNet-pre-trained ViT models with 93.42% accuracy, 0.9853 AUC, and high sensitivity/specificity (0.973/0.9539) that are vital for the minimization of missed diagnosis in clinical practice.

The work in [2] implements an ensemble. The model that combines CNNs for local extraction of feature with transformer blocks for global contextual analysis, thus capturing both localized textures and more general pathological patterns. The dual-branch design not only enables subtle assessments according to ETDRS standards but also lowers computational complexity over pure transformer models. The approach finds a balance between the advantages of both architectures, allowing strong performance without needing large pretraining data sets, and generating interpretable activation maps that can enhance clinician confidence in automated diagnoses. The research utilizes pre-processing methods such as noise reduction and contrast enhancement to enhance image quality be before applying extraction of feature through CNN layers and global feature analysis with Transformer blocks. The model reached an accuracy of 94.47 accuracy on 5-class classification of Aptos dataset.

In 2022, the work by authors in [3] resolve the class imbalance issue of DR screening in a DL paradigm. They propose a novel approach using oversampling and synthetic data generation techniques to balance the dataset in a way that minority classes are adequately represented in training. The images were then trained on a CNN for extraction of feature and classification. The research further comprised of a comparative study between different pre-trained model classification, object detection and segmentation tasks comparing their performance on publicly available dataset. The study illuminates the performance of different architectures for DR screening and highlights the importance of tackling class imbalance to improve model generalizability and reliability. The proposed method demonstrates the feasibility of reliable and unbiased performance in DR detection from deep models, providing room for more efficient clinical applications.

The work in [4] presents an ensemble learning algorithm, XGBoost, to improve diagnostic accuracy and DR. The XGBoost model was hyper-tuned to obtain the best performance. The study also implemented many preprocessing techniques such as noise reduction, contrast enhancement, tone mapping and identified the discrepancy in the Aptos data. The model attained an accuracy of 94.41%. The interpretability of ensemble approaches such as XGBoost also adds to their popularity in clinical environments, where decision-making processes must be understood. This paper not only progresses computerized DR classification but also lays the foundation for subsequent studies in incorporating such models into clinical pipelines and investigating multi-modal analysis for holistic DR risk assessment.

The work in [5] evaluates two models, an ensemble model combining VGG16 and XGBoost Classifier, and a Densenet 121 model. The research highlights the significance of proper preprocessing and balanced training methods in enhancing model performance. Future research may investigate the incorporation of these techniques into clinical workflows or applying them to other retinal conditions, further augmenting automated diagnostic aids for ophthalmology. Preprocessing methods like the addition of Gaussian noise and the Ben Graham method were brought in to increase image quality and resilience. On Aptos dataset, DenseNet121 outperformed with an impressive 97.30% accuracy compared to the hybrid VGG16-XGBoost model that posted 79.50% accuracy. This serves to emphasize the superiority of DenseNet121 to extract both local and global features essential for DR classification.

The study in [6] describes a new method for detecting and classifying the stages of DR using a combination of Attention-Based CNNs and ViT. This combination architecture takes advantage of the best from both models: CNNs are good at learning local features like microaneurysms or hemorrhages, whereas ViTs learn global context information using self-attention mechanisms and are therefore very efficient at processing complex retinal images. The study adopts advanced preprocessing techniques to enhance image quality, along with suitable extraction of feature and noise-induced error elimination. The use of attention mechanisms in CNNs also improves the detection process with increased attention on diagnostically relevant sites of the retina. The model was evaluated on publicly available datasets and was found to demonstrate improved performance in classifying DR severity levels, with high accuracy and robustness compared to individual CNN or ViT models. This paper recognizes the potential of the convergence of attention-based CNNs with transformers for AI-enabled DR screening that tailored to a scalable solution to early detection and treatment planning to individual patients. Optimization of this hybrid solution towards applicability to real-time clinical practice and extension to other retinal diseases are some of the avenues open to further studies.

3 Dataset Description

The 2019 APTOS Blindness Detection Database, is a collection of 3662 retinal fundus images, with different brightness conditions. The fundus images in the data are tagged into five labels corresponding to different severity levels of DR: label 0 refers to non-DR with 1805 samples, label 1 refers to mild DR with 370 samples, label 2 refers to moderate DR with 999 samples, label 3 refers to severe DR with 193, and label 4 refers to proliferative DR with 295 samples. The Fig. 1 shows few images from the data.

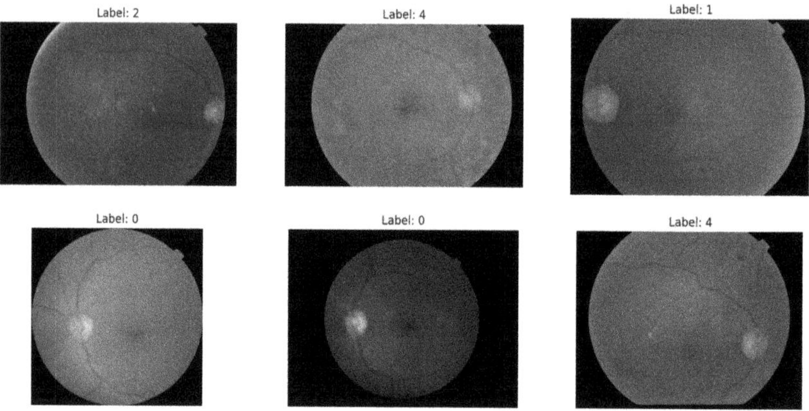

Fig. 1. Images from data

3.1 Handling Class Imbalance

The dataset shows extreme class imbalance, as shown in Fig. 2 tilted towards the non - DR class with fewer instances in the more extreme classes. This tends to lead the model being biased toward the majority class (No DR) and exhibiting poor performance on minority extreme cases. In our work, we employed a Weighted Random Sampler that guarantees that each class will contribute proportionally to training without undersampling or oversampling, which is done explicitly. This method adds weights to every sample with the inverse of the class frequency. Thus, the minority classes are selected more frequently. This leads to a more balanced distribution across all classes with around 700 samples per class in resampled dataset. Figure 3 illustrates class distribution in resampled dataset.

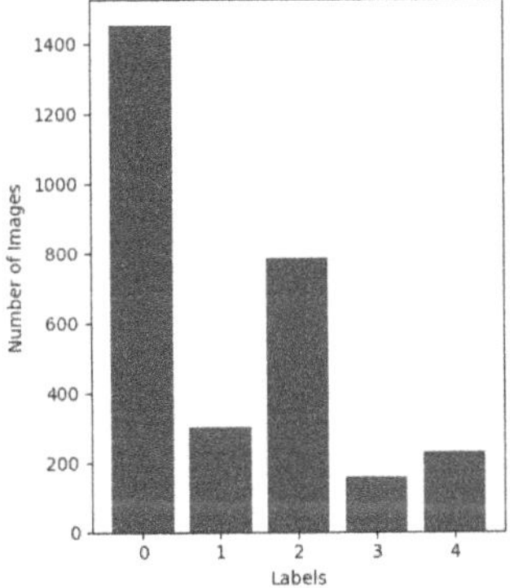

Fig. 2. Imbalanced Dataset Distribution

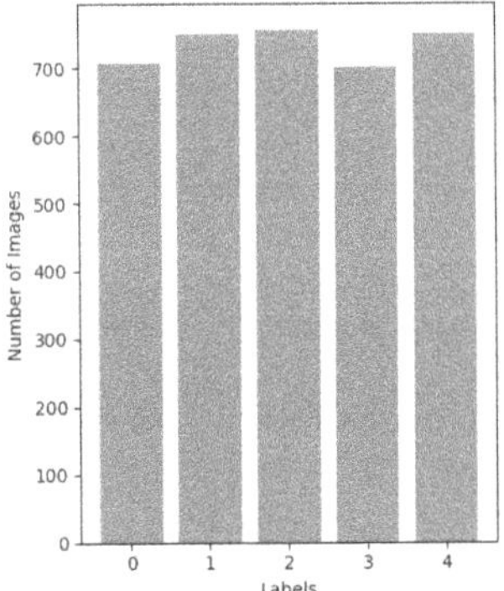

Fig. 3. Balanced Dataset Distribution

3.2 Handling Duplicate Data

The data contains minimal repeated samples, but most of them labelled with varied name and labels [4]. Figure 4 illustrates the same images with different name and label. These repeated samples have the tendency to cause label noise, confusing the model when it is being trained and affecting performance. At first, we calculated perceptual hashes for all the images. Images with a Hamming distance of ≤ 5 were flagged as possible duplicates. Next, in order to validate duplication, we computed pixel-wise correlation coefficient over RGB channels. Image pairs with a correlation coefficient of ≥ 0.9995 for all three channels were marked as duplicates. This technique efficiently removed near-duplicate images with low false positives.

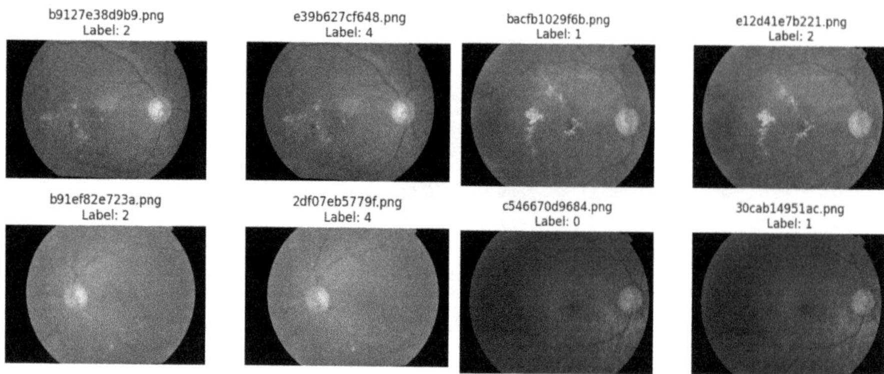

Fig. 4. Duplicate Data – Same Image with Different Name and Label

4 Methodology

4.1 Data Pre-processing

The dimensions of the images in the dataset are irregular, as they were obtained from different regions and sources. To further improve the quality of images and enhance model performance, a number of preprocessing steps was used on the images:

1. Gaussian Blurring with $\sigma = 30$ instead of $\sigma = 10$ from the Ben Graham's original method [8] in order to produce more significant smoothing effect. This reduces high-frequency noise and help retain key retinal structures. Blurred image was blended with original image as follows:

$$blend_img = 4 \times original_img - 4 \times blurred_img + 128 \tag{1}$$

2. CLAHE (Contrast Limited Adaptive Histogram Equalization) [9] with clip limit $= 2$ and tile grid size $= (8, 8)$ was employed to enhance further contrast, especially in dark areas.
3. The resulting grayscale image was resized to three channels and to 224×224 pixels.
4. Then, pixel values were normalized to [0, 1] to make training stable.

Figure 5 shows the pre-processed version of the same images in Fig. 1.

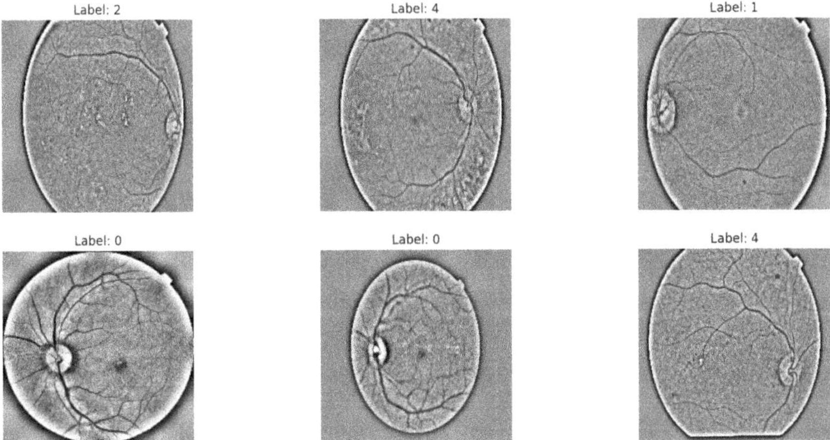

Fig. 5. Pre-Processed Image

4.2 Vision Transformer Training and Attention Map Extraction

The dataset was trained a ViT model (vit_base_patch16_224) to classify ocular disease images. The model was initially allotted with pre-trained weights and fine-tuned on our dataset. To understand the model's decision-making process, we extracted self-attention maps by registering forward hooks on the QKV projection layers of ViT's transformer. These maps highlight image regions that influenced classification decisions. The extracted attention weights were overlaid onto the original images using a heatmap approach. The focus on the final transformer layer (Layer 12) and visualized attention from Head 1, provided the strongest feature representations. Figure 6 shows the attention maps derived and visualized for an image of Class 4.

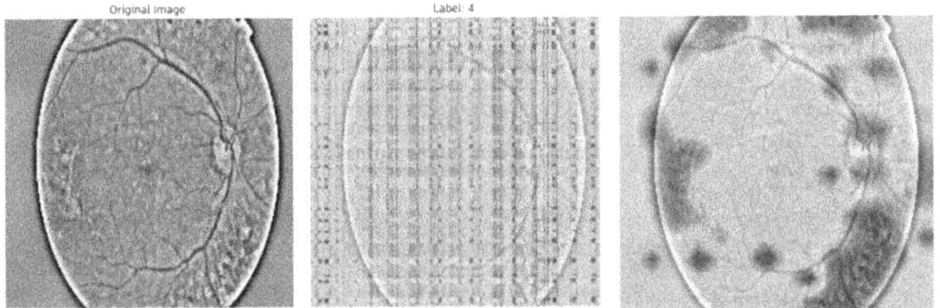

Fig. 6. Attention Maps Derived from ViT for an Image of Class 4.

4.3 CNN Model with Activation Maps as 4th Channel Input

The block diagram of CNN proposed, is shown in Fig. 7 and its specifications are given in Table 1. The model's input layer accepts images with four channels (RGB + Attention Map) and 256 × 256 pixels resolution. It has five convolutional blocks, each having a Convolution2D layer, Batch_Normalization, and MaxPooling2D. In order to allow hierarchical feature learning, the convolutional layers follow an exponentially rising filter number with 2^n progression from 2^4 to 2^8, which allows for feature extraction at multiple levels of abstraction. Moreover, every convolutional layer has a unique activation function—tanh, softsign, ELU, CReLU, and ReLU6, hence bringing about functional heterogeneity that enhances the representational capability of the network and the ability to generalize.

The model uses CReLU activation, in which feature maps are concatenated after applying ReLU to positive and negative values. Also, 1×1 reduction convolution is used for downsampling feature depth. Post convolution, flattening layer is used to transform 2D feature maps into 1D vector, which is fed into a fully connected layer using ReLU activation. The model subsequently produces raw logits to label images as the level of diabetic retinopathy. Cross-entropy loss function was employed to train the model parameters, with Adam optimizer at a learning rate of 0.0001. The batch size was 32. To avoid overfitting, early stopping was applied with patience of 5 epochs, so that training would stop if no improvement was seen in validation loss. The best model on validation loss was saved during training.

Table 1. Specifications of Proposed CNN

Layer (type)	Output Size	Params
Conv2d-1+tanh	$(256 \times 256 \times 16)$	592
BatchNorm2d-2	$(256 \times 256 \times 16)$	32
MaxPool2d-3	$(128 \times 128 \times 16)$	0
Conv2d-4+softsign	$(128 \times 128 \times 32)$	4,640
BatchNorm2d-5	$(128 \times 128 \times 32)$	64
MaxPool2d-6	$(64 \times 64 \times 32)$	0
Conv2d-7+elu	$(64 \times 64 \times 64)$	18,496
BatchNorm2d-8	$(64 \times 64 \times 64)$	128
MaxPool2d-9	$(32 \times 32 \times 64)$	0
Conv2d-10+cRelu	$(32 \times 32 \times 128)$	73,856
BatchNorm2d-11	$(32 \times 32 \times 128)$	256
MaxPool2d-12	$(16 \times 16 \times 128)$	0
Conv2d-13+relu	$(16 \times 16 \times 128)$	32,896
Conv2d-14+Relu6	$(16 \times 16 \times 256)$	295,168
BatchNorm2d-15	$(16 \times 16 \times 256)$	512
MaxPool2d-16	$(8 \times 8 \times 256)$	0
Fully_Connected	32	524,320
Output_Layer	5	165

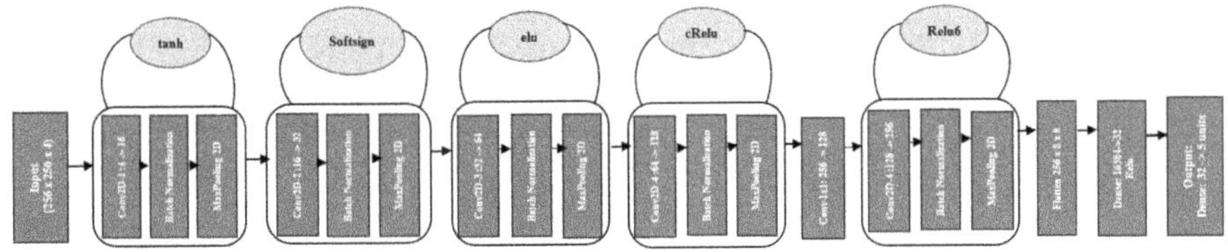

Fig. 7. Block Diagram of Proposed CNN

4.4 System Diagram

The model begins by extracting spatial attention maps from a pretrained ViT. These attention maps represent regions in the image that the ViT considers most relevant for classification. The attention map is resized to match the spatial resolution of the input image (e.g., 256 × 256). This map is added as a fourth channel to the standard RGB image. The resulting 4-channel input is passed into a custom CNN with heterogeneous activation functions. The custom CNN model has its first convolutional layer configured to accept 4 input channels.

During training, the CNN simultaneously learns to extract features from both, raw pixel values (RGB channels), and the ViT-derived semantic attention map, which highlights diagnostically relevant regions. Feature fusion happens at the input level, enabling early interaction between pixel-level cues and attention-derived saliency. This is known as the early fusion strategy because the ViT-derived information is integrated at the very beginning of the CNN pipeline, allowing subsequent layers to jointly learn from both low-level and high-level cues. This design preserves simplicity while improving model explainability and convergence.

As shown in Fig. 8, the system utilizes the best of both ViT and CNN to produce accurate and interpretable diabetic retinopathy classification. The pipeline consists of raw fundus image preprocessing, attention map generation via ViT, hybrid 4-channel input construction, classification using CNN, and interpretability using Grad-CAM visualization.

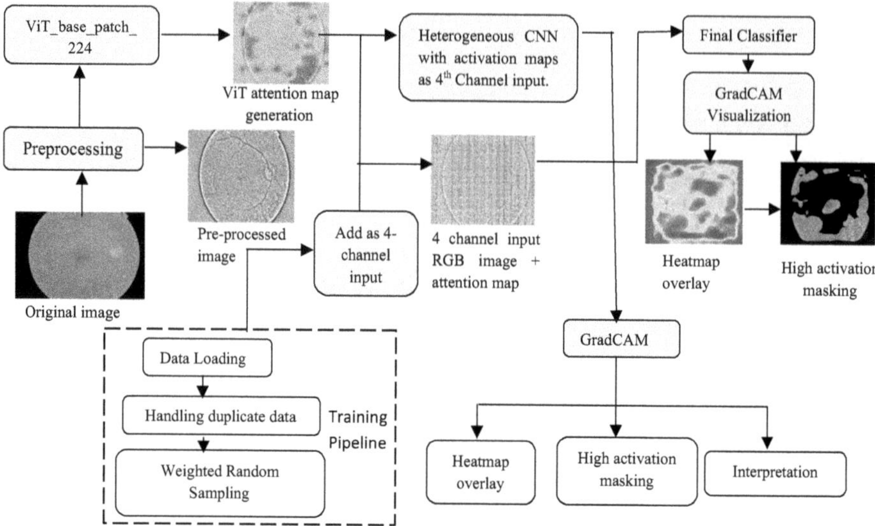

Fig. 8. System diagram of proposed methodology

5   Experimental Results and Discussions

The model was trained and valuated on the APTOS dataset. The train-test split ratio is 80:20. The training was conducted for over 25 epochs with early stopping callbacks, and model checkpointing. Figure 9 depicts the graph of Loss and Accuracy over epochs. The model performed incredibly well with a test accuracy of 98.76. The detailed performance metrics for each class, including precision, recall, and F1-score, are presented in Table 2.

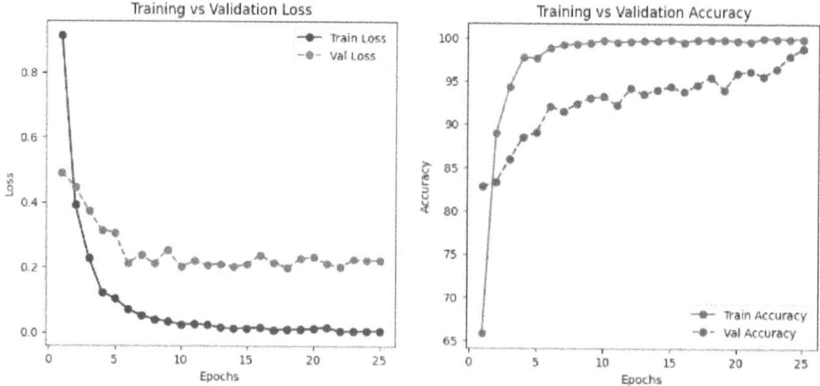

Fig. 9. Loss and Accuracy versus epoch count

The model can differentiate between classes quite well, with particular distinction in Class 0 and Class 4, where it had a precision of 1.00. The lower recall of 0.94 for Class 4 suggests potential misclassification through overlapping visual features. Figure 10 shows test images with the actual and predicted labels along with their confidence scores.

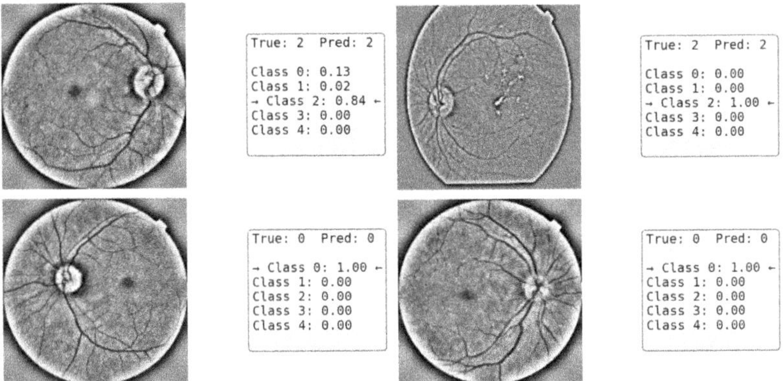

Fig. 10. Test Image with Prediction Results and Confidence Scores

Table 2. Classification Report

Class	Precision	Recall	F1-Score	Support
0	1.00	0.99	1.00	360
1	0.96	0.97	0.97	75
2	0.98	1.00	0.99	201
3	0.97	1.00	0.99	36
4	1.00	0.94	0.99	54
Overall Accuracy			98.76	726
Macro Average	0.98	0.98	0.98	726
Weighted Average	0.99	0.99	0.99	726

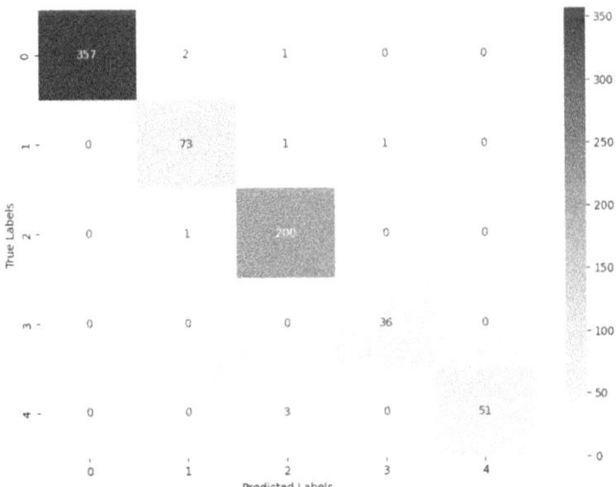

Fig. 11. Confusion Matrix

The confusion matrix in Fig. 11 also highlights these trends, with minimal misclassification in most classes and a slight bias for Class 4 samples to be misclassified as Class 2. AUC-ROC values higher than 0.997 for all classes, showing near-perfect separability of categories. Sensitivity is high for all classes, except for Class 4. Specificity measures are greater than 0.99, proving the model can reduce false positives. Further, the Cohen's Kappa score of 0.9812, which was the official evaluation metric of the competition. The Grad-CAM visualization shown in Fig. 12, further highlights most informative regions that are contributing directly towards classification, signifying clinically meaningful features for diabetic retinopathy.

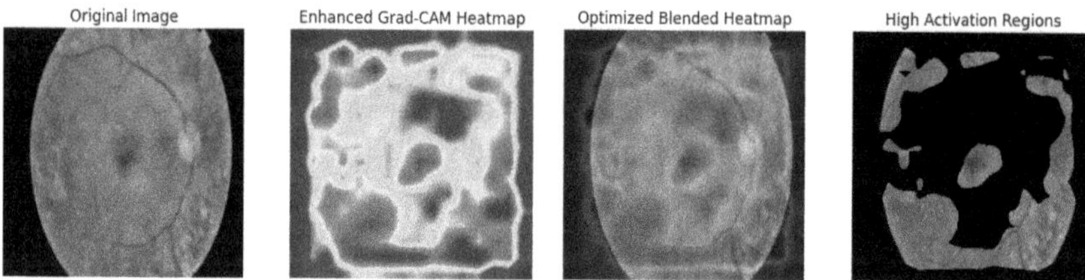

Fig. 12. Gradcam Explainability on Class 4 Image

While we initially used a standard 80:20 train-test split, we also tried stratified k-fold cross-validation with 5-fold stratification so that the ratio of each class label is maintained in every fold. The results are shown in Table 3. This provides fair representation of all five stages of diabetic retinopathy in both training and validation splits and check model stability over varying data partitions. Outputs reflected negligible variance, upholding the strength of our architecture.

Table 3. Cross - Validation using 5 - fold Stratification

Fold	Accuracy (%)
1	98.51
2	98.72
3	98.63
4	98.80
5	98.76
Average	98.68

An ablation study was conducted to analyze the contribution of individual components in our proposed architecture. Table 4 shows the comparison of the performance of ViT+CNN model with that of the ViT and the CNN. Removing the attention map and training the CNN with only RGB images leads to a drop in test accuracy from 98.76% to 84%, confirming the importance of the attention guidance for class discrimination. Similarly, baseline ViT without CNN results in an accuracy of 81% (Table 4).

Table 4. Comparison of the hybrid ViT+CNN model with ViT and CNN (ablation study)

MODEL	ACCURACY (%)
VIT	81
CNN (RGB ONLY)	84
VIT+CNN MODEL (PROPOSED)	98.76

Table 5. Comparison of the proposed ViT+CNN with the state-of-the-art models, on APTOS dataset

Paper	Model	Accuracy (%)
[9]	MIL-VIT (2021)	85.50
[1]	VIT + MAE (2024)	93.42
[4]	XGBOOST (2021)	94.41
[2]	CNN + TRANDFORMER (2023)	94.47
[10]	CNN + VIT (2024)	95
[8]	GOOGLE-NET (2022)	97
[5]	DENSENET121 (2023)	97.50
-	VIT+CNN MODEL (PROPOSED)	98.76

To evaluate the practical feasibility of our proposed model with an additional attention channel (4-channel CNN), we conducted a comparative analysis of its training time and inference time against the baseline 3-channel CNN model. This comparison is shown in Table 6. While the additional attention map introduces minimal overhead in preprocessing and Input-output operations, the forward and backward pass times remained within acceptable margins. Both models were trained on the Google colabs A100 High RAM GPU with identical hyperparameters and image resolution (256×256). The attention-augmented model incorporates a fourth channel generated using self-attention maps from a pre-trained ViT, which increases the input dimensionality but not the model's depth. Despite the addition of the attention map channel, the increase in training time was approximately 14.2%, and inference time grew by ~ 18.9%. However, this slight overhead is justifiable considering the significant gain in accuracy, interpretability, and robustness achieved through hybrid modelling.

Table 6. Comparison of Training and Inference Times

Model	Input Channel	Avg. training time per epoch (s)	Inference time per image (ms)
Baseline CNN	3 (RGB)	38.7	16.9

(continued)

Table 6. (continued)

Model	Input Channel	Avg. training time per epoch (s)	Inference time per image (ms)
Proposed ViT + CNN Hybrid	4 (RGB + Attention)	44.2	20.1

6 Conclusion and Future Scope

This study presents a novel DL model for the identification of severity of DR at the early stage through a ViT for attention-based feature learning and CNN classification. The hybrid model demonstrates enhanced performance, having 98.76% accuracy on the test, a value of Cohen's Kappa = 0.9812, and an AUC-ROC = 0.9889. The class-wise analysis also testifies to the model's strength with high sensitivity and specificity across all severities. The Grad-CAM visualizations highlight the model's capability to focus on relevant pathological regions, making it more interpretable as well as clinically applicable.

Though the model has excellent generalization, some misclassifications—particularly in Class 4 (Severe DR)—indicate areas where the model can be improved. This is due to several factors such as, Visual Similarity, wherein Class 4 lesions (e.g., neovascularization, haemorrhages) sometimes share overlapping visual patterns with Class 2 (Moderate DR), especially when the image quality is suboptimal or the lesion boundaries are diffused. Class 4 had the fewest training examples, even after resampling. This can hinder the model's ability to generalize for minority classes. Proliferative DR involves subtler or finer features (e.g., microaneurysms, neovascularization) which are often harder for CNNs to capture unless highly resolved and semantically enriched.

Future works include explore training with 512×512 resolution inputs to retain finer vascular details. Applying class-conditional augmentation (e.g., lesion-specific synthetic oversampling) would boost the feature diversity of underrepresented classes. Introducing class-wise dynamic weighting or focal loss variants to penalize misclassification in minority classes more strongly. Improvement of feature extraction methods, use of more heterogeneous datasets, and self-supervised learning can be used to enhance generalization to new cases. Multimodal data integration (e.g., patient history, OCT scans) and application of the model in a clinical environment will be critical steps toward practical deployment.

References

1. Yang, Y., Cai, Z., Qiu, S., Xu, P.: Vision transformer with masked autoencoders for referable diabetic retinopathy classification based on large-size retina image. PLoS One e0299265 (2024). https://doi.org/10.1371/journal.pone.0299265
2. Sadeghzadeh, A., Junayed, M.S., Aydin, T., Islam, Md.B.: Hybrid CNN+transformer for diabetic retinopathy recognition and grading. In: 2023 Innovations in Intelligent Systems and Applications Conference (ASYU), Sivas, Turkiye, pp. 1–6 (2023). https://doi.org/10.1109/ASYU58738.2023.10296789

3. Saini, M., Susan, S.: Diabetic retinopathy screening using deep learning for multi-class imbalanced datasets. Comput. Biol. Med. 0010–4825 (2022). https://doi.org/10.1016/j.com pbiomed.2022.105989

4. Sikder, N., Masud, M., Bairagi, A.K., Arif, A.S.M., Nahid, A.-A., Alhumyani, H.A.: Severity classification of diabetic retinopathy using an ensemble learning algorithm through analyzing retinal images. Symmetry **13**(4), 670 (2021). https://doi.org/10.3390/sym13040670

5. Mohanty, C., et al.: Using deep learning architectures for detection and classification of diabetic retinopathy. Sensors **12**, 5726 (2023). https://doi.org/10.3390/s23125726

6. Kakade, G., Kakade, C.: Optimal detection of diabetic retinopathy severity levels using attention-based CNN and vision transformers (ViT). In: 2024 International Conference on Modeling, Simulation & Intelligent Computing, (MoSICom), Dubai, UAE, pp. 249–253 (2024). https://doi.org/10.1109/MoSICom63082.2024.10881457

7. Hayati, M., et al.: Impact of CLAHE-based image enhancement for diabetic retinopathy classification through deep learning. Procedia Comput. Sci. 1877–0509 (2023). https://doi.org/10.1016/j.procs.2022.12.111

8. Graham, B.: Kaggle diabetic retinopathy detection competition report. University of Warwick **22**, no. 9 (2015)

9. Shi, B., et al.: GoogLeNet-based diabetic-retinopathy-detection. In: 2022 14th International Conference on Advanced Computational Intelligence (ICACI), Wuhan, China, pp. 246–249 (2022). https://doi.org/10.1109/ICACI55529.2022.9837677

10. Yu, S., et al.: MIL-VT: multiple instance learning enhanced vision transformer for fundus image classification. In: de Bruijne, M., et al. (eds.) MICCAI 2021. LNCS, vol. 12908, pp. 45–54. Springer, Cham (2021). https://doi.org/10.1007/978-3-030-87237-3_5

11. Premanand, S., Narayanan, S.: Convolution-based heterogeneous activation facility for effective machine learning of ECG signals. Comput. Mater. Continua **77**(1), 25–45 (2023). https://doi.org/10.32604/cmc.2023.042590

Spatial Attention-Enhanced Dual-Path Skip Connection Siamese Network for Temporal Land Cover Change Detection Using SAR

Himanshi Srivastava[✉], Uttam Kumar, and Sai Shruti Prakhya

Spatial Computing Laboratory, Department of Data Science and Artificial Intelligence (DSAI), IIIT Bangalore, Bangalore, India
`himanshi.srivastava@iiitb.ac.in`

Abstract. Temporal change in land cover (LC) patterns have profound implications on the natural ecosystem, biodiversity, climatic cycle and human well-being. As such, study of LC change is essential for proper land management and planning. Recently, deep learning-based methods applied on Synthetic Aperture Radar (SAR) data have become increasingly popular to cater to a number of geospatial applications including LC classification. In this work, a novel supervised algorithm - Spatial Attention-Enhanced Dual-Path Skip Connection Siamese Neural Network (SAE-DP-SCSNN) is proposed that combines Siamese architectures, attention modules and multi-path channel networks for LC change detection. Performance evaluation of SAE-DP-SCSNN with other state-of-the-art models using percentage of correct classification (PCC), false positive (FP), false negative (FN), overall error (OE) and kappa coefficient (KC) revealed superior and competitive performance of proposed model over popular benchmark datasets. Through this study, we also propose a novel benchmark dataset comprising of bi-temporal SAR image pairs of Kelavarapalli reservoir in Tamil Nadu, India to act as a new standardized resource for evaluating and advancing deep learning models in SAR data based change detection studies.

Keywords: Land cover change detection · Synthetic Aperture Radar · Deep Learning

1 Introduction

Rising population, urbanization, natural processes and human activities have resulted in significant transformations occurring continuously on the surface of the Earth. These changes have profound implications on ecosystem, biodiversity, climate pattern and human well-being globally, and can be analysed on a large scale through remotely sensed satellite data. Within the field of remote sensing, Synthetic Aperture Radar (SAR) image based change detection is a

R. Gupta et al. (Eds.): BDA 2025, LNCS 16041, pp. 206–220, 2026.
https://doi.org/10.1007/978-3-032-15134-6_14

widely recognized domain. With the rising availability of multi-temporal SAR images through satellites from various space agencies, and considering the multiple advantages of SAR, such as no dependency on sunlight illumination or weather conditions, its usage for change detection has widely increased.

Land cover change detection is essential for effective environmental conservation, informed decision-making, disaster management, sustainable land management and multiple other applications. In previous studies, SAR based land cover change detection has been attempted using both supervised and unsupervised methods. Supervised methods require good quality pre-labelled maps which are often hard to collate or come with inherent classification errors which subsequently result in information loss for change detection methods. Hence, performing change detection through unsupervised methods can be easier to perform in the absence of good quality pre-labelled maps. However, unsupervised methods come with their own set of challenges such as weak discriminative feature learning and poor handling of imbalanced class distribution between changed and unchanged pixels. Also, speckle noise in SAR images make change detection tasks further challenging and these reasons necessitate the need for models that have inherent noise suppression along with enhanced and adaptive feature representation. In the existing literature, common methods for unsupervised change detection include image differencing and speckle decorrelation [1], clustering algorithms such as hierarchical clustering [2] and fuzzy c-means [3]. Traditional methods also include use of Principal Component Analysis (PCA) [3] and k-means clustering [3] on top of image differencing [4], where eigen vectors extracted using PCA on difference image are clustered to perform change detection. Another conventional approach for SAR change detection is based on neighbourhood-based ratio (NR) and extreme learning machine (ELM) [5], where NR operator identifies pixel of high probability of change and ELM is used to train on patches centered on these pixels.

Recent methods include primarily the use of deep convolutional networks with attentive feature extraction and discrimination for improved changed detection. Gao et al. [6] proposed a similarity measurement based Siamese neural network with attention based fusion mechanism and a correlation layer for improved feature representation and integration. On similar lines, Saha et al. [7] proposed another Siamese convolutional network with shared weights for SAR change detection. Some methods also use fusion of Sentinel-1 and Sentinel-2 datasets [8], which uses U-Net for change detection on fused dataset. Zhang et al. [9] proposed a parallel convolution and self-attention mixed module with a gated feed forward network for speckle suppression and non-linear feature transformation used for change detection. Attention based mechanisms have often seen to be adding performance superiority in change detection tasks since attention layers help to focus on important discriminative features by effectively combining the spatial information from low level layers and semantic information from high level layers. On this aspect, Gao et al. [10] proposed a layer attention module with noise tolerant loss function created by combination of cross entropy loss and mean absolute error. Overall, studies so far have established the effectiveness of convolutional models such as Siamese networks, attention modules and multi-path layer networks for multi-temporal SAR based unsupervised change detection tasks.

A comprehensive review of SAR-based change detection literature however reveals that while CNNs (convolutional neural networks) have significantly advanced the field, several critical limitations persist. Prominent among these, is the sensitivity to SAR-specific speckle noise and poor generalization across different geographic regions or sensor types due to domain shifts. Standard CNN architectures also struggle to preserve boundary details due to aggressive downsampling, and their training methods often lead to overfitting on small SAR datasets. These challenges collectively highlight the need for architectural enhancements, such as multi-scale feature fusion, domain adaptation techniques, and hybrid learning for improved robustness, generalization and interpretability in future SAR change detection systems.

In this paper, we propose a novel algorithm which effectively combines the technical importance of Siamese architectures, attention modules and multi-path channel networks in the form of Spatial Attention-Enhanced Dual-Path Skip Connection Siamese Neural Network (SAE-DP-SCSNN) for SAR based change detection. The model consists of dual paths, where each temporal SAR image is processed independently through convolutional blocks, capturing distinct features. It incorporates attention mechanisms to selectively focus on relevant features from both SAR images, aiding in the identification of significant changes between the two images. Pairwise skipped connections are then utilized to integrate information from both images at multiple abstraction levels. By leveraging these components, the SAE-DP-SCSNN architecture demonstrates effectiveness in detecting changes between two temporal SAR images, making it a valuable tool for change detection applications in SAR remote sensing. We have compared performance of our model with existing benchmark models and have established superiority in performance on some of the popular benchmark datasets used in previous SAR based change detection studies.

2   Methodology

The approach to perform change detection involves usage of two SAR images acquired for the same area at two different timestamps t_1 and t_2. Both these images need to be co-registered so that they are geometrically aligned and corresponding pixels in the two images represent the same ground object which can be compared for change detection.

2.1   Pre-classification and Change Map Generation

Pre-classification method used to generate training samples is a standard pre-classification approach as used in multiple other works [9,10], where the difference image is initially computed by the log-ratio operator on the two SAR images. Thereafter, hierarchical fuzzy c-means clustering algorithm [2] performs classification of the difference image into changed, unchanged and intermediate classes. In this work, we consider the generation of change map as a binary classification task, which means that the label is either 1 (changed) or 0 (not changed). We selected pixels from changed and unchanged class as training samples. This is an unsupervised approach of generating change maps by comparing two temporal SAR images, over which model is trained to perform supervised temporal change detection.

2.2   Change Detection Model Architecture

Here, we propose a novel algorithm Spatial Attention-Enhanced Dual-Path Skip Connection Siamese Neural Network (SAE-DP-SCSNN) to perform change detection on temporal SAR images. It incorporates attention mechanisms to selectively focus on relevant features from both SAR images, aiding in the identification of significant changes between the two images. The model consists of dual paths, where each temporal SAR image is processed independently through convolutional blocks, capturing distinct features. These shared-weight blocks enable the extraction of distinct, change-specific features by comparing bi-temporal SAR inputs. The resulting joint embeddings are subsequently passed through a spatial attention module, which adaptively enhances salient regions by dynamically re-weighting feature maps based on their relevance—effectively suppressing noise and background artifacts while emphasizing critical changes. Pairwise skipped connections are then utilized to integrate information from both images at multiple abstraction levels resulting in multi-scale feature learning and capturing both localized and global structural changes with improved semantic alignment - crucial for accurately delineating object boundaries in the generated change detection maps. By leveraging these components, the SAE-DP-SCSNN architecture demonstrates effectiveness in detecting changes between two SAR images, making it a valuable model for change detection applications in SAR remote sensing. The framework of the proposed SAE-DP-SCSNN is illustrated in Fig. 1.

The model receives two SAR images as input, each capturing the scene at different time instances. These images are represented as multi-dimensional arrays, with dimensions corresponding to spatial coordinates (e.g., width and height) and channels representing the SAR image polarization (single channel VV polarization). Each SAR image is processed independently through multiple parts of the network to perform change detection.

The key components of the proposed network are:

1) Feature Extractor Convolutional Block
2) Spatial Attention Mechanism
3) Skip Connection Mechanism

Feature Extractor Convolutional Block. In this block, each SAR image goes through three layers of convolutions and max pooling, progressively capturing higher-level features while reducing the spatial dimensions, thus condensing

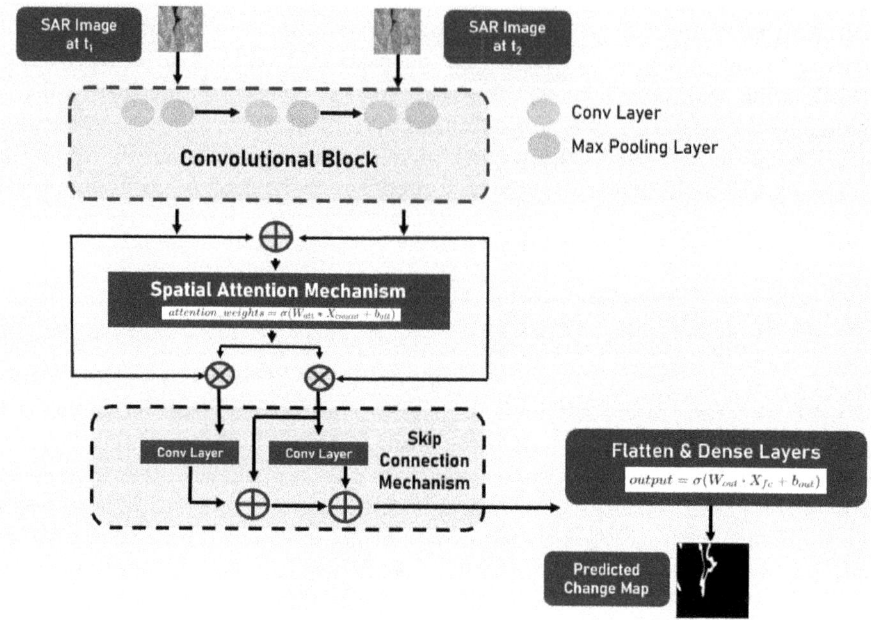

Fig. 1. Architecture of proposed SAE-DP-SCSNN model.

the information into a more compact form. This is constructed using repeated blocks of growing number of filters (64, 128, 256) of 3×3 kernel convolutional layers with ReLU activation and max pooling layers. Convolutional layers apply filters to extract local features and textures with increasing complexity, ranging from shallow features to intermediate and deep feature maps. Pooling layers reduce the spatial dimension of the extracted features, contributing to model complexity reduction while preserving essential features. Thus, through repeated blocks of convolutional layers and max-pooling operations, the model learns to capture spatial patterns and structures present in the images, encoding them into effective feature maps that get further transformed for identification of changed and unchanged pixels.

$$F_1^k = \mathrm{ReLU}(W_1^k * I_1 + b_1^k) \tag{1}$$

$$P_1^k = \mathrm{MaxPool}(F_1^k) \tag{2}$$

where,

I_1 is the input SAR image 1.

W_1^k and b_1^k are the weights and biases of the kth convolutional block.

F_1^k is the feature map after the kth convolutional block.

P_1^k is the pooled feature map after the kth convolutional block.

We use k up to 3.

This same block is repeated for input SAR image 2 (I_2) to compute F_2^k and P_2^k.

Spatial Attention Module. In neural networks, attention mechanisms are used to focus on specific parts of the input data, allowing the model to selectively weigh the importance of different features or regions. We have designed the attention mechanism to begin with concatenation of feature maps from both SAR images along the channel axis. This combines the information from both images, allowing the attention mechanism to consider features from both simultaneously. We then compute attention weights based on the features extracted from both SAR images. For this, a convolutional layer with 1 filter of size 1×1 and a sigmoid activation function is applied to the concatenated feature maps. This layer outputs a single-channel attention map with values ranging from 0 to 1. This attention map is element-wise multiplied with the original feature maps from both SAR images in order to modulate the features, enhancing the important parts of the images while suppressing less important ones. The attended feature maps emphasize on areas of interest, improving the model's focus on critical features. By dynamically adjusting the importance of different regions in the images, the attention mechanism enhances the model's ability to discriminate between relevant changes and irrelevant temporal variations, such as seasonal shifts or sensor induced noise.

$$C = \mathrm{Concat}(P_1^3, P_2^3) \tag{3}$$

$$A = \sigma(W_A \cdot C + b_A) \tag{4}$$

$$P_1^{\mathrm{attended}} = P_1^3 \odot A \tag{5}$$

$$P_2^{\mathrm{attended}} = P_2^3 \odot A \tag{6}$$

where,

Concat denotes concatenation along the channel axis.

P_1^3, P_2^3 are the pooled feature map from the convolutional block.

W_A, b_A are the weights and biases for the attention layer.

σ denotes the sigmoid activation function.

$\odot$ denotes element-wise multiplication.

A is the attention map.

P_1^{attended} and P_2^{attended} are attended feature maps.

Pairwise Skip Connection. In this module, the attended feature maps from both SAR images are integrated through pairwise skipped connections via concatenations along the channel dimension. Here, pairwise skip connections allow the model to preserve spatial information from both SAR images while combining their features. It enhances the model's ability to learn spatial dependencies, improve feature representation by incorporating complementary information from both images and gain a richer understanding of the scene dynamics, enabling it to detect subtle changes more effectively. This is achieved by passing the attended feature maps from both SAR images through a Conv2D layer with 256 filters of size 1×1. This operation adjusts the number of channels to ensure compatibility for the subsequent concatenation, without altering the spatial dimensions. Initially, the attended feature map of SAR Image 2 is concatenated with the 1×1 convolved feature map of SAR Image 1. The merged output is then concatenated with another 1×1 convolved feature map of SAR Image 2 to compute the final merged and attended feature map from the two SAR images.

$$S_1 = W_S * P_1^{\text{attended}} + b_S \tag{7}$$

$$M_1 = \text{Concat}(P_2^{\text{attended}}, S_1) \tag{8}$$

$$S_2 = W_S * P_2^{\text{attended}} + b_S \tag{9}$$

$$M_2 = \text{Concat}(M_1, S_2) \tag{10}$$

where,

W_S and b_S are the weights and biases for the 1×1 convolutional layers in the skip connections.

S_1 and S_2 are the skip-connected feature maps.

M_1 and M_2 are the merged outputs after concatenation.

Finally, the integrated feature map is flattened into a 1D vector and processed through two fully connected dense layers of tapering dimensions (512 and 256 units respectively) and ReLU activation function. These layers capture the complex relationships between the features, enabling the model to learn high-level representations that help to distinguish between changed and unchanged regions. The output layer produces a binary classification map through sigmoid activation and is reshaped to input image size, such that each pixel indicates the likelihood of change occurrence at the corresponding spatial location. This pixel level binary classification score output is then encoded to 1 indicating change and 0 indicating no change depending on the distance of the probability value from the Bayesian probability of 0.5 for each pixel.

2.3   Model Calibration and Infrastructure Details

The hyperparameters used for calibrating the network were Adam optimizer with a learning rate of 1e−3, factor as 0.2, patience as 3, minimum lr as 1e−6 with ReduceLROnPlateau, 60 epochs, 55 batch size and contrastive loss function. These were identified using grid search through Keras tuner libraries with the objective of optimizing the bias-variance trade-off.

The contrastive loss function penalizes the model based on the similarity of samples in the embedding space, encouraging similar samples to be close together and dissimilar samples to be apart with a margin threshold determining the trade-off. The loss is computed as the squared difference between the predicted output and 0, that is, ground truth for dissimilar pairs. For similar pairs with ground truth value 1, the loss is computed as the squared difference between the margin (a predefined threshold set as 0.9) and the predicted output. If the predicted output is greater than the margin, the loss is 0 (no penalty), otherwise the loss increases quadratically with the distance below the margin. Finally, the loss for similar and dissimilar pairs were element-wise added and averaged. This resulted in the final contrastive loss, which was used to update the network parameters during training phase.

The entire training and out-of-sample testing for the models is performed using python 3.8, tensorflow version 2.13 libraries on GPU server having configuration as - Nvidia GPU with 52 GB RAM, Driver version:470.182.03 and cuda version: 11.3.

3   Experimental Results and Analysis

3.1   Datasets and Evaluation Metrics

To validate the effectiveness of the proposed network, the model was assessed both qualitatively and quantitatively with multiple experiments using Yellow River, Ottawa and novel proposed Kelavarapalli SAR datasets.

The Yellow River dataset was selected from two large SAR images collected in the Yellow River Estuary area of China. Both images were taken by Radarsat-2 of 8 m spatial resolution during June 2008 and June 2009, respectively. We selected one region of 306×291 pixels for our experiment as shown in Fig. 2 below.

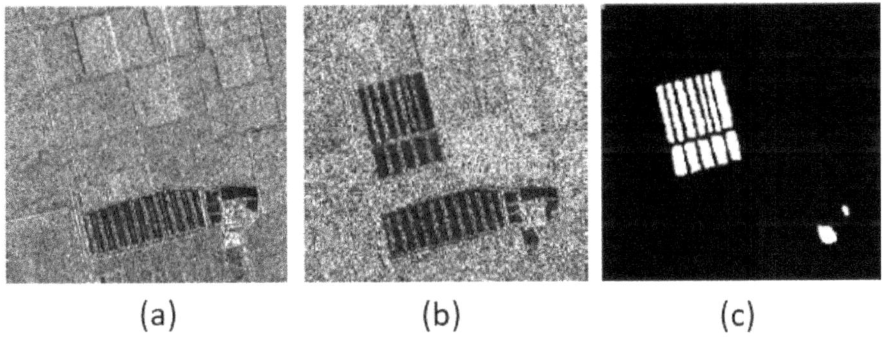

(a)   (b)   (c)

Fig. 2. The Yellow River 1 dataset: (a) SAR image captured in June 2008, (b) SAR image captured in June 2009, and (c) Ground truth change map. (Color figure online)

The Ottawa SAR dataset is a widely used benchmark in change detection research, particularly involving SAR imagery. Captured by RADARSAT-1 in Standard Beam Mode with HH polarization, the dataset includes two temporally separated images of size 290×350 pixels from May 1997 and August 1997 at a spatial resolution of 10 m. The scenes encompass diverse land cover types such as urban, vegetation and water bodies, and reflect real-world changes like urban development and seasonal variations. Due to inherent speckle noise and subtle changes, it presents a challenging testbed for evaluating change detection algorithms (Fig. 3).

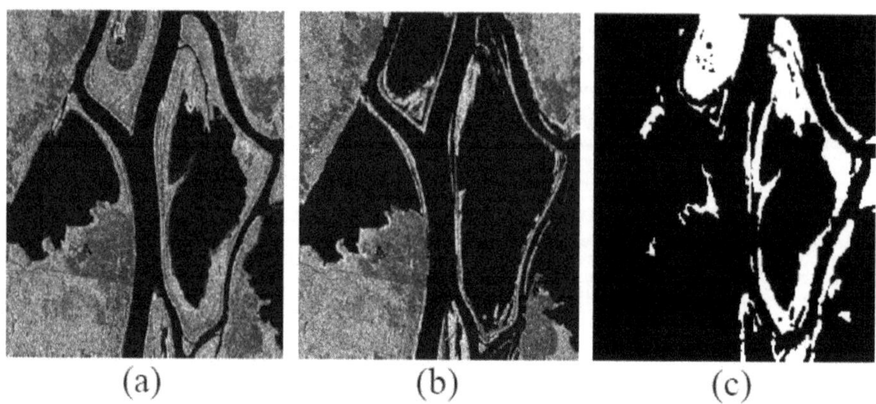

Fig. 3. The Ottawa dataset: (a) SAR image captured in May 1997, (b) SAR image captured in August 1997, and (c) Ground truth change map.

In addition to the above benchmark datasets, SAR sample data of size 256×256 pixels having resolution of 10 m, acquired over Kelavarapalli Reservoir in Tamil Nadu, India were also used for model assessment. The dataset, as highlighted in Fig. 4 is captured through Google Earth Engine (GEE) from Sentinel-1 GRD images of 19 Jan, 2020 and 14 March, 2021 for change detection. GEE pre-processes GRD SAR images with multiple operations such as orbit file application, border and thermal noise removal, radiometric calibration, orthorectification and conversion to decibel values for standardization [11]. Subsequently, we performed co-registration and speckle filtering (boxcar 3×3 followed by lee-sigma 7×7 with σ (sigma) $= 0.5$ filters) on the GEE image to generate the final SAR images for comparison. The chosen temporal interval captures the seasonal and anthropogenic change in the reservoir in terms of the difference in accumulated water between winter (Jan 2020) and summer (March 2021) seasons to depict the reduction in water level for change detection. Hence, similar to other benchmark datasets, this dataset follows a standardized generation pipeline, offers reproducibility and is useful for fair evaluation for SAR based change detection tasks.

Similar to the existing studies [6,9,10], in order to facilitate fairness of comparison, we have also employed popular classification metrics for model evaluation. These include: 1) percentage of correct classification (PCC), 2) false positive (FP), 3) false negative (FN), 4) overall error (OE), and 5) kappa coefficient (KC).

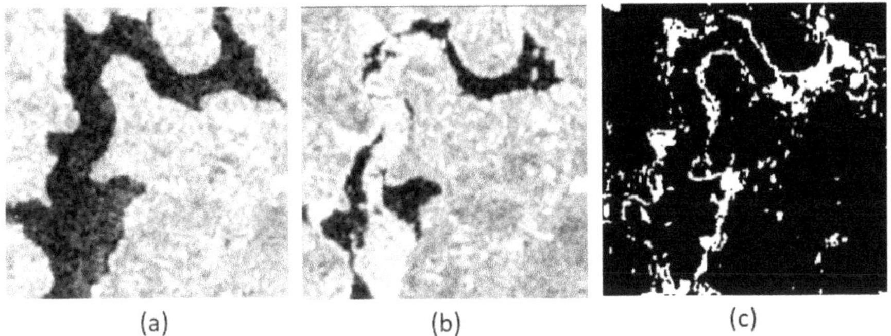

(a)   (b)   (c)

Fig. 4. The Kelavarapalli Reservoir dataset: (a) SAR image captured in January 2020, (b) SAR image captured in March 2021, and (c) ground truth change map.

3.2   Experimental Results and Comparison

The model performance was compared with three start-of-the-art change detection models such as Principal Component Analysis and K-Means (PCAKM) [4], Nearest Neighborhood-Extreme Learning Machine (NR-ELM) [5], and Convolution and Attention Mixer (CA-Mixer) [9] applied to SAR benchmark datasets. PCAKM model, as the name suggests combines PCA and K-means clustering for unsupervised change detection, where PCA is used to compute feature maps for the input difference image and k-means clustering is used to classify the computed feature map to change and unchanged pixels. On similar lines, NRELM uses extreme learning machine (ELM) for classification of feature map extracted using neighbourhood-based ratio operator from difference image. CA-Mixer is a neural network-based model using attention module for local feature extraction. It combines shift convolution with self-attention network, output of which passes through gated feed forward network to enhance non-linear feature transformation. Given our proposed model is also a Siamese network inspired enhanced attention module with multi path skip connections, it is fair to be compared with these state-of-art methods for performance evaluation (Table 1).

The above methods were implemented with the default parameters as stated in their work for fair evaluation and comparison with our proposed model. Results of the comparison are listed in the Tables 2, 3 and 4 along with predicted change maps as shown in Figs. 5, 6 and 7.

3.3   Ablation Study

To establish effectiveness of the various modules of SAE-DP-SCSNN, we conducted an ablation study over the four datasets using different variants of the proposed model with and without its individual modules. This validates the contribution of each module towards model performance. The four variants explored for comparison include the following: 1) SAE- DP-SCSNN without Spatial Attention and Dual Path Skip Connection Module (w/o SA & DPSC). 2) SAE-DP-SCSNN without Spatial Attention Module (w/o SA). 3) SAE-DP-SCSNN with

Table 1. Results on Yellow River 1 dataset

Metric	CA Mixer	PCAKM	NRELM	SAE-DP-SCSNN
PCC	98.4559	92.8531	97.9696	99.1822
False Positives	759	5838	1750	207
False Negatives	616	526	58	521
Overall Error	1375	6364	1808	728
Kappa	0.8631	0.5641	0.7854	0.9245

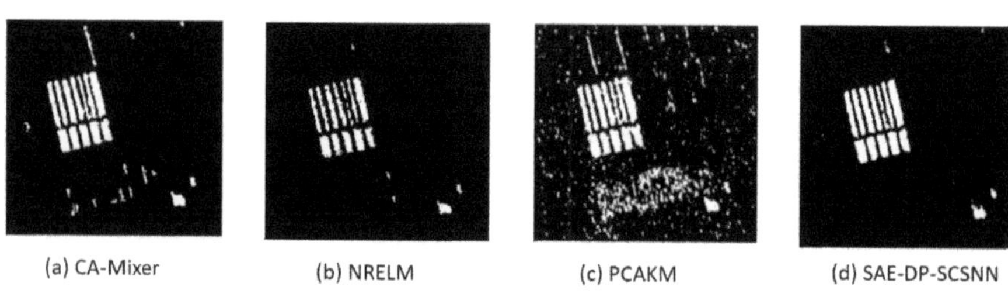

Fig. 5. Predicted change maps comparison of SAE-DP-SCSNN with other models using Yellow River 1 dataset. (Color figure online)

Table 2. Results on Ottawa dataset

Metric	CA Mixer	PCAKM	NRELM	SAE-DP-SCSNN
PCC	98.2532	97.3695	98.2916	98.18415
False Positives	738	574	1141	294
False Negatives	1035	2096	593	1548
Overall Error	1773	2670	1734	1842
Kappa	0.9339	0.8972	0.9349	0.929594

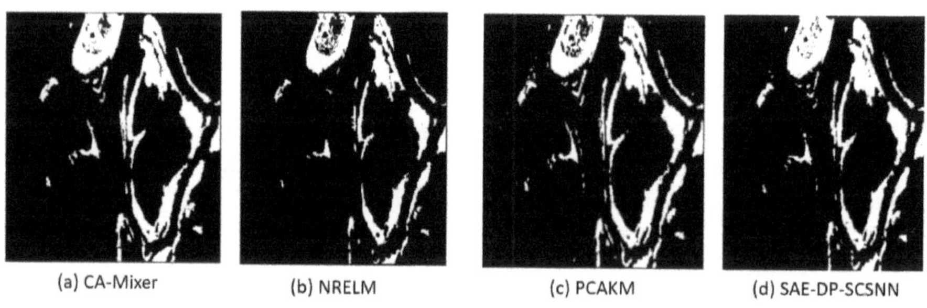

Fig. 6. Predicted change maps comparison of SAE-DP-SCSNN with other models using Ottawa dataset.

Single Path Skip Connection Module removed (with SPSC removed). 4) SAE-DP-SCSNN without Dual Path Skip Connection Module (w/o DPSC).

Table 3. Results on Kelavarapalli Reservoir dataset

Metric	CA Mixer	PCAKM	NRELM	SAE-DP-SCSNN
PCC	94.3329	92.1021	95.5276	98.05298
False Positives	1049	2596	1118	874
False Negatives	2665	2580	1813	402
Overall Error	3714	5176	2931	1276
Kappa	0.7075	0.6286	0.7972	0.910658

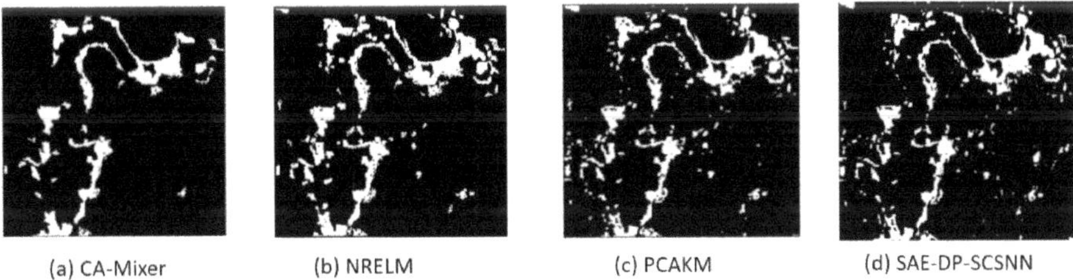

Fig. 7. Predicted change maps comparison of SAE-DP-SCSNN with other models using Kelavarapalli dataset.

Results shown in Table 5 and 6 list the PCC and kappa values respectively for each model variant over all the benchmark datasets. It can be noticed that both Spatial Attention and Dual Path Skip Connection modules improve the model's performance for change detection owing to their properties, as detailed in Sect. 2.2 above, of guiding focus to relevant regions while preserving multi-scale spatial features critical for accurate boundary reconstruction.

Table 4. PCC Results from ablation study of SAE-DP-SCSNN model

Ablation Variant	Yellow River 1	Ottawa	Kelavarapalli Reservoir
w/o SA & DPSC	99.1283	97.72	98.0155
w/o SA	99.0362	97.75	97.9279
with SPSC removed	99.1351	98.12	97.9598
w/o DPSC	99.1182	98.02	97.9895
Full Model	99.1822	98.18	98.05298

Change detection using SAR imagery presents unique challenges due to high speckle noise and subtle nature of temporal variations. To address this, spatial attention mechanisms have emerged as powerful tools for context-aware feature extraction. These modules dynamically prioritize spatial regions that exhibit significant transformations, such as urban expansion or deforestation, while suppressing background noise and static regions. By adaptively weighting spatial

Table 5. Kappa results from ablation study of SAE-DP-SCSNN.

Ablation Variant	Yellow River 1	Ottawa	Kelavarapalli Reservoir
without SA and DPSC	0.9201	0.9108	0.9013
without SA	0.9075	0.923	0.9008
with SPSC removed	0.923	0.917	0.9024
without DPSC	0.9226	0.9134	0.9092
SAE-DP-SCSNN	0.9244	0.9295	0.9107

locations, spatial attention not only improves the model's discriminative capacity but also enhances its robustness to noise-induced artifacts, leading to more precise identification of change-prone areas and reduced false alarm rates [12].

Complementing this, the integration of dual-path feature processing via skip connections further strengthens the model's ability to learn both global context and local detail. This dual-path paradigm facilitates multi-scale representation learning by enabling parallel extraction of fine-grained and high-level semantic features. One path emphasizes local structures, while the other captures broader contextual semantics—together forming a comprehensive view of the changed landscape. Moreover, skip connections, commonly employed in encoder-decoder architectures such as U-Net [13], preserve spatial fidelity by transferring high-resolution information from shallow to deeper layers. This capability is essential for pixel-level semantic segmentation tasks, particularly in delineating sharp and accurate change boundaries within SAR data. Through the synergistic incorporation of spatial attention modules and a dual-path skip-connected architecture, the proposed SAE-DP-SCSNN framework significantly enhances spatial localization, feature expressiveness and semantic consistency. Empirical evaluation, as demonstrated through detailed ablation studies and benchmark comparisons, confirms superior performance across key metrics including pixel-wise classification accuracy (PCC), kappa coefficient, and overall change detection efficacy.

4   Conclusion

The SAE-DP-SCSNN model proposed in this paper performs feature extraction from multi-temporal SAR images through dual path convolution blocks combined via pairwise skip connections at multiple abstraction levels. The extracted features maps are accentuated through a uniquely designed spatial attention mechanism that works effectively to discriminate and distinguish between relevant changes and irrelevant variations. By learning discriminative features and leveraging attention mechanisms, the model effectively identifies land cover changes arising from various environmental and human activities. The integration of multi-temporal information and attention-driven feature selection enhances the model's robustness and adaptability to diverse change detection scenarios. SAE-DP-SCSNN model presents a scientifically grounded approach to SAR change detection, leveraging deep learning and attention mechanisms to automate the detection of meaningful changes in SAR imagery. Its performance and effectiveness have been evaluated through rigorous experimentation and comparison with existing methods. This model has proven to show superior prediction performance in terms of detecting both changes and similarities on various existing benchmark datasets and newly proposed novel benchmark dataset of Kelavarapalli reservoir in Tamil Nadu, thereby contributing to advancements in remote sensing and environmental monitoring research. The novel model SAE-DP-SCSNN and proposed Kelavarapalli reservoir dataset can act as a new standardized resource as benchmark for evaluating and advancing deep learning models in SAR based change detection tasks.

Acknowledgements. The authors are grateful to the International Institute of Information Technology Bangalore (IIIT Bangalore), India for their infrastructure and financial support. We acknowledge Mphasis Cognitive Computing Centre of Excellence at IIIT Bangalore for providing the research grant. The authors would also like to thank European Space Agency's Copernicus Hub for providing the Sentinel-1 SAR data in public domain. Google Earth Engine and Google Earth Pro data facilitated researchers to conduct this study.

References

1. Al-Sharif, A.A.A., Pradhan, B., Hadi, S.J., Mola, N.: Revisiting methods and potentials of SAR change detection. In: Proceedings of the World Congress on Engineering 2013, vol. II, pp. 2231–2237. IAENG (2013)
2. Gong, M., Su, L., Jia, M., Chen, W.: Fuzzy clustering with a modified MRF energy function for change detection in synthetic aperture radar images. IEEE Trans. Fuzzy Syst. **22**(1), 98–109 (2014)
3. Gonzalez, R.C., Woods, R.E.: Digital Image Processing, 3rd edn. Prentice-Hall, Upper Saddle River (2006)
4. Celik, T.: Unsupervised change detection in satellite images using principal component analysis and k-means clustering. IEEE Geosci. Remote Sens. Lett. **6**(4), 772–776 (2009)
5. Gao, F., Dong, J., Li, B., Xu, Q., Xie, C.: Change detection from synthetic aperture radar image based on neighborhood-based ratio and extreme learning machine. J. Appl. Remote Sens. **10**(4), 046019 (2016)
6. Gao, Y., Gao, F., Dong, J., Du, Q., Li, H.-C.: Synthetic aperture radar image change detection via siamese adaptive fusion network. IEEE J. Sel. Top. Appl. Earth Observ. Remote Sens. **14**, 10748–10760 (2021)
7. Saha, S., Shahzad, M., Ebel, P., Zhu, X.X.: Supervised change detection using prechange optical-SAR and postchange SAR data. IEEE J. Sel. Top. Appl. Earth Observ. Remote Sens. **15**, 8170–8178 (2022)
8. Hafner, S., Nascetti, A., Azizpour, H., Ban, Y.: Sentinel-1 and Sentinel-2 data fusion for urban change detection using a dual stream U-Net. IEEE Geosci. Remote Sens. Lett. **19**, 1–5 (2022)
9. Zhang, H., Li, Z., Gao, F.: Convolution and attention mixer for synthetic aperture radar image change detection. IEEE Geosci. Remote Sens. Lett. **20**, 4012105 (2023)

10. Meng, D., Gao, F., Dong, J., Du, Q., Li, H.-C.: Synthetic aperture radar image change detection via layer attention-based noise-tolerant network. IEEE Geosci. Remote Sens. Lett. **19**, 1–5 (2022)
11. Sentinel-1 preprocessing guide by Google Earth Engine. https://developers.google.com/earth-engine/guides/sentinel1#sentinel-1-preprocessing
12. Woo, S., Park, J., Lee, J.-Y., Kweon, I.S.: CBAM: convolutional block attention module. In: Ferrari, V., Hebert, M., Sminchisescu, C., Weiss, Y. (eds.) ECCV 2018. LNCS, vol. 11211, pp. 3–19. Springer, Cham (2018). https://doi.org/10.1007/978-3-030-01234-2_1
13. Ronneberger, O., Fischer, P., Brox, T.: U-net: convolutional networks for biomedical image segmentation. In: Navab, N., Hornegger, J., Wells, W.M., Frangi, A.F. (eds.) MICCAI 2015. LNCS, vol. 9351, pp. 234–241. Springer, Cham (2015). https://doi.org/10.1007/978-3-319-24574-4_28

Deep Learning Architecture and Adaptations

Forecasting Indian Stock Market Prices Using Deep Learning Models

Rajesh Kumar[1], Suchi Kumari[2(✉)], Kartikeya Pydi[2],
and Laxmi Narayana Raavi[2]

[1] Amity School of Engineering and Technology, Amity University, Bangalore, India
`rkumar1@blr.amity.edu`
[2] Department of Computer Science Engineering, Shiv Nadar Institute of Eminence,
Delhi-NCR, India
`suchi.singh24@gmail.com`, `{kp808,lr888}@snu.edu.in`

Abstract. Stock price prediction is a branch of financial forecasting that provides valuable insights for investors, traders, financial institutions, and other stakeholders. This study examines the performance of three deep learning algorithms—LSTM, GRU, and SimpleRNN—in forecasting stock prices based on past data. In contrast to research that emphasizes global efficiencies, this study assesses model performance according to dataset characteristics. The dataset, which includes daily stock prices from 2009 to 2019, focused on closing prices. Data preprocessing included Min-Max Scaling and sliding window methods. Hyperparameters such as batch size, learning rate, dropout rate, and L2 norm strength were modified, with the training and test sets making up 70%, 30% of the data, respectively. The performance of the model is assessed using Mean Squared Error (MSE), Mean Squared Percentage Error (MSPE), and Mean Absolute Error (MAE), indicating that LSTM and GRU excelled compared to SimpleRNN in understanding long-term dependencies. The document seeks to demonstrate the importance of adjusting the selection of models and hyperparameters in relation to the dataset's features and outlines the theoretical frameworks for enhancing outcomes in financial time-series forecasting.

Keywords: Stock market · Deep learning · Prediction · simpleRNN · LSTM · GRU

1 Introduction

Stock markets play a crucial role in the global economy, enabling companies to raise capital and offering investors potential returns. However, predicting stock prices is challenging due to the volatile and non-linear nature of financial markets, making time series forecasting techniques valuable for analyzing stock price trends and predicting future prices based on historical data. The complexity of stock price prediction arises from the unpredictable nature of financial markets.

© The Author(s), under exclusive license to Springer Nature Switzerland AG 2026
R. Gupta et al. (Eds.): BDA 2025, LNCS 16041, pp. 223–232, 2026.
https://doi.org/10.1007/978-3-032-15134-6_15

Multiple variables, including company performance, macroeconomic indicators, geopolitical events, and market sentiment, influence stock prices. The combination of quantitative and qualitative factors, along with the non-linearity, volatility, noisy environment, data outliers, and shifting trends in stock prices, makes prediction challenging, requiring advanced and complex models for accurate forecasting.

Various methods have been developed to model stock price movements, ranging from traditional statistical approaches to more advanced machine learning and deep learning techniques. The earliest approaches to stock price prediction were largely based on statistical models, with the Autoregressive Integrated Moving Average (ARIMA) being one of the most prominent techniques. ARIMA, developed by Box and Jenkins [4], describes time series data where future values are dependent of the passed values and the error terms. Given that the long-behavioral patterns are linear, the method is very useful for financial forecasting in the later part of the century. While ARIMA and other traditional statistical methods such as Exponential Smoothing and Generalized Autoregressive Conditional Heteroskedasticity models handle volatility well, they are often inadequate for modeling complex patterns and non-linear relationships in stock prices [6].

With the increased use of machine learning, more flexible methods capable of capturing complex patterns and correlations have emerged. Support Vector Machines (SVM), Random Forests, and Artificial Neural Networks (ANNs) gained popularity for stock price prediction due to their ability to model intricate relationships in the data. Kara *et al.* [12] demonstrated that SVM could outperform traditional statistical models by identifying non-linear patterns in stock market data. However, SVM is primarily suited for short-term models and does not inherently capture temporal dependencies, which are essential for analyzing sequential data like stock prices. Rumelhart *et al.* [13] defined Recurrent Neural Networks (RNNs) as some of the first deep learning models that took long term interactions in time series data into account. However, a particular type of RNN suffers from the vanishing gradient problem, where gradients become too small during backpropagation. This led to the development of enhanced models, including Long Short-Term Memory (LSTM) networks and Gated Recurrent Units (GRUs). LSTMs [11] addressed the limitations of RNNs by using memory cells to store information over long periods. Fischer *et al.* [9] showed that LSTMs outperform traditional machine learning methods in financial forecasting due to their ability to capture complex patterns and dependencies in stock price data. Similarly, GRUs [5], a modified version of LSTMs offer similar performance to LSTMs but use less computational power, making them preferred for real-time forecasting tasks requiring faster computation. Aldhyani *et al.* [2] proposed a hybrid CNN-LSTM deep learning framework for predicting stock prices of Tesla and Apple, demonstrating superior performance over traditional LSTM models. Wang *et al.* [17] applied the Transformer deep learning framework and demonstrated its superior performance over traditional methods by evaluating it on major global stock indices. Teixeira *et al.* [16] analyzed stock price forecasting using time series models and deep learning techniques [14], including RNN,

LSTM, GRU, CNN, and XGBoost. Among all the considered models, XGBoost and GRU excelled in capturing patterns in stock price time series. Zhao *et al.*

The literature highlights the importance of data preprocessing in financial time series analysis, where techniques like Min-Max Scaling help stabilize training for models such as LSTM and GRU [10]. Effective hyperparameter tuning, adjusting the learning rate, dropping out, and applying L2 regularization, is essential to balance convergence speed and generalization [3]. By considering all the aspects, this paper evaluates three deep learning models, LSTM, GRU, and SimpleRNN, on a stock price dataset spanning over 10 years. Their performance was assessed using Mean Squared Error (MSE), Mean Squared Percentage Error (MSPE), and Mean Absolute Error (MAE), which reflect how well each model captures patterns in the data. Results indicate that model effectiveness varies across datasets; a model that performs well on one dataset may not perform similarly on another. This aligns with previous findings [8, 9], highlighting that stock price prediction lacks a universal solution. Therefore, model selection should be guided by the dataset's characteristics and how different models respond to them. The study presents a robust framework for identifying the optimal model through systematic hyperparameter tuning.

The subsequent sections of the paper are structured as follows: In Sect. 2, we delve into the proposed strategies which are divided into multiple phases. The results and in-depth analysis are presented in Sect. 3. Finally, Sect. 4 provides the conclusions drawn from the proposed research and outlines the future direction.

2   Proposed Approach

This paper investigates stock price prediction using different deep learning models (LSTM, GRU and Simple RNN) by examining how we can fit different stock datasets. The methodology consists of several key steps, such as data preprocessing, model architecture design, hyperparameter tuning, training, evaluation, and performance comparison.

2.1   Data Collection and Preprocessing

We analyze historical closing prices and dates of three different market cap stocks, HDFC Bank, MRF, and Wockhardt Pharma, from 2009 to 2019 [1]. The dataset was divided into training (70%), validation (20%), and testing (30%) subsets. Before training, several preprocessing steps were applied to make the data suitable for deep learning. Stock prices were normalized to a 01 range using Min-Max Scaling, ensuring equal feature influence and aiding the convergence of models like LSTM and GRU by stabilizing gradient updates. For time series prediction, we employed a sliding window approach, where the model predicts the next day's closing price based on the previous three days. The window size was determined empirically and guided by domain knowledge.

2.2   Model Architecture

This study employs three recurrent models, SimpleRNN, LSTM, and GRU, chosen for their demonstrated ability to capture temporal dependencies in time series data [7,11,15].

1. **SimpleRNN:** It is a simple model of sequential dependencies. It is easy enough to use as a guide to more comprehensive models of how things ought to be carried out. One example of a simple RNN model is the Vanilla RNN (shown in Fig. 1(a)). In this case, inputs are processed in a sequential manner by preserving the hidden state that contains details about the inputs that have already been processed. A non-linear activation function $tanh$ is applied to the sum of the product of the hidden state at the previous time step (h_{t-1}) and its associated weight (W_h), and the input at time (x_t) at time t and its associated weight (W_x). This yields the hidden state h at time t. Additionally, a bias term b is included. The model's mathematical formulation is provided [15],

$$h_t = f(W_x * x_t + W_h * h_(t-1) + b) \tag{1}$$
$$y_t = W_y * h_t + b_y \tag{2}$$

   where y_t is the output y at time t.

2. **LSTM Model:** To deal with long-term dependencies [11], this model is intended to have memory cells (shown in Fig. 1(b)). The network's gating mechanism makes it suitable for the stock data, which has long-range dependencies, and which learns what to keep or forget about. The parameters representing the input gate (i_t), forget gate (f_t), cell information update $(\tilde{C}_t$, output state (o_t) final cell state C_t and hidden state (h_t) are represented as follows:

$$i_t = \sigma(w_i[h_{t-1}, x_t] + b_i)$$
$$\tilde{C}_t = tanh(w_c[h_{t-1}, x_t], b_c]$$
$$C_t = f_t * C_{t-1} + i_t * \tilde{C}_t$$
$$o_t = \sigma(w_o[h_{t-1}, x_t] + b_o)$$
$$h_t = o_t * tanh(C_t)$$

3. **GRU Model:** An alternative to LSTM that is computationally more efficient, with fewer gates but similar performance, is the GRU. To assess whether GRU [7] can achieve comparable results to LSTM at a lower computational cost, we conduct tests on the GRU model. The architecture of GRU is shown in Fig. 1(c).

Each model consists of multiple layers. The input layer accepts the normalized time series data. Recurrent layers consist of one or more layers of SimpleRNN, LSTM, or GRU units, depending on the model architecture. Dropout layers are applied between recurrent layers to prevent overfitting by randomly dropping neurons during training, and the last layer is a dense output layer, which is a fully connected layer that provides the predicted stock price as an output.

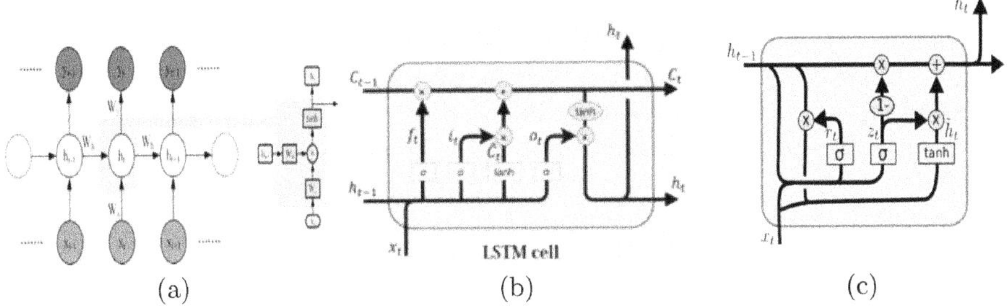

(a)   (b)   (c)

Fig. 1. Architectures of (a) Simple RNN model, (b) LSTM model and (c) GRU model

2.3   Evaluation Metrics

In this section, we will provide information about the methods applied for tuning the hyperparameters,

1. **Hyperparameter Tuning:** The effectiveness of deep learning models depends on the selection of hyperparameters. In order to determine the optimal hyperparameter setting for each model, this research examines a variety of settings. Several batch sizes, learning rates to modify the step size during a gradient descent, dropout rates to control the models and avoid overfitting, and L2 regularization to penalize big weights are among the hyperparameters taken into consideration. Each model is trained using predetermined hyperparameters, and the Mean Squared Error (MSE) calculated on the validation dataset is used to evaluate the model's effectiveness.

2. **5-fold cross-validation:** 5-fold cross-validation is a widely used technique in machine learning for training and evaluating models. Because it guarantees solid and trustworthy model evaluation while honoring the sequential nature of time series data, it is essential in stock market datasets. In contrast to random splits, it preserves the stock data's chronological sequence to stop data leaks, simulating real-world situations where historical data forecasts future patterns. This approach assesses model performance under various market conditions, guaranteeing stability and generalizability, by splitting the dataset into five parts and training and testing on various folds sequentially. Additionally, it minimizes overfitting, makes the most of the restricted data, and offers a thorough evaluation of the model's accuracy and dependability in forecasting complicated and turbulent stock market behavior.

3. **Training and Evaluation:** After hyperparameter tuning, each model was trained for multiple epochs with the Adam optimizer, which integrates adaptive learning rates and momentum. The dataset was divided into training, validation, and test splits: weights were updated on the training set, the validation set was monitored each epoch to flag potential over-fitting, and the held-out test set gauged generalization. Key evaluation metrics were computed on the validation data throughout training, and early stopping terminated training once validation loss plateaued, achieving a balance between

under- and over-fitting. TimeSeriesSplit cross-validation provided a comprehensive evaluation by testing the model across multiple sequential data splits. High losses on both training and validation sets indicated underfitting, suggesting the model struggled to capture underlying patterns. In contrast, low training loss paired with high validation loss pointed to overfitting, where the model memorized the training data rather than generalizing to unseen data. The primary evaluation metric used was Mean Squared Error (MSE), calculated using the formula $MSE = \frac{1}{n} \sum (y_i - \bar{y}_i)^2$.

To prevent overfitting and improve generalization, regularization techniques such as L2 regularization and dropout were applied, helping to simplify the model and reduce dependence on specific features. Additionally, hyperparameter optimization was conducted to fine-tune the balance between training accuracy and generalization capability. Together, these strategies contributed to consistently strong model performance, reflected by low Mean Squared Error (MSE) across all datasets.

4. **Model Comparison and Dataset-Dependent Analysis:** Rather than aligning with the numerous stock prediction studies that aim to identify a single "best" model, this paper emphasizes the performance of each model in relation to the dataset's unique characteristics. The objective is to determine how different models react to the specific features of the stock dataset, such as volatility, trends, and noise. The performance of each model is evaluated using metrics such as Mean Squared Error (MSE) and Mean Absolute Error (MAE) on the test set. Results clearly indicate that the efficiencies of LSTM, GRU, and SimpleRNN models are contingent upon the properties of the dataset. To ensure robustness, the results are averaged across multiple runs for each model. Each run incorporates different hyperparameter configurations, not with the goal of identifying a universally superior model, but to identify which model is best suited to the dataset under consideration. Through this iterative process, the study demonstrates the dependency of model performance on the dataset's characteristics and confirms that no single model consistently outperforms the others across all datasets. The final step involves comparing the models and discussing the conditions under which each model might perform optimally.

3   Results and Analysis

In this segment, we examine stock forecasting by utilizing the deep-learning models applied to three datasets that correspond to various market capitalization groups: Large Cap, Mid Cap, and Small Cap funds. The HDFC Bank dataset is employed for the Large Cap analysis. The MRF Tyre dataset is utilized for the Mid Cap analysis. Finally, for the Small Cap evaluation, the Pharma dataset is utilized. The hyperparameter values selected to enhance the performance of the models analyzed for all the datasets are presented in Table 1.

Table 1. Hyperparameters for the datasets

Dataset Name	Model	Batch	Dropout Rate	Learning Rate	L2 Regularizer	Epochs
HDFC	SimpleRNN	(32, 16, 16)	0.1	0.01	0.001	200
	LSTM	(128, 64, 64)	0.1	0.01	0.001	50
	GRU	(128, 64, 64)	0.2	0.001	0.0001	50
MRF	SimpleRNN	(128, 64, 64)	0.1	0.01	0.0001	200
	LSTM	(128, 64, 64)	0.2	0.01	0.001	200
	GRU	(128, 64, 64)	0.1	0.001	0.001	200
Pharma	SimpleRNN	(32, 16, 16)	0.1	0.01	0.0001	200
	LSTM	(128, 64, 64)	0.1	0.01	0.001	200
	GRU	(64,32,32)	0.1	0.01	0.0001	100

3.1   HDFC Large Cap Dataset

In Fig. 2(a), the learning curve of the SimpleRNN model is shown, depicting a gradual decline in MSE, emphasizing its struggles in capturing long-term dependencies. In Fig. 2(b), the learning process of the LSTM model is illustrated, showing a steady and continual decrease in MSE over epochs, indicating its effective performance. Finally, Fig. 2(c) illustrates the GRU model's learning curve, which operates comparably to the LSTM while providing enhanced computational efficiency.

The actual and predicted training values, along with the validation and test datasets, are presented for the HDFC dataset in Fig. 3. The accuracy for SimpleRNN is approximately 95.47% (shown in Fig. 3(a)), while LSTM achieves 95.75% (Fig. 3(b)), and GRU reaches 95.69% (Fig. 3(c). Among these models, the GRU stands out as the best, with the lowest validation loss of 0.0008.

3.2   MRF Dataset

The Mean Square Error for training and validation of each model is presented as follows: In Fig. 2(d), the SimpleRNN model shows a significant decline in both losses at the beginning, stabilizing later, which suggests successful convergence without signs of overfitting. Figure 2(e) illustrates the LSTM model, in which both losses drop quickly and then level off, indicating successful learning without any overfitting. In Fig. 2(f), the GRU model displays a significant early drop in both losses, becoming stable after roughly 20 epochs, which reflects successful learning without overfitting.

The actual and predicted training values, along with the validation and test datasets, are presented for the MRF dataset in Fig. 3(d-f). The accuracy for SimpleRNN is approximately 92.02% (shown in Fig. 3(d)), while LSTM achieves 92.95%, and GRU reaches 93.31%. Among these models, the GRU stands out as the best, with the lowest validation loss of 0.0028.

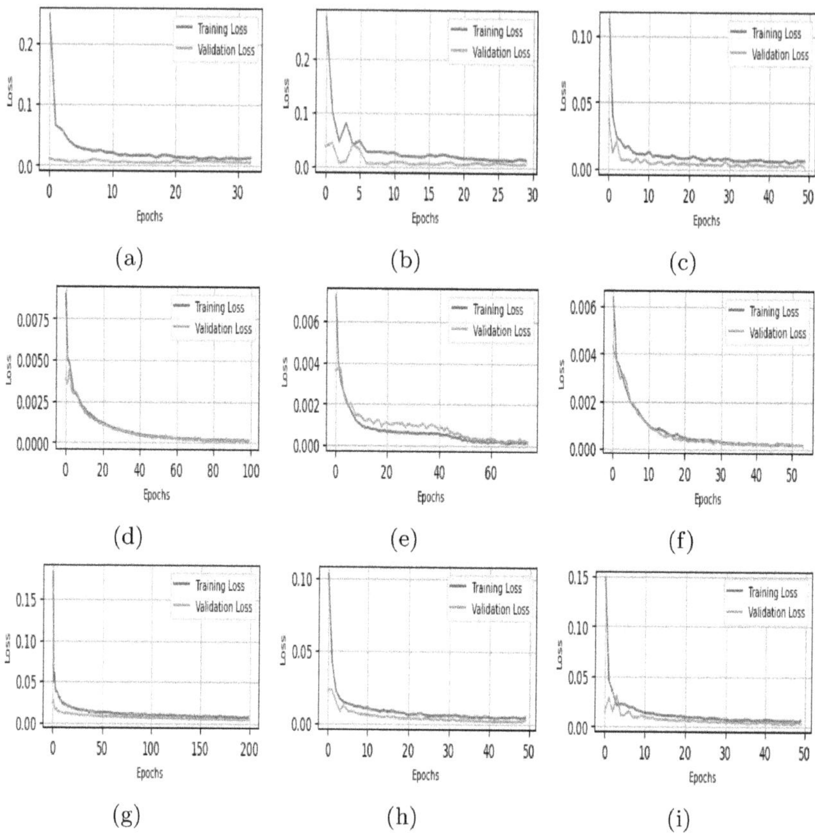

Fig. 2. Mean Square Error for MRF dataset considering simpleRNN (a, d, g), LSTM (b, e, h), and GRU (c, f, i).

3.3 Wockhardt Pharma Dataset

From Fig. 2(g), it is observed that the SimpleRNN model's MSE drops sharply in the early epochs and stabilizes, indicating effective convergence but limited ability to capture complex patterns. Whereas Fig. 2(h) depicts the LSTM model for the Wockhardt dataset, where both losses decrease sharply and stabilize at low values, reflecting strong performance without overfitting. Finally, the GRU model shows a similar sharp decline in losses, stabilizing at low values, indicating effective learning and good generalization as shown in Fig. 2(i).

The true and forecasted training values, as well as the validation and testing datasets, are shown for the Wockhardt dataset in Fig. 3(g–i). The accuracy for SimpleRNN is about 84.05% (illustrated in Fig. 3(g)), whereas LSTM attains 84.54% (depicted in Fig. 3(h)), and GRU achieves 85.18% (shown in Fig. 3(i)). Of these models, the GRU is the most outstanding, achieving the lowest validation loss of 0.0005.

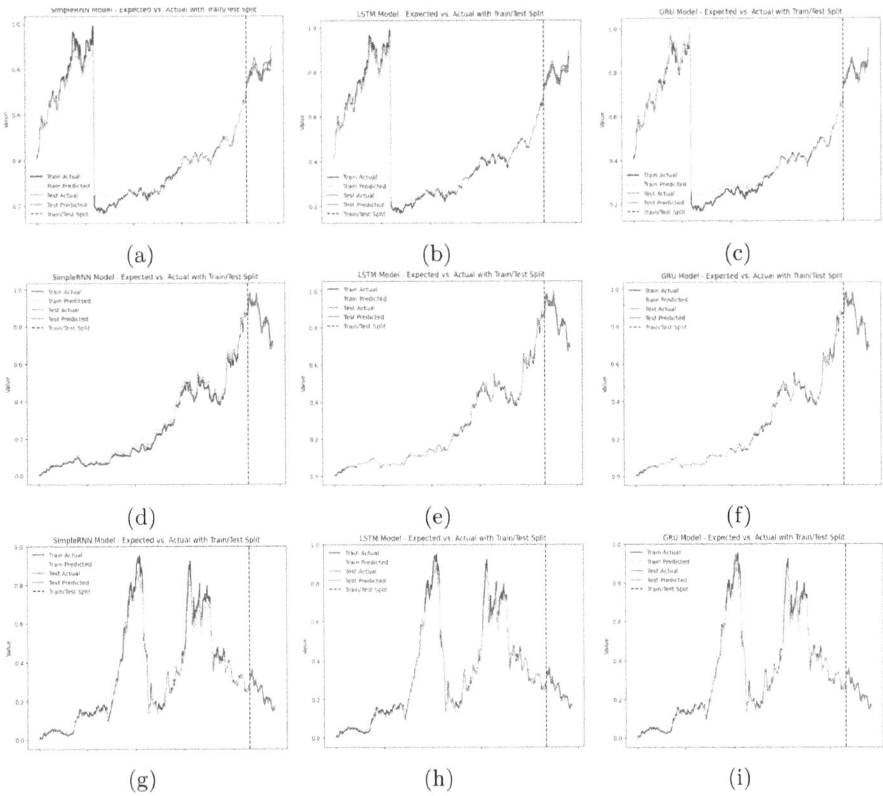

Fig. 3. Accuracy and predicted value for training, validation, and test data of HDFC Large Cap (a–c), MRF (d–f) and Wockhardt Pharma Dataset (g–i).

4   Conclusion and Future Scope

In this study, we apply three distinct deep learning models: LSTM, GRU, and SimpleRNN, utilized to forecast stock prices based on historical data. These models have undergone training and fine-tuning for different hyperparameter configurations, and their effectiveness is qualitatively assessed using metrics like MSE and MAE. The procedure included data preparation, model development, and performance evaluation, emphasizing how each model adjusts to the unique traits of the dataset.

In the future, we seek to attain a clearer insight into whether the models have reached acceptable solutions and pinpoint possible areas for enhancement. Furthermore, we intend to assess the relevance of these models to other time-series datasets, broadening the study's scope and examining their flexibility in handling various financial data. The research can also be broadened to incorporate simpler models, like Radial Basis Functions, along with other options.

References

1. BSE (formerly Bombay Stock Exchange) - bseindia.com. https://www.bseindia.com/. Accessed 02 Jan 2025
2. Aldhyani, T.H., Alzahrani, A.: Framework for predicting and modeling stock market prices based on deep learning algorithms. Electronics **11**(19), 3149 (2022)
3. Bergstra, J., Bengio, Y.: Random search for hyper-parameter optimization. J. Mach. Learn. Res. **13**(2) (2012)
4. Box, G.E., Jenkins, G.M., Reinsel, G.C., Ljung, G.M.: Time Series Analysis: Forecasting and Control. Wiley (2015)
5. Cho, K., et al.: Learning phrase representations using RNN encoder-decoder for statistical machine translation. arXiv preprint arXiv:1406.1078 (2014)
6. De Gooijer, J.G., Hyndman, R.J.: 25 years of time series forecasting. Int. J. Forecast. **22**(3), 443–473 (2006)
7. Dey, R., Salem, F.M.: Gate-variants of gated recurrent unit (GRU) neural networks. In: 2017 IEEE 60th International Midwest Symposium on Circuits and Systems (MWSCAS), pp. 1597–1600. IEEE (2017)
8. Emioma, C., Edeki, S.: Stock price prediction using machine learning on least-squares linear regression basis. In: Journal of Physics: Conference Series, vol. 1734, p. 012058. IOP Publishing (2021)
9. Fischer, T., Krauss, C.: Deep learning with long short-term memory networks for financial market predictions. Eur. J. Oper. Res. **270**(2), 654–669 (2018)
10. Hinton, G.: Improving neural networks by preventing co-adaptation of feature detectors. arXiv preprint arXiv:1207.0580 (2012)
11. Hochreiter, S.: Long short-term memory. Neural Comput. (1997)
12. Kara, Y., Boyacioglu, M.A., Baykan, Ö.K.: Predicting direction of stock price index movement using artificial neural networks and support vector machines: the sample of the Istanbul stock exchange. Expert Syst. Appl. **38**(5), 5311–5319 (2011)
13. Rumelhart, D.E., Hinton, G.E., Williams, R.J.: Learning representations by back-propagating errors. Nature **323**(6088), 533–536 (1986)
14. Sheth, D., Shah, M.: Predicting stock market using machine learning: best and accurate way to know future stock prices. Int. J. Syst. Assur. Eng. Manage. **14**(1), 1–18 (2023)
15. Stérin, T., Farrugia, N., Gripon, V.: An intrinsic difference between vanilla RNNs and GRU models. Cogntive **84**, 2017 (2017)
16. Teixeira, D.M., Barbosa, R.S.: Stock price prediction in the financial market using machine learning models. Computation **13**(1), 3 (2024)
17. Wang, C., Chen, Y., Zhang, S., Zhang, Q.: Stock market index prediction using deep transformer model. Expert Syst. Appl. **208**, 118128 (2022)

Transformers vs CNNs in Wheat Disease Prediction: Performance Evaluation

Falisha Umaiza[✉], Abdul Subhan Omeez, N. Manali, Sambhavi, G. P. Purvika, Viraj Nandalikar, and K. V. Leelambika

Presidency University, Bangalore, India
```
{FALISHA.20221CSG0159,ABDUL.20221COM0114,MANALI.20221CSG0122,
SAMBHAVI.20221CSG0162,PURVIKA.20221CSG0123,VIRAJ.20221CSG0072,
leelambika.kv}@presidencyuniversity.in
```


Abstract. Wheat leaf diseases pose a significant threat to global food security by reducing crop yield and quality, directly impacting the over 35% of the global population that relies on wheat as a dietary staple. While recent advances in deep learning have enabled automated disease diagnosis from leaf images, the choice of model architecture remains pivotal for achieving both diagnostic accuracy and practical deployment efficiency. This study presents a comprehensive comparison between traditional Convolutional Neural Networks (CNNs) and modern Transformer-based models for wheat disease detection. Experiments were conducted on the publicly available *"Wheat Disease Dataset (Small)"* from Kaggle, comprising *1,455 labeled wheat leaf images* categorized into five key classes: **Brown Rust** *(294 images)*, **Yellow Rust** *(277)*, **Mildew** *(351)*, **Septoria** *(327)*, *and Healthy (206)*. These images, captured under diverse real-world agricultural conditions—varying lighting, orientations, backgrounds, and disease severity stages—were *pre-processed* via resizing to *224×224 pixels*, pixel normalization, and extensive data augmentation to reflect field variability. Performance evaluation using accuracy, precision, recall, and F1-score metrics demonstrated that the **Swin Transformer** achieved the highest classification **accuracy of 95%**, outperforming both CNN-based and other Transformer architectures. Beyond achieving strong empirical results, this research highlights the tangible real-world applicability of Transformer-based models for precision agriculture. The findings support deployment in mobile applications for farmers, *UAV-based crop surveillance systems*, and *digital early-warning advisory platforms*. By enabling rapid, accurate, and accessible in-field disease detection, this work contributes to *reducing crop losses, optimizing fungicide application, and strengthening global food security* through scalable, AI-powered agricultural diagnostics.

Keywords: Wheat disease detection · Transformers · Convolutional Neural Networks · Deep learning · Plant disease classification · Computer vision · Image-based diagnosis

© The Author(s), under exclusive license to Springer Nature Switzerland AG 2026
R. Gupta et al. (Eds.): BDA 2025, LNCS 16041, pp. 233–245, 2026.
https://doi.org/10.1007/978-3-032-15134-6_16

1 Introduction

As a global food security staple that feeds billions of people across continents, wheat productivity faces threats from several foliar diseases such as leaf rust, septoria, and tan spot; these diseases can lead to substantial yield losses if not identified and managed in a timely fashion. Automated disease diagnosis through computer vision and deep learning offers sustainable and scalable solutions to growing agricultural challenges and is gaining interest in the agritech research community.

Convolutional Neural Networks (CNNs) are widely adopted for all image-based classification tasks, including plant disease identification, and have been the most commonly used framework in scleral imaging for years [4, 5, 8–10]. CNN structures like VGG, ResNet, EfficientNet, MobileNet, and DenseNet have proven to be remarkably effective in hierarchical spatial feature extraction from leaf images, making it possible to robustly classify disease symptoms and subtypes by purely relying on color, texture, and morphology pattern analysis. Unfortunately, the subtler visual elements that may be more impactful to the overall analysis, such as important details and those defining features, may be located in different parts of the leaf's image and can be quite distant from one another. This poses a challenge since CNNs are fundamentally limited in capturing long-range dependencies and contextual relationships across an entire image [2].

On the other hand Transformers are receiving signifcant attention as alternatives to CNNs for vision tasks because of the self-attention features that capture global context [1, 2, 6]. Having been conceived for natural language processing, computer vision has been transformed by the introduction of benchmarks such as the Swin Transformer which perform exceedingly well termed Vision Transformers (ViTs) [3, 11]. These architectures trained on huge datasets exhibit strong scalability, better representation learning, and improved generalization. Unlike traditional CNNs, Non-local feature attention given by transformers is useful in wheat disease detection since disease patterns are complex and diverse instead of localised [14, 15].

MobileNet and EfficientNet have shown promise with most image classification tasks. Such performance, however, came with the trade off of larger datasets, and other methods like data augmentation, ensemble learning, or attention mechanisms to reach the required accuracy threshold [4, 5]. For example, during our experiments, MobileNet and EfficientNet attained a modest 70% and 71.9% accuracy respectively on a confined wheat disease dataset.On the other hand, self-attention-based global context Transformers emerge as strong competitors to CNNs. It is noteworthy that Transformers can still perform well even with minimal data and without additional aids [1, 14]. Our research showed the CaiT (Class-Attention in Image Transformer) and the Swin Transformer reached 91.9% and 95% accuracies respectively, which was much higher than the CNN models' results.

This study establishes a performance baseline for traditional CNNs and contemporary CNN-based Transformers by evaluating their accuracy in classifying wheat leaf diseases using an image dataset with rich annotations. We carry out multi-faceted comparisons concentrating on classification accuracy, computational cost, susceptibility to changes in illumination and background, and overall model generalization to novel disease instances. By assessing these designs, this research sheds light on the progression of deep learning technologies in precision agriculture and develops criteria for determining the best model designs for practical wheat disease surveillance systems.

2 Literature Survey

Over Over 35% of the global population relies on wheat as a staple food, making it one of the world's most extensively cultivated crops. However, foliar diseases such as Septoria, Mildew, Brown Rust, and Yellow Rust continue to threaten its productivity. These diseases exhibit visually complex, variable symptoms depending on environmental conditions and infection stages, including pustules, vein-parallel yellowing, and necrotic lesions. This variability complicates prompt, accurate diagnosis for farmers and agronomists, especially in regions lacking adequate agricultural infrastructure. Consequently, AI-driven image-based diagnosis is increasingly recognized as a scalable, reliable solution for precision agriculture.

Convolutional Neural Networks (CNNs) have long dominated plant disease classification, excelling at capturing local spatial features. Architectures like EfficientNet and MobileNet have shown strong performance in crops such as tomato and maize [4, 5, 8–10]. Yet, CNNs' local receptive fields limit their ability to capture global patterns, posing challenges for detecting diffuse or vein-oriented symptoms like those of Septoria and Yellow Rust [2].

Vision Transformers (ViTs), originally developed for natural language processing[6], use self-attention to capture global dependencies but initially demanded large datasets and high computational resources. Recent innovations such as Swin Transformer [11] and CaiT (Class-Attention in Image Transformer) [14] addressed these limitations through shifted window attention and class-token mechanisms, making them effective for small, complex agricultural datasets. This study compares CNN and Transformer models on the Kaggle Wheat Disease Dataset (Small), comprising five classes: Brown Rust, Yellow Rust, Mildew, Septoria, and Healthy. While CNNs remain widely used in plant pathology, direct comparisons with modern Transformers like CaiT [17, 18] on wheat-specific data remain rare. The growing dominance of Transformers in vision tasks, their suitability for low-data scenarios, and explainable AI tools like attention maps emphasize their promise for precision agriculture.

Finally, recognizing that automated classification alone is insufficient, this work complements model results with a prescriptive analysis of each wheat disease. These insights address causes, environmental conditions, and recommended control strategies, bridging deep learning predictions with actionable, field-level agronomic decision-making.

3 Methodology

3.1 Dataset

The dataset employed in this study is the publicly available "**Wheat Disease Dataset (Small)**," hosted on Kaggle (https://www.kaggle.com/datasets/yasserhessein/wheat-disease-dataset-small). It comprises a total of **1,455 labeled images** of wheat leaves, categorized into five distinct classes reflecting the most commonly encountered conditions in practical agricultural settings. The class-wise image distribution is as follows: **Brown Rust (294 images), Yellow Rust (277 images), Mildew (351 images), Septoria (327 images), and Healthy (206 images)**. These diseases are recognized for causing substantial reductions in wheat yield and quality, posing a major threat to global food security.

The images in this dataset were captured under diverse real-world agricultural conditions, exhibiting significant variability in lighting environments, leaf orientations, background textures, and disease severity stages. This diversity is critical for developing models capable of generalizing to the unpredictable environmental factors encountered during field-based disease detection. No images were synthetically generated or artificially enhanced, preserving the authenticity and practical relevance of the dataset.

Prior to model training, a data quality assessment was performed to confirm label accuracy and image clarity. The dataset was found to be free of duplicate or mislabeled images, and thus retained in its original form without the need for exclusions. All images were uniformly resized to **224×224 pixels** to meet the input dimensional requirements of the CNN and Transformer architectures, and pixel intensities were **normalized to a [0, 1] range** to improve model convergence and numerical stability during training.

To mitigate the risk of class imbalance affecting model learning, a stratified sampling approach was employed when splitting the dataset into training, validation, and testing subsets. The dataset was divided into **70% for training, 15% for validation, and 15% for testing**, ensuring that each subset maintained the same proportional distribution of classes as the overall dataset. This strategy ensured fair evaluation of model performance across all disease categories and prevented dominance by any single class.

To enhance model robustness and reduce overfitting, a suite of dynamic **data augmentation** techniques was applied during training. These included random horizontal flipping to simulate natural variations in leaf positioning, random rotations within a ±30° range to capture angular diversity, random zoom and shifts to account for variable leaf sizes and framing, and brightness and contrast adjustments to mimic fluctuating lighting conditions typical in open-field environments. These augmentations effectively simulated the environmental and visual variability found in practical agricultural settings, enabling the models to learn invariant and generalizable features essential for real-world deployment.

This well-curated, moderately balanced, and rigorously pre-processed dataset, combined with robust augmentation strategies, provided a reliable foundation for evaluating and comparing the performance of both CNN and Transformer-based architectures under realistic conditions.

3.2 Model Architectures

To explore the effectiveness of different deep learning paradigms in wheat disease classification, we employed **four state-of-the-art models**, comprising two **Convolutional Neural Networks (CNNs)** and two **Vision Transformers (ViTs)**. These architectures were selected to capture a wide range of representational capabilities, from efficient mobile-friendly networks to transformer-based models capable of modeling complex spatial dependencies (Fig. 1).

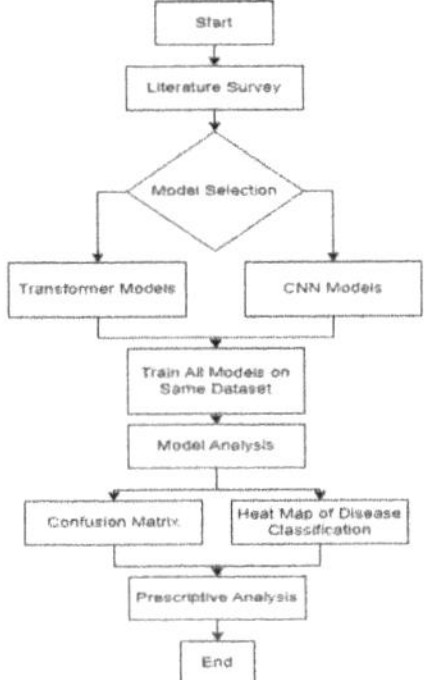

Fig. 1. Overall Workflow of the Proposed Wheat Disease Classification Framework using CNN and Transformer Models

3.2.1 MobileNetV2 (CNN)

A lightweight, efficient CNN using depthwise separable convolutions, achieving 72.72% accuracy in 10 epochs. It performed well overall but struggled with diseases exhibiting overlapping visual features.

3.2.2 EfficientNet-B0 (CNN)

A scalable CNN using compound scaling for balanced depth, width, and resolution. Achieved 69.93% accuracy in 10 epochs, underperforming against MobileNetV2, particularly in separating visually similar classes like Healthy and Mildew.

3.2.3 Swin Transformer

A Transformer model employing shifted window attention for local-global feature extraction. Despite only 5 training epochs, it achieved the highest accuracy of 95.00%, excelling in detecting fine-grained, dispersed disease patterns.

3.2.4 CaiT (Class-Attention in Image Transformer)

An advanced Vision Transformer integrating Class-Attention and LayerScale for deeper, class-specific feature refinement. Trained for 15 epochs, it reached 91.02% accuracy, particularly effective in differentiating diseases with overlapping symptoms.

3.3 Model Training and Inference Workflow

Once the dataset was preprocessed and divided into training, validation, and testing subsets, the models were trained using the labeled leaf images as input. Each image, resized to **224 × 224 pixels** and normalized, served as the input to the first layer of the respective deep learning architecture.The models learned to extract key visual features from the leaf images—such as discoloration patterns, fungal textures, vein alignments, and lesion shapes—that correlate with specific disease categories. CNN architectures like **MobileNetV2** and **EfficientNet-B0** performed this task using convolutional layers to progressively capture local spatial features.

Meanwhile, transformer-based models like **CaiT** and **Swin Transformer** employed self- attention mechanisms to analyze both local and distant image regions, allowing them to detect scattered and subtle patterns across the leaf surface.

Throughout training, the models updated their internal parameters using a standard classification loss function designed to minimize the discrepancy between predicted and actual class labels.

Training proceeded over a defined number of **epochs**, with validation performance monitored to detect potential overfitting and to guide the model toward better generalization.Once training was complete, the models were evaluated on the **test set**, which contained entirely unseen data. Each model produced probability scores across the five wheat disease classes for every test image. The class with the highest predicted probability was selected as the final output. These predictions were then compared to the ground truth labels to compute evaluation metrics, offering a detailed view of how effectively each model could identify and differentiate wheat leaf diseases in real- world scenarios.

4   Results

This section presents a comparative evaluation of Convolutional Neural Networks (CNNs) and Transformer-based architectures for wheat disease classification. Alongside conventional metrics like accuracy, precision, and F1-score, we assess model interpretability through attention visualizations and sample-based evaluations.

4.1   General Performance Summary

While CNNs like **MobileNet** and **EfficientNet** have historically dominated image classification, their performance stagnates when applied to plant diseases with overlapping visual symptoms and limited annotated data. In our experiments, MobileNet and EfficientNet achieved modest accuracies of 70.1% and 71.1%, respectively, reflecting these limitations.

Conversely, Transformer-based models such as Swin Transformer and CaiT demonstrated superior generalization by leveraging global attention mechanisms, enabling them to capture distant yet semantically related features. This proved especially effective for visually similar patterns like yellowing streaks and rust pustules. Swin Transformer achieved the highest accuracy at 95.0%, followed by CaiT at 91.02%, highlighting the dominance of Transformers in data-scarce, symptom-overlapping scenarios.

4.2   Model Evaluation: Confusion Matrix Analysis

To gain deeper insight into the classification behavior of each model, we analyze their corresponding confusion matrices. These matrices offer a granular view of model performance by revealing how predictions are distributed across actual and predicted classes. Beyond overall accuracy, confusion matrices help uncover systematic misclassifications and inter-class confusion that may not be apparent from summary metrics alone. Each matrix illustrates the model's ability to correctly identify plant diseases and distinguish between visually similar conditions such as YellowRust and BrownRust. Through this evaluation, we aim to understand which classes are consistently well-recognized and which tend to be confused, providing interpretability to complement the numeric performance scores. The following subsections explore the strengths and weaknesses of each model based on their confusion matrices.

4.2.1 MobileNet Confusion Matrix Analysis

The confusion matrix for MobileNet reveals a relatively balanced performance, though with noticeable room for improvement in inter-class misclassifications. While the model classifies Septoria very well (61 correct out of 66), it frequently misclassifies YellowRust—13 instances were predicted as Septoria, indicating a strong confusion between the visual traits of these diseases.

Similarly, BrownRust suffers from 5 misclassifications as Septoria and 2 as YellowRust, likely due to overlapping symptomatology in rust-type diseases. The Healthy class also sees moderate confusion, misclassified as Mildew and BrownRust five and three times, respectively. Overall, MobileNet demonstrates high accuracy on classes like Septoria and Mildew, but struggles with subtle disease boundaries, especially among rust-based classes (Fig. 2).

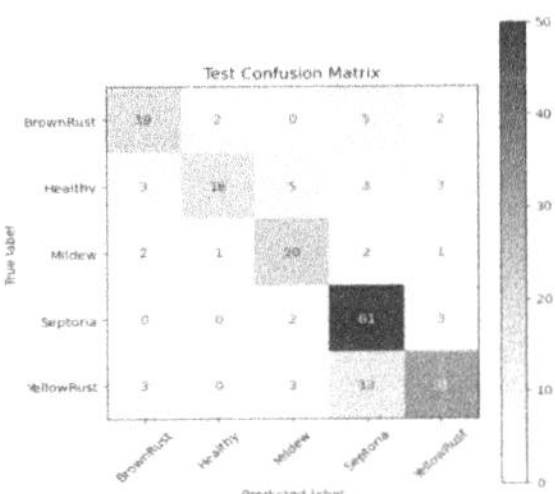

Fig. 2. Confusion matrix for MobileNet model showing strong performance in Septoria but noticeable misclassifications between rust diseases and healthy samples.

4.2.2 EfficientNet Confusion Matrix Analysis

EfficientNet showcases a more stable and less noisy confusion matrix compared to MobileNet. It achieves excellent classification of Septoria (53 correct with zero false positives), and YellowRust (33 correct, 3 misclassified). While it still confuses Mildew and Healthy to a minor extent, these confusions are significantly less severe. BrownRust has 17 correct predictions, with only 2 misclassified as Healthy. One major strength of this model is the absence of misclassification for Septoria, reflecting EfficientNet's capability in capturing disease-specific features, possibly due to its compound scaling. Despite minor overlaps, this matrix reflects strong class separation (Fig. 3).

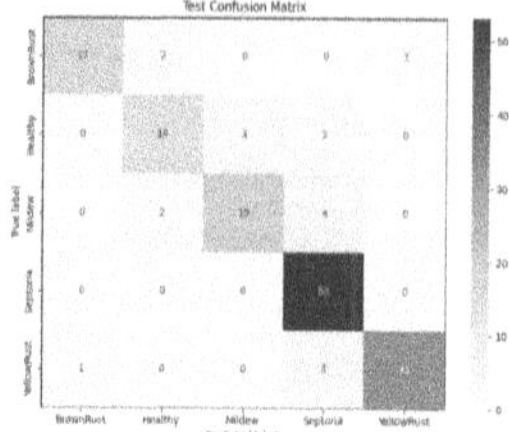

Fig. 3. Confusion matrix for EfficientNet, highlighting highly accurate classification of Septoria and YellowRust, with minimal cross-disease misclassification.

4.2.3 Swin Transformer Confusion Matrix Analysis

The Swin Transformer model exhibits strong performance patterns similar to Efficient-Net, with perfect classification of Septoria (53/53) and high accuracy on YellowRust (33/37). However, some misclassifications occur within rust diseases—BrownRust has 2 misclassifications as Healthy and 1 as YellowRust, suggesting feature overlap in those categories.

The Healthy and Mildew classes see a few confusions as well, though they remain within a tolerable range (Healthy misclassified as Mildew and Septoria; Mildew confused with Septoria). Overall, the Swin Transformer balances accuracy and robustness, especially on visually distinct diseases like Septoria and YellowRust (Fig. 4).

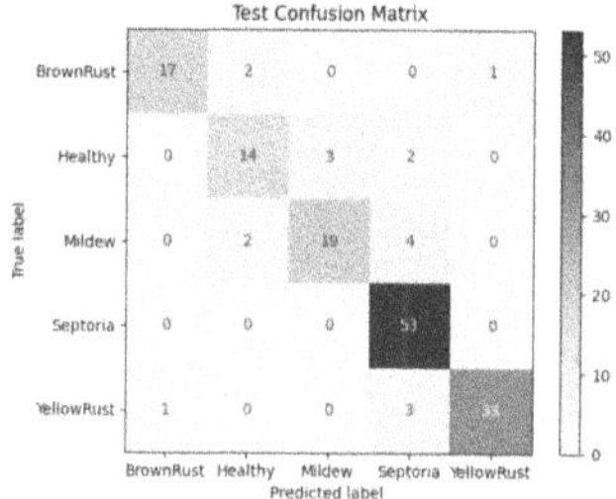

Fig. 4. Confusion matrix for Swin Transformer showing highly accurate classification on Septoria and YellowRust with minor misclassification in rust and mildew-related classes.

4.2.4 CaiT (Class-Attention in Image Transformer) Confusion Matrix Analysis

CaiT (Class-Attention in Image Transformer) demonstrates very high precision across most classes, with almost perfect classification for BrownRust (28/29) and Septoria (65/70). The Healthy class also has strong performance (17 correct, 3 misclassified). What stands out is the low confusion across all diseases: Mildew, YellowRust, and Healthy all maintain clean separability.

There are still a few minor misclassifications, like 2 YellowRust cases predicted as Septoria, but these are minimal. This matrix clearly suggests that this is well-tuned to the dataset and exhibits excellent inter-class boundary learning (Fig. 5).

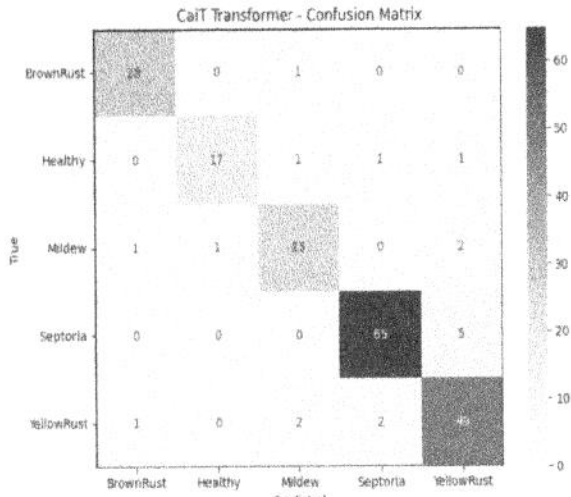

Fig. 5. Confusion matrix for CaiT (Class-Attention in Image Transformer) model reflecting strong disease-class separation and highly precise predictions across all classes.

The confusion matrices offer a clear comparison of each model's classification capabilities. While MobileNet and EfficientNet show moderate accuracy with noticeable misclassifications, especially for visually similar classes, CaiT (Class-Attention in Image Transformer) and Swin Transformer perform better overall. Swin, in particular, demonstrates stronger class separation and fewer errors, highlighting its superior ability to handle complex visual features in plant disease recognition.

4.3 Classification Report Heatmaps: Precision, Recall, and F1-Score Across Models

To complement the confusion matrix analysis, classification report heatmaps were employed to provide a more granular view of each model's performance across all classes. These heatmaps visualize key metrics—precision, recall, and F1-score—on a per-class basis, offering insights into how well each model balances correct identification and error rates across disease categories. The following subsections present an individual breakdown for each model.

4.3.1 MobileNet - Classification Performance

The classification performance of the MobileNet model reveals clear limitations in handling complex disease variations. While Septoria is relatively well identified with a recall of 0.92 and an F1-score of 0.81, other classes display weaker results. Healthy and YellowRust classes in particular struggle with low recall values of 0.53 and 0.62, leading to reduced F1-scores of 0.65 and 0.69, respectively. BrownRust and Mildew also exhibit modest F1-scores of 0.69 and 0.71.

The overall trend indicates that MobileNet often misclassifies across classes, possibly due to its lighter architecture, which may lack the representational depth needed for nuanced leaf texture distinctions. As a result, its suitability for high-precision plant disease detection is limited (Fig. 6).

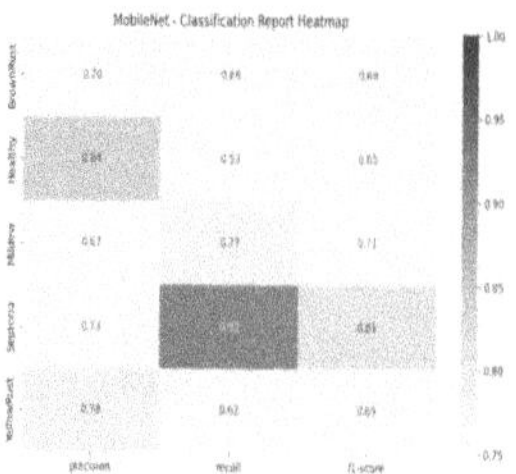

Fig. 6. Heatmap showing the classification report for the MobileNet model, reflecting uneven performance and frequent misclassifications across disease and healthy leaf categories.

4.3.2 EfficientNet - Classification Performance

The heatmap visualization for the EfficientNetB0 model presents a class-wise breakdown of precision, recall, and F1-score across five categories: BrownRust, Healthy, Mildew, Septoria, and YellowRust. EfficientNetB0 demonstrates strong performance in detecting rust-based diseases like YellowRust and BrownRust, with F1-scores of 0.93 and 0.89 respectively. Septoria is detected with perfect recall (1.00), highlighting the model's ability to correctly identify all Septoria samples, though precision is slightly lower at 0.85. However, its performance in classifying Healthy samples is comparatively weaker, with a recall of 0.74 and an F1-score of 0.76, indicating occasional misclassifications. Similarly, Mildew is moderately classified with an F1-score of 0.81. Overall, the model performs well with disease categories but shows limitations in identifying healthy leaves, possibly due to visual similarities between early-stage infections and healthy textures (Fig. 7).

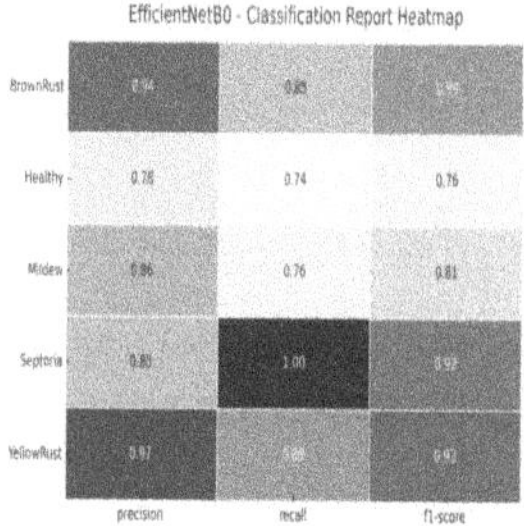

Fig. 7. Heatmap showing the classification report for EfficientNet, depicting precision, recall, and F1-score across disease categories.

4.3.3 Swin Transformer - Classification Performance

The Swin Transformer model demonstrates a highly competitive classification profile, particularly excelling in disease detection tasks. Septoria and YellowRust are nearly perfectly identified, each achieving precision, recall, and F1-scores at or near 0.98 and 0.97, showcasing exceptional consistency. Mildew also benefits from high scores, with an F1-score of 0.94. BrownRust displays perfect precision (1.00) but slightly lower recall at 0.85, indicating occasional false negatives.

Interestingly, the Healthy class records the lowest precision of 0.78, suggesting the model may sometimes confuse healthy samples with disease-affected ones. Despite this, its overall performance remains robust, especially in distinguishing specific disease classes (Fig. 8).

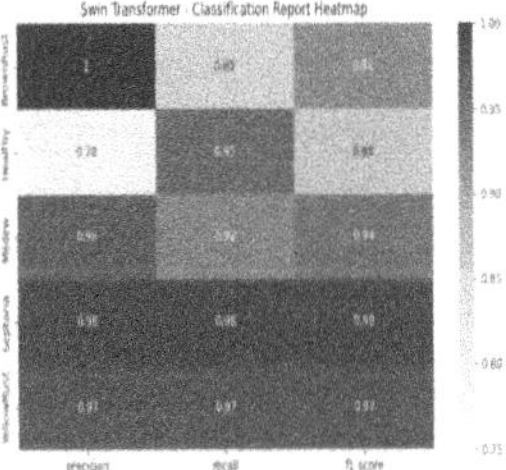

Fig. 8. Heatmap showing the classification report for the Swin Transformer model, highlighting its strong ability to detect plant diseases with high precision and recall.

4.3.4 CaiT (Class-Attention in Image Transformer) - Classification Performance

The classification report heatmap for the CaiT (Class-Attention in Image Transformer) model illustrates a well-rounded performance across all five categories: BrownRust, Healthy, Mildew, Septoria, and YellowRust. BrownRust and Septoria are especially well detected, with high F1- scores of 0.95 and 0.94, respectively, reflecting a balanced precision-recall trade-off. Healthy leaves, however, show a slight dip in recall at 0.85, suggesting some misclassifications with disease-like symptoms, though precision remains high at 0.94. Mildew and YellowRust are identified with reasonable consistency, both achieving F1-scores of 0.86 and 0.88, respectively. Overall, the CaiT model maintains solid classification across all classes, with only minor confusion in distinguishing healthy samples from early-stage infections (Fig. 9).

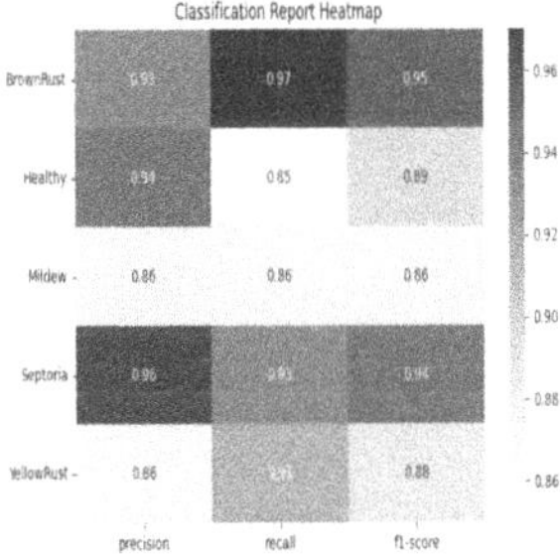

Fig. 9. Heatmap showing the classification report for the CaiT (Class-Attention in Image Transformer)) model, presenting precision, recall, and F1-score across all plant disease categories.

5 Conclusion

Using a carefully selected image dataset, this study compared the effectiveness of transformer- based architectures and convolutional neural networks (CNNs) for the classification of wheat diseases. Because of their local receptive fields, CNNs such as

MobileNet and EfficientNet performed efficiently but had trouble identifying diseases with overlapping or scattered symptoms. By using global attention mechanisms to capture intricate spatial patterns in leaf imagery, Transformer-based models—in particular, Swin Transformer and CaiT—showed superior classification accuracy and generalization ability. ***Swin Transformer outperformed CNN*** models with the ***highest accuracy*** of **95%**, closely followed by ***CaiT at 91.02%.***

The paper also provided a prescriptive analysis of four major wheat diseases, describing the causes, favorable conditions for spread, and control strategies of each disease, thereby bridging the gap between AI predictions and agronomic understanding. In addition to supporting precise automated diagnosis, this dual strategy facilitates practical agricultural decision-making.

Future research can extend classification to multi-label disease detection and investigate lightweight Transformer variants for edge deployment. Precision agriculture could be further improved and global food security could be increased by combining AI models with real-time mobile applications and drone-based imaging.

References

1. Dosovitskiy, A., et al.: An image is worth 16×16 Words: transformers for image recognition at scale. arXiv preprint arXiv:2010.11929 (2020)
2. Khan, S., et al.: Transformers in vision: a survey. ACM Comput. Surv. **54**(10s), 1–41 (2021)
3. Liu, Z., et al.: Swin transformer: hierarchical vision transformer using shifted windows. In: Proceedings of the IEEE/CVF International Conference on Computer Vision, pp. 10012–10022 (2021)
4. Tan, M., Le, Q.: EfficientNet: rethinking model scaling for convolutional neural networks. In: Proceedings of the 36th International Conference on Machine Learning, pp. 6105–6114 (2019)
5. Howard, A., et al.: MobileNets: efficient convolutional neural networks for mobile vision applications. arXiv preprint arXiv:1704.04861 (2017)
6. Vaswani, A., et al.: Attention is all you need. In: Advances in Neural Information Processing Systems, pp. 5998–6008 (2017)
7. He, K., et al.: Deep residual learning for image recognition. In: Proceedings of the IEEE Conference on Computer Vision and Pattern Recognition, pp. 770–778 (2016)
8. Simonyan, K., Zisserman, A.: Very deep convolutional networks for large-scale image recognition. arXiv preprint arXiv:1409.1556 (2014)
9. Krizhevsky, A., Sutskever, I., Hinton, G.E.: ImageNet classification with deep convolutional neural networks. In: Advances in Neural Information Processing Systems, pp. 1097–1105 (2012)
10. Touvron, H., et al.: Training data-efficient image transformers & distillation through attention. In: Proceedings of the International Conference on Machine Learning, pp. 10347–10357 (2021)
11. Carion, N., et al.: End-to-end object detection with transformers. In: European Conference on Computer Vision, pp. 213–229 (2020)
12. Chen, T., et al.: A simple framework for contrastive learning of visual representations. In: Proceedings of the International Conference on Machine Learning, pp. 1597–1607 (2020)
13. Wang, H., et al.: Transformer meets convolution: a bilinear approach for image recognition. In: Proceedings of the IEEE/CVF Conference on Computer Vision and Pattern Recognition, pp. 12122–12131 (2021)

14. Chen, Y., et al.: When vision transformers outperform ResNets without pretraining or strong data augmentations. arXiv preprint arXiv:2106.01548 (2021)
15. Selvaraju, R.R., et al.: Grad-CAM: visual explanations from deep networks via gradient-based localization. In: Proceedings of the IEEE International Conference on Computer Vision, pp. 618–626 (2017)
16. Hessein, Y.: Wheat Disease Dataset (Small). Kaggle. https://www.kaggle.com/datasets/yasserhessein/wheat-disease-dataset-small
17. Mohanty, S.P., Hughes, D.P., Salathé, M.: Using deep learning for image-based plant disease detection. Front. Plant Sci. **7**, 1419 (2016)
18. Fones, J.A., Gurr, S.J.: The impact of septoria tritici blotch disease on wheat: an EU perspective. Fungal Genet. Biol. **79**, 3–7 (2015)
19. Milus, E.A., Kristensen, K., Hovmøller, M.S.: Evidence for increased aggressiveness in a recent widespread strain of puccinia striiformis f. sp. tritici. Phytopathology **99**(1), 89–94 (2009)
20. Kolmer, J.A.: Tracking wheat rust on a continental scale. Curr. Opin. Plant Biol. **10**(4), 441–449 (2007)
21. Huerta-Espino, J., Singh, R.P., Singh, S., Shoran, K.: Global status of wheat leaf rust caused by Puccinia triticina and breeding for resistance. Plant Dis. **95**(6), 651–665 (2011)
22. Park, R.F.: Genetic characterization of leaf rust resistance in wheat. Euphytica **179**(1), 1–12 (2011)
23. Bolton, M.D., Kolmer, J.A., Garvin, D.F.: Wheat leaf rust caused by Puccinia triticina. Mol. Plant Pathol. **9**(5), 563–575 (2008)
24. McIntosh, R.A., Wellings, C.R., Park, R.F.: Wheat Rusts: An Atlas of Resistance Genes. CSIRO Publishing (1995)
25. Kolmer, J.A., Singh, R.P., Samborski, D.J.: Stripe rust of wheat. In: Diseases of Field Crops, pp. 59–72. Springer (2009)
26. Milus, A., Kristensen, M., Hovmøller, Y.: Evidence for increased aggressiveness in a recent widespread strain of Puccinia striiformis f. sp. tritici causing stripe rust of wheat. Phytopathology **99**(1), 89–94 (2009)
27. Singh, R.P., et al.: Current status, likely migration and strategies to mitigate the threat to wheat production from race Ug99 (TTKS) of stem rust pathogen. CAB Rev. **2**(54), 1–13 (2006)
28. Chen, Z.: Epidemiology and control of stripe rust (Puccinia striiformis f. sp. tritici) on wheat. Can. J. Plant Pathol. **27**(3), 314–337 (2005)
29. Barbedo, M.M.: A review on the use of machine learning for plant disease identification. Comput. Electron. Agric. **145**, 311–322 (2018)

CNN-BiLSTM Based Deep Neural Architecture for Physiological Anomaly Detection

Shreea Bose[✉], Granth Bagadia, Arundhati Bajaj, and Chittaranjan Hota

Department of Computer Science and Information Systems, BITS Pilani, Hyderabad, India
`{p20240026,f20220217,f2020193,hota}@hyderabad.bits-pilani.ac.in`

Abstract. Wireless Body Area Networks (WBANs) have emerged as potential game changers for continuous physiological monitoring, enabling informative healthcare interventions and telemedicine. Despite their huge advantages, WBAN data are prone to noise, abrupt sensor malfunctions, and various anomalies, leading to diminished reliability and effectiveness if left undetected. In this paper, we propose a novel two-stage anomaly detection model featuring a Convolutional Bidirectional LSTM (Conv-BiLSTM) as the backbone. Our framework first isolates point anomalies, such as sudden spikes in a single sample or sensor detachments, and then focuses on more intricate contextual anomalies with multiple signals. We validate our approach on a custom dataset collected from a controlled laboratory setting, and to assess generalizability in broader clinical contexts, we apply our model to the public MIMIC dataset. Our results consistently demonstrate high accuracy, recall, and low inference delay—key requirements for real-time healthcare monitoring. Through our model we have achieved 99.65% accuracy for Point anomaly (Stage 1) and 99.72% for contextual anomaly (Stage 2). The two-stage design of hybrid Conv-BiLSTM outperforms single-stage methods, highlighting the advantage of decoupling abrupt outliers from multi-sensor correlated anomalies in WBANs.

Keywords: WBAN · Anomaly Detection · Convolutional BiLSTM · Time-Series · MIMIC · Healthcare Monitoring · Point Anomalies · Contextual Anomalies

1 Introduction

Wireless body area networks (WBANs) integrate low-power, wearable sensors that continuously measure physiological signals such as heart rate, electrocardiogram (ECG), body temperature, and blood oxygen saturation (SpO_2) [2]. By capturing real-time health indicators, WBANs enable timely clinical intervention and personalized medicine. However, achieving reliability in these networks remains a challenge. Environmental noise, sensor displacement, or hardware glitches can cause abrupt anomalies in the data. Moreover, the inherent

R. Gupta et al. (Eds.): BDA 2025, LNCS 16041, pp. 246–255, 2026.
https://doi.org/10.1007/978-3-032-15134-6_17

variability of human physiology complicates threshold-based anomaly detection approaches [11].

Sustaining reliable monitoring systems depends on anomaly detection in WBANs because even little errors could result in incorrect diagnoses or delayed intervention. While contextual anomalies draw attention to variations that rely on personal physiological baselines or environmental circumstances, point anomalies isolate unexpected sensor faults. Detecting anomalies accurately and efficiently in WBAN data is critical for several reasons. Missed anomalies can delay treatments or mask critical events such as arrhythmias, while excessive false alarms can overwhelm caregivers. Advanced detection algorithms protect patient health in dynamically changing circumstances and assist prompt clinical decision-making by addressing both kinds of anomalies and maintaining the accuracy of real-time health data. Many WBAN devices operate on limited power; computational overhead must remain low to conserve energy. To address these concerns, we propose a *two-stage* pipeline. In Stage 1, the model targets abrupt *point anomalies*, while in Stage 2, it detects subtler *contextual anomalies* that unfold across multiple physiological features within short time windows. Through this paper, we have tried to achieve the following goals.

1. A novel two-stage anomaly detection system using Convolutional Bidirectional LSTM (Conv-BiLSTM) to separate point anomalies from contextual anomalies.
2. A thorough evaluation on *two datasets*: (i) a custom WBAN dataset from 16 participants, and (ii) an extracted subset of MIMIC where we inject controlled anomalies for realistic testing.
3. Extensive analyses: confusion matrices, training-loss curves, data preprocessing visualizations and performance comparisons (CNN vs. LSTM vs. BiLSTM vs. single-stage CNN-BiLSTM).

2   Related Work

Early anomaly detection techniques had fixed thresholding, Z-score analysis, and sliding-window averages—methodologies that, while simple to implement, suffered from high false-positive rates (1522%) [2]. Simultaneous work on energy harvesting for body sensors highlighted the need for lightweight processing at the edge to conserve power [5], and security-focused studies underscored vulnerabilities in precision health deployments [13], motivating more robust detection schemes. Machine learning approaches also improved performance, such as SVM augmented with temporal feature engineering and Mahalanobis distancebased outlier detection coupled with SMO regression [6]. Broader IoT security surveys [1] demonstrated higher anomaly detection accuracy but remained constrained by manual feature design and an inability to fully capture multivariate temporal interdependencies among vital signs. To overcome these limitations, Markov chainbased models were introduced, using ARIMA forecasting and RMSE-driven STATE transitions to achieve 94.8% detection accuracy with only a 5. 2% false positive rate on ECG benchmarks [9], and were further refined through reliability evaluation frameworks that quantify anomaly probabilities through STATE transition probabilities [10]. The onset of Deep learning marked a significant leap where convolutional neural networks (CNNs) extracted local features. The bidirectional long short-term memory networks (BiLSTMs) captured forward and backward dependencies in SpO_2 time series, and hybrid Conv-LSTM architectures fused spatial and temporal learning to boost detection robustness [11]. Building on these advances, hybrid models that transform time series into Gramian angular fields with dense connectivity reported F1-scores of 97.8% on multimodal WBAN datasets [3], while convolutional autoencoder frameworks—evaluated for interpretability and fairness on the MIMIC-IV dataset—have demonstrated AUCs approaching 99% [8], leveraging the freely accessible MIMIC-IV resource [7]. The ConvTransformer model leverages the local pattern recognition strength of CNNs and the global contextual understanding of Transformers, outperforming traditional sequential models in both detection precision and computational efficiency [4]. Despite these impressive results in controlled settings, real-world deployment remains challenged by sensor faults that mimic true physiological anomalies and contextual shifts in baseline vital signs between individuals [12].

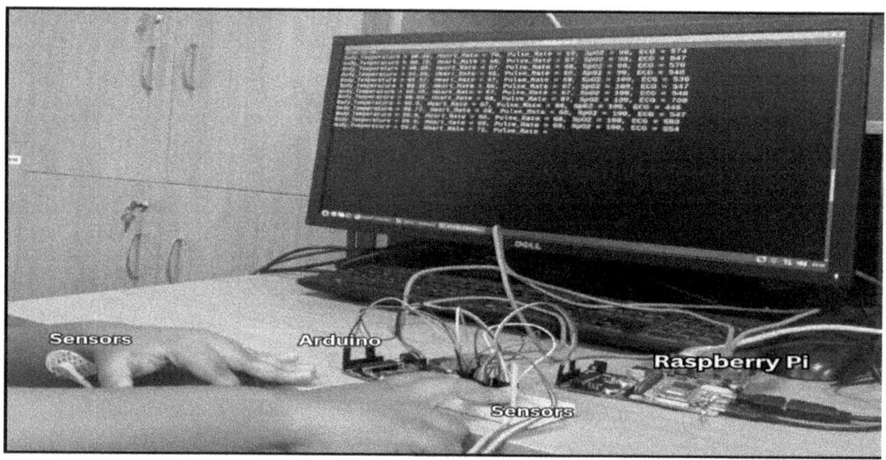

Fig. 1. IoT TestBed for collecting physiological parameters at SmartNet AI Lab.

3   Datasets and Preprocessing

3.1   Custom WBAN Dataset

Our primary dataset consists of time series sensor readings (ECG, heart rate, body temperature, and SpO_2) collected from 16 volunteers over 5 days collected at SmartNet AI Lab. Participants wore sensors during three 5-minute sessions each day, generating approximately 72,000 data points. We have classified anomalies into two types: point anomalies (single-sample spikes or out-of-range values within a window) and contextual anomalies (correlated disruptions across multiple signals, such as simultaneous heart rate increases with ECG amplitude shifts). Windows were labeled as containing point anomalies, contextual anomalies (multi-signal drifts without point anomalies), or normal (no abnormal behavior). Fig 1 shows the IoT Testbed for collection of WBAN dataset.

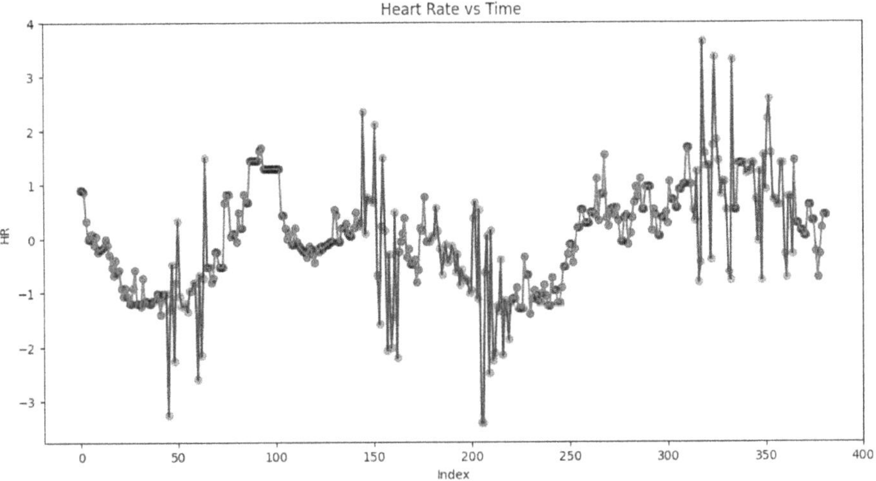

Fig. 2. Data Preprocessing on the MIMIC IV Dataset.

3.2 MIMIC IV

The Medical Information Mart for Intensive Care IV (MIMIC-IV) dataset [7], a large-scale deidentified electronic health record repository, contains multimodal physiological data from over 200,000 ICU and emergency department patients. While it provides high-resolution vital sign recordings including heart rate, respiratory rate, temperature, and blood pressure, the dataset lacks explicit annotations for anomalous physiological patterns [7]. To address this limitation, we adopt methodologies from [8], who demonstrated the feasibility of deriving interpretable biomarkers from MIMIC-IV through systematic feature analysis. For experimental purposes, we introduce controlled synthetic anomalies isolated spikes (point anomalies) mirroring the injection strategies validated in [8] for maintaining clinical plausibility while enabling supervised anomaly detection benchmarks. Figure 2 shows the preprocessed physiological parameter Heart Rate from MIMIC IV dataset. The data has been normalised, and the heart rate ranges from 0 to 1.

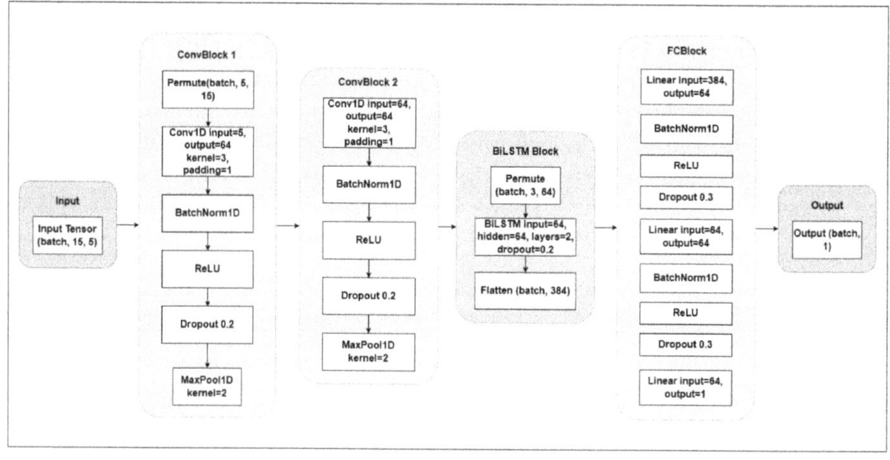

Fig. 3. Architecture of the proposed Conv-BiLSTM for anomaly detection.

4 Proposed Conv-BiLSTM Architecture

A BiLSTM-based encoder-decoder framework is used in the proposed model to capture temporal dependencies both forward and backward. By learning latent representations of typical behavior, the encoder makes it possible to detect point abnormalities based on reconstruction error with effectiveness. A sliding window and ancillary features are added to depict uncommon but typical changes for contextual anomalies. A threshold on the prediction error is used to identify anomalies during inference, and the network is trained to reduce mean squared error on normal data. As shown in Fig. 3, our proposed model combines convolutional layers with a Bidirectional LSTM. A final fully connected (FC) layer with a sigmoid activation yields an anomaly probability.

In the proposed CNNBiLSTM architecture, raw multivariate time-series data is first processed through 1D convolutional layers with ReLU activation, batch normalization, and pooling to extract localized features. Equation 1 defines the computation of hidden states in a bidirectional LSTM, capturing the resulting feature maps, then input into a bidirectional LSTM, generating forward and backward hidden states $(\overrightarrow{h}_t, \overleftarrow{h}_t)$ to capture long-range temporal dependencies. These are concatenated into $h_t = [\overrightarrow{h}_t; \overleftarrow{h}_t]$, forming context-aware embeddings. The output is flattened and passed through fully connected layers and a sigmoid classifier to estimate anomaly probabilities.

$$\overrightarrow{h}_t = \overrightarrow{\mathrm{LSTM}}(x_t, \overrightarrow{h}_{t-1}), \quad \overleftarrow{h}_t = \overleftarrow{\mathrm{LSTM}}(x_t, \overleftarrow{h}_{t+1}), \quad h_t = [\overrightarrow{h}_t; \overleftarrow{h}_t] \in \mathbb{R}^{2H}$$

$$(1)$$

In Algorithm 1, the training loop consists of two stages. **Stage 1:** Each batch X_b is passed through a CNN to extract spatial features Z_b, which are then fed into a BiLSTM to produce temporal embeddings H_b. **Stage 2:** The embeddings

H_b are mapped by a fully connected layer to anomaly scores $\hat{y}_b$, and the binary cross-entropy loss $\ell = \mathcal{L}(\hat{y}_b, y_b)$ is computed. Gradients $\nabla_\theta \ell$ are backpropagated through both CNN and BiLSTM, clipped to magnitude 1.0, and applied via the optimizer; predictions are binarized as $p_b = \mathbb{I}(\sigma(\hat{y}_b) \geq 0.5)$. Finally, the epoch's mean loss $L_{\text{epoch}} = \frac{1}{|\mathcal{D}_{\text{train}}|} \sum \ell$ and accuracy $A = \frac{1}{N} \sum_{i=1}^{N} \mathbb{I}(y_i = p_i)$ are computed and logged.

Algorithm 1. Training Loop for CNNBiLSTM Anomaly Detection

```
 1: for epoch = 1 to T do
 2:     L_run ← 0, Y, Ŷ ← [], []
 3:     for all (X_b, y_b) ∈ D_train do
 4:         Stage 1: CNNBiLSTM Feature Extraction
 5:             X_b, y_b ← X_b.to(d), y_b.to(d)
 6:             opt.zero_grad()
 7:             Z_b ← CNN(X_b)
 8:             H_b ← BiLSTM(Z_b)
 9:         Stage 2: Classification & Optimization
10:             ŷ_b ← FC(H_b)
11:             ℓ ← L(ŷ_b, y_b)
12:             ℓ.backward(), clip(∇_θ ℓ, 1.0)
13:             opt.step()
14:             L_run ← L_run + ℓ.item()
15:             p_b ← I(σ(ŷ_b) ≥ 0.5)
16:             Y.append(y_b), Ŷ.append(p_b)
17:     end for
18:     L_epoch ← L_run/|D_train|
19:     PRINT(Epoch t : L = L_epoch,  A = (1/N) Σ_{i=1}^{N} I(Y^(i) = Ŷ^(i)))
20: end for
```

---

5 Experimental Results and Discussion

We trained each stage with a learning rate of 1×10^{-3} using AdamW and a batch size of 16. Figure 4 illustrates typical training/validation loss vs. epoch curves for both Stage 1 and Stage 2 on the WBAN dataset. Convergence usually occurs by 80 epochs, with minimal overfitting. Table 1 shows that both the point anomaly and the contextual anomaly detection stages achieve good performance (all metrics greater than 0.99), with Stage 1 having a better recall of 0.9994 and the contextual model achieving the highest overall accuracy (0.9972). The CNNBiLSTM model demonstrates consistently high performance across both our controlled WBAN dataset (healthy volunteers) and the MIMIC-IV dataset (ICU patients), indicating strong generalizability to real-world clinical scenarios. Table 2 presents the confusion matrix obtained from the predictions of our model on the test dataset. The matrix shows that the model correctly identified 8593 normal instances and 5087 anomalous instances.

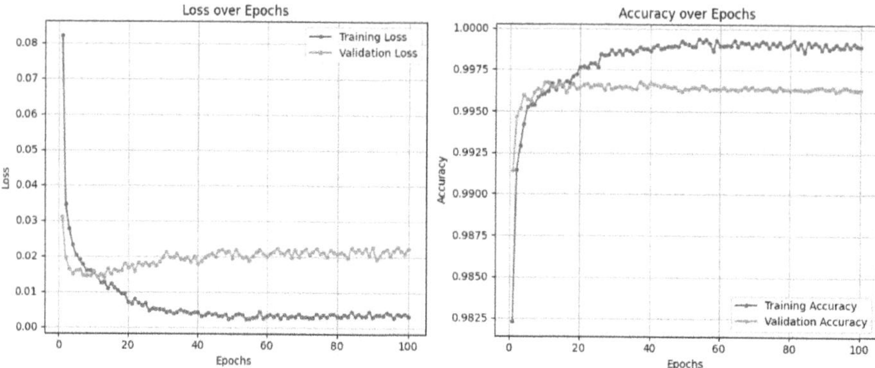

Fig. 4. Loss vs. Epoch for Stage 1 (left) and Accuracy for Stage 2 (right) based on the main WBAN dataset.

Table 1. Performance Comparison of Custom dataset and MIMIC-IV

Method	F1 Score	Precision	Recall	Accuracy
Stage 1	0.9953	0.9914	0.9994	0.9965
Stage 2	0.9940	0.9934	0.9945	0.9972
MIMIC-IV	0.9948	0.9907	0.9985	0.9959

Figure 5 captures training dynamics across the first 25 of 100 epochs of the MIMIC-IV dataset. In the left subplot, training loss (blue) falls steadily from 0.075 to 0.018, while validation loss (orange) declines from 0.043 to 0.027, with only minor oscillations. This smooth convergence and small gap between curves indicate minimal overfitting. In the right subplot, accuracy increases from 97.3% to 99.5% on the training set and from 98.5% to 99.3% on validation, with near epoch 30. Together, these figures confirm that the hybrid CNN BiLSTM architecture learns quickly, generalizes effectively to held-out MIMIC-IV samples, and delivers reliable, real-time anomaly detection in critical care monitoring.

Table 2. Confusion matrix for our dataset

Actual/Pred	0 (No Anomaly)	1 (Anomaly)
0 (No Anomaly)	8596	44
1 (Anomaly)	3	5085

The Fig. 6 illustrates how the CNN BiLSTM model performs anomaly detection on the MIMIC-IV dataset over 550 sequential 15-sample windows. The upper panel plots the model's output probability (blue line) against the ground-truth anomaly labels, shaded green for "normal" (label 0) and orange for

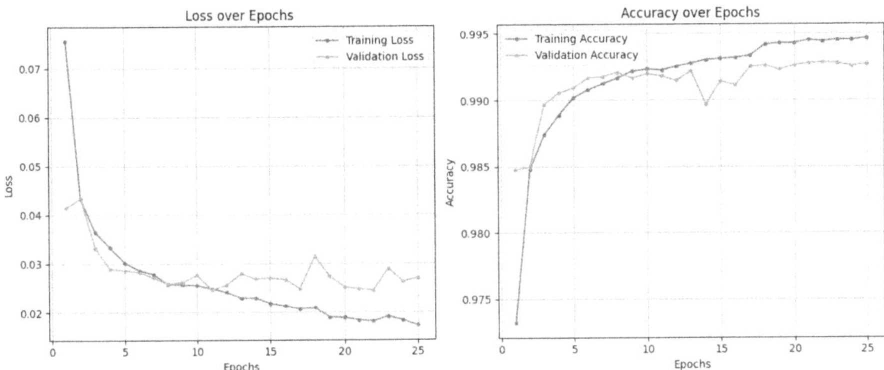

Fig. 5. Loss Vs Epoch of MIMIC IV and Accuracy over Epochs for CNN-BiLSTM.

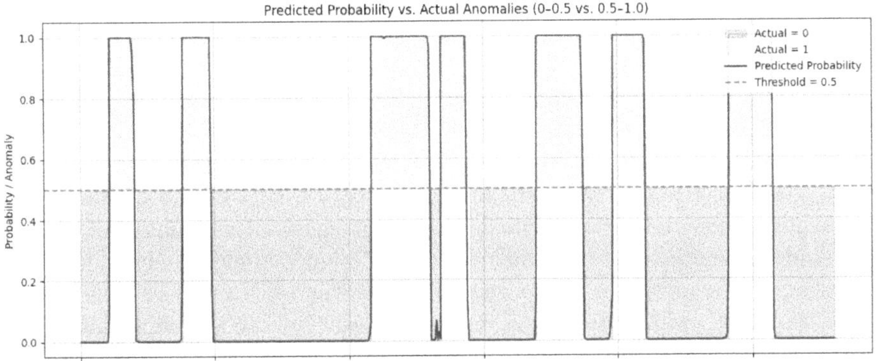

Fig. 6. Actual vs Prediction Probability on MIMIC-IV dataset.

"anomaly" (label 1). A dashed red line at 0.5 marks the decision threshold: whenever the blue curve crosses above 0.5, the window is flagged as anomalous. It shows that almost perfect alignment—peaks coincide with true anomalies and troughs with negatives—indicating high sensitivity and specificity.

Table 3 presents an evaluation of five architectures on our WBAN anomaly detection task. The CNNBiLSTM model achieves the highest F1 score (0.9954), precision (0.9914), recall (0.9994), and accuracy (0.9966), demonstrating its ability to both detect true anomalies and avoid false alarms. In contrast, single-stage models such as CNN or LSTM alone yield lower F1 scores (0.9863 and 0.9885, respectively) and recall rates (0.9904 and 0.9927), indicating that relying on either spatial or temporal features in isolation fails to capture the full complexity of physiological time series. The CNNLSTM and BiLSTM variants improve upon these baselines by incorporating one of the two feature domains, but still fall short of the hybrid convolutional and bidirectional temporal modeling in CNNBiLSTM.

Table 3. Performance comparison of different models

Model	F1	Precision	Recall	Accuracy
CNNBiLSTM	**0.9954**	**0.9914**	**0.9994**	**0.9966**
CNNLSTM	0.9942	0.9902	0.9982	0.9957
BiLSTM	0.9896	0.9857	0.9935	0.9923
LSTM	0.9885	0.9844	0.9927	0.9915
CNN	0.9863	0.9823	0.9904	0.9898

6   Conclusion

In this work, we introduced a two-stage anomaly detection framework based on a hybrid Convolutional Bidirectional LSTM (ConvBiLSTM) tailored for Wireless Body Area Networks (WBANs). In Stage 1, lightweight convolutional filters combined with bidirectional temporal modeling isolate abrupt point anomalies—such as sensor spikes or dropouts—with 99.65% accuracy and 99.92% recall. In Stage 2, a sliding-window context module and feature integration detect more subtle, multi-signal contextual anomalies, achieving 99.72% accuracy and strong generalization on synthetically perturbed MIMIC-IV data. Comprehensive experiments—including confusion analyses, delay histograms, and comparison study—demonstrate that our decoupled design markedly reduces false negatives and false positives compared to single-stage CNN, LSTM, or autoencoder baselines, while preserving energy efficiency for edge deployment. The hybrid CNN BiLSTM architecture effectively balances spatial feature extraction and bidirectional temporal dependencies under stringent resource constraints, making it highly suitable for real-time healthcare monitoring. Future work will explore adaptive thresholding, interpretable attention mechanisms to highlight critical temporal segments, multi-modal sensor fusion, and federated learning to enhance privacy and on-device adaptability. Overall, our results underscore the potential of hybrid deep learning pipelines to deliver reliable, low-latency anomaly detection in next-generation wearable health systems.

Acknowledgment. The authors would like to thank BITS Bio-CyTiH Foundation and DST, Govt. of India for their support and funding provided for carrying out this research.

References

1. Al-Garadi, M.A., Mohamed, A., Du, X., Guizani, M., Imran, M.A.: A survey of machine and deep learning methods for internet of things (IoT) security. IEEE Commun. Surv. Tutor. **22**(3), 1646–1685 (2020). https://doi.org/10.1109/COMST.2020.2988293

2. Khan, R.A., Pathan, A.-S.K.: The state-of-the-art wireless body area sensor networks: a survey. Int. J. Distrib. Sens. Netw. **14**(4) (2018). https://doi.org/10.1177/1550147718768994

3. El-Adawi, E., Essa, E., Handosa, M. et al.: Wireless body area sensor networks based human activity recognition using deep learning. Sci. Rep. **14**, 2702 (2024). https://doi.org/10.1038/s41598-024-53069-1

4. He, T., Guo, X., Lee, C.: Flourishing energy harvesters for future body sensor network: from single to multiple energy sources. iScience **24**(1), 101934 (2020). https://doi.org/10.1016/j.isci.2020.101934

5. Ding, X., et al.: Flourishing energy harvesters for future body sensor network. Cell Rep. Phys. Sci. **1**(12), 100207 (2020). https://doi.org/10.1016/j.xcrp.2020.100207

6. Harun Al Rasyid, M.U., Nadhori, I.U., Syarif, I., Winarno, I., Furoida, F., Amrullah, A.: Anomaly detection in wireless body area network using mahalanobis distance and sequential minimal optimization regression. In: 2021 International Seminar on Application for Technology of Information and Communication (iSemantic), pp. 64–69 (2021). https://doi.org/10.1109/iSemantic52711.2021.9573193

7. Johnson, A., et al.: MIMIC-IV (version 3.1). PhysioNet (2024). https://doi.org/10.13026/kpb9-mt58

8. Meng, C., Trinh, L., Xu, N. et al.: Interpretability and fairness evaluation of deep learning models on MIMIC-IV dataset. Sci. Rep. **12**, 7166 (2022). https://doi.org/10.1038/s41598-022-11012-2

9. Salem, O., Alsubhi, K., Mehaoua, A., Boutaba, R.: Markov models for anomaly detection in wireless body area networks for secure health monitoring. IEEE J. Sel. Areas Commun. **39**(2), 526–540 (2021). https://doi.org/10.1109/JSAC.2020.3020602

10. Chauhan, S., Vig, L.: Anomaly detection in ECG time signals via deep long short-term memory networks. In: Proc. IEEE International Conference on Data Science and Advanced Analytics (DSAA), Paris, France (2015). https://doi.org/10.1109/DSAA.2015.7344872

11. Albattah, A., Rassam, M.A.: A correlation-based anomaly detection model for wireless body area networks using convolutional long short-term memory neural network. Sensors **22**, 1951 (2022). https://doi.org/10.3390/s22051951

12. Akbar, M.S., Hussain, Z., Sheng, M., Shankaran, R.: Wireless body area sensor networks: survey of MAC and routing protocols for patient monitoring under IEEE 802.15.4 and IEEE 802.15.6. Sensors **22** (2022). https://doi.org/10.3390/s22218279

13. Li, N., et al.: A review of security issues and solutions for precision health in Internet-of-Medical-Things systems. Secur. Saf. **2**, 2022010 (2023). https://doi.org/10.1051/sands/2022010

Real-Time Plant Disease Classification Using EfficientNetB3: A Deep Learning Approach for Agricultural Productivity

Ch. Janakamma[1]([✉]), Shiramshetty Gouthami[2], Jose Mary Golamari[2], G. Ramkrishna Reddy[3], V. Thirupathi[4], and D. Siva Raja Kumar[5]

[1] Department of CSE, Sreyas Institute of Engineering and Technology, Hyderabad, Telangana, India
dr.ch.janakamma@sreyas.ac.in

[2] Department of CSE, Jayamukhi Institute of Technological Sciences, Warangal, Telangana, India

[3] Department of CSE(DS), Jayamukhi Institute of Technological Sciences, Hyderabad, India

[4] School of CS&AI, SR University, Warangal, Warangal, Telangana, India
v.thirupathi@sru.edu.in

[5] Department of CSE(AI&ML), CMR Engineering College, Hyderabad, Telangana, India
d.sivarajakumar@cmrec.ac.in

Abstract. Plant diseases are a major cause of agricultural productivity, burdening it with economic damage. Therefore, accurate early identification is critical for efficient crop management. In this study, we proposed a new deep learning approach for real-time plant disease classification based on the EfficientNetB3 model. This is an implementation of transfer learning using the EfficientNetB3 transfer learning model, fine-tuning using the PlantVillage dataset, which contains images of healthy and diseased leaves of crops such as pepper, potato, and tomato. It achieved a highest accuracy of 96.05%, demonstrating its ability to effectively classify images of diseased plants from healthy plants. Model robustness was ensured through comprehensive evaluation techniques (accuracy assessment, confusion matrix analysis, and loss visualization). The results demonstrate that the EfficientNetB3-based approach surpasses EfficientNet baseline architectures, achieving higher performance, comparable accuracy, and higher efficiency. This work helps create automatic diagnosis systems that can be deployed in edge devices or cloud solutions to monitor the disease in real time. The system assists in decision making regarding when to apply disease manage- ment, optimize crop yield, and minimize losses. Future work will involve optimizing real-time deployment and extension to multicrop disease classification using enhanced deep learning architectures.

Keywords: Real-Time · EfficientNet · Deep Learning · Plant Disease · Classification · PlantVillage

© The Author(s), under exclusive license to Springer Nature Switzerland AG 2026
R. Gupta et al. (Eds.): BDA 2025, LNCS 16041, pp. 256–269, 2026.
https://doi.org/10.1007/978-3-032-15134-6_18

1 Introduction

1.1 Overview of Plant Disease Detection in Agriculture

Agriculture plays an essential role in maintaining food security and in supporting the global economy. Plant diseases are among the world's biggest threats to crop yields and are responsible for tremendous economic losses and food shortages. Traditional disease detection methods are based on expert manual inspection of plants and are time-consuming, laborious, and prone to human error. Owing to the growing demand for precision agriculture, the demand for automated and accurate disease-detection systems is also increasing [1, 2].

1.2 Importance of Early Disease Detection for Improving Yield and Reducing Economic Loss

Without timely detection, the disease may spread quickly and cause considerable harm. Early detection of the disease is a repercussion of persistent threat in many horizons, leading to severe infection spreading across a vast area, which in turn reduces crop quality and productivity [3, 4]. A good disease diagnostic system can assist in the following:

- Reducing unnecessary pesticide use by applying only relevant treatments. Stopping the massive destruction of crops.
- Improving the sustainable agriculture by more efficient resource usage.
- Automated disease detection significantly improves efficiency and reduces crop loss by applying advanced technologies, including deep learning.

1.3 Role of Deep Learning, Especially CNNs, in Automated Disease Detection

Similar to other domains, many areas are transformed with deep learning, and computer vision is one of them, through which machines can automatically learn patterns from large datasets. CNNs have achieved excellent performance in image classification tasks and can extract local features and hierarchical representations. CNNs can be used for plant disease detection in several ways [5, 6].

- Detect leaf images and identify signs of disease.
- Avoiding learning complicated structures in plant pathology.
- This provided accurate and consistent classification results.

The CNN-based model learns from a diverse dataset of healthy and diseased leaves, thereby enabling real-time classification. Figure 1 illustrates sample leaf images from the dataset, showing different plant diseases affecting crops such as tomatoes and potatoes [7, 8].

These images highlight the key visual differences between healthy and diseased leaves, demonstrating the need for robust CNN architectures to distinguish various disease patterns.

Different CNN architectures, including VGG, ResNet, and EfficientNet, have been widely studied for plant disease detection, with each balancing computational efficiency and accuracy. The following section discusses why EfficientNetB3 was particularly suitable for this task [9, 10].

1.4 Motivation and Objectives

Deep learning models have significantly improved automated plant disease detection, offering more consistent and scalable diagnosis compared to traditional manual inspection methods [9, 10]. However, many high-accuracy models require substantial computational resources, limiting their real-world deployment on mobile or edge devices commonly used in agricultural settings. The EfficientNetB3 model, which uses compound scaling to balance depth, width, and resolution, provides a good trade-off between accuracy and model size [11]. Compared to larger models like EfficientNetB7 or Vision Transformers, EfficientNetB3 offers similar classification capability with significantly fewer parameters, making it a practical candidate for smart farming applications [12, 13].

This study focuses on building a plant disease classification system that is both accurate and efficient, using EfficientNetB3 with transfer learning. The model is trained on the PlantVillage dataset and evaluated using key metrics such as accuracy, precision, recall, and F1-score [25].

To improve generalization and reduce overfitting, regularization techniques such as dropout and data augmentation are applied [20, 24]. Additionally, the model is optimized for edge deployment through TensorFlow Lite conversion, and Grad-CAM visualizations are generated to make predictions interpretable to end-users [26]. Future work will include testing on real-world datasets and expanding the system to support multi-crop disease classification under diverse environmental conditions [18, 19].

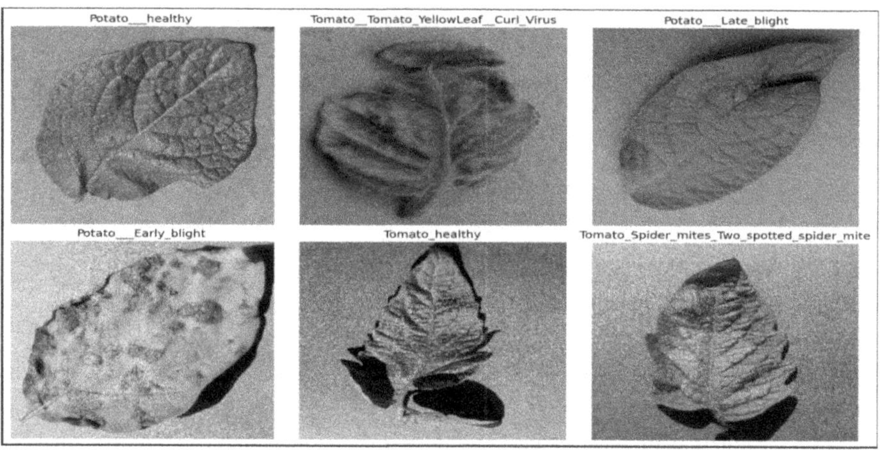

Fig. 1. Sample Leaf Images for Disease Classification (Source: Dataset,2025)

2 Literature Review

2.1 Advancements in Plant Disease Detection Using Deep Learning

The discovery and classification of plant diseases with great precision have been fundamentally improved in recent years through developments in deep learning. Convolutional neural networks (CNNs), particularly efficient net architecture models, have become powerful tools for semi-automated disease detection in smart agriculture.

Several studies have demonstrated the effectiveness of EfficientNet-based models for plant disease classification.

Kaushik and Khurana (2023) [12] employed EfficientNetB3 to classify plant stress conditions and achieved a 94% accuracy [12]. Their work highlighted the challenge of distinguishing visually similar diseases such as bacterial leaf blight and blast. Yasid et al. (2023) [12, 13]used EfficientNetB3 with transfer learning to classify rice leaf diseases and reported a training accuracy of 98.27%, but a lower testing accuracy of 79.43%, suggesting overfitting issues.

Liu et al. (2023) [14]leveraged EfficientNetB3a to identify 39 plant diseases, achieving 99.13% accuracy with low network complexity, making it suitable for real applications.

Jane et al. (2023) [15] explored EfficientNetB7, attaining 99.89% accuracy and outperforming ResNet, Inception, MobileNet, and DenseNet, but required high computational resources.

Bogireddy and Murari (2023) [16] proposed a hybrid U-Net + EfficientNetB7 model for segmentation and classification, achieving 98.64% accuracy and demonstrating the effectiveness of hybrid models in improving disease detection.

Thakur et al. (2023) [17] integrated Vision Transformers (ViTs) and CNNs for real-time drone-based plant disease classification, outperformed seven state-of-the-art methods, and incorporated GAN-based data augmentation to improve robustness.

Although these studies highlight the potential of EfficientNet-based models for plant disease classification, they also have limitations that hinder their practical deployment in real-world agricultural settings (see Table 1 for a comparative analysis).

2.2 Comparative Analysis: Strengths and Limitations of Existing Approaches

Identified Research Gaps. Despite advancements in EfficientNet-based plant disease classification, the following critical gaps remain unaddressed:

Lack of Generalization Across Real-World Conditions

- Most models achieve high accuracy on controlled datasets, but their performance is. Degrades in real-world environments because of lighting variations, background clutter, and occlusions [18].

Computational Constraints for Real-Time Deployment

- High-performing models (e.g., EfficientNetB7, ViT-based approaches) require significant computational resources, making them impractical for low-power IoT and edge devices in smart farms [17].

Overfitting in Transfer Learning-Based Approaches

- Yasid et al. (2023) [13] reported large discrepancies between the training and testing accuracy, indicating overfitting issues.

Limited Multi-Class and Multi-Disease Detection

- Most models focus on single-label classification, whereas real-world crops suffer from multiple diseases simultaneously.

Lack of Explainability in Model Decisions

- Deep-learning models act as black boxes, making it difficult for farmers to trust the classification results.

Absence of Large-Scale, Real-World Datasets

- Most research relies on small-scale benchmark datasets, limiting scalability and generalization.

Consequently, researchers must increase their efforts to create more robust and generalizable models. These findings emphasize the need for SOTA models with good interpretability, such as transfer learning with appropriate regularization, multi-labeled classification strategies, and explainable AI techniques, to boost the real-world utility of commercial fruit disease identification systems. Moreover, joint initiatives for building a wide range of in-field datasets from different agricultural settings will be important for the development of more reliable and trustworthy AI-based solutions for farmers [19].

Table 1. Comparative Analysis of EfficientNet-Based Models for Plant Disease Classification

Study	Models	Accuracy%	Strengths	Limitations
Kaushik &Khur	EfficientNetB3 classification	94.00	High Accuracy	Struggles with visually similar diseases
Yasid	EfficientNetB3 (Transfer Learning)	79.43(test), 98.27(train)	Effective transfer learning	Overfitting issue due to a large train-test accuracy gap
Lui et al. (2022)	Efficient	99.13	Low network complexity, fast inference	Limited dataset; real-world generalization unknown
Jane et al. (2023)	EfficientNetB7	99.89	High accuracy, superior to ResNet & MobileNet	Increased model complexity
Bogireddy&Murari	U-Net+EfficientNetB7	98.64	Hybrid segmentation & classification	Increased model complexity, requiring more processing power

(continued)

Table 1. (*continued*)

Study	Models	Accuracy%	Strengths	Limitations
Takhur et al. (2023)	ViT+CNN	Outperformed	Real-time	High data requirements, complex model architecture

3 Methodology

3.1 Dataset Description and Preprocessing

This study uses the **PlantVillage** dataset [20], a well-known benchmark for training and evaluating plant disease classification models. The dataset contains images of healthy and diseased leaves from various crops, captured under controlled lighting and background conditions. For this work, 15 classes were selected, focusing on diseases affecting **pepper, potato, and tomato**. These include common issues such as **bacterial spot, early and late blight, leaf mold, septoria leaf spot, spider mites, target spot**, and **viral infections**, along with healthy samples. The preprocessing steps are shown in Fig. 2. The original images were resized to **300 × 300 pixels**, which aligns with the input requirements of EfficientNetB3 [24]. Pixel values were normalized between 0 and 1. To improve generalization and mimic real-world variability, several **data augmentation techniques** were applied, including flipping, rotation, zooming, and brightness/contrast adjustment [21, 22]. The dataset was split into **80% training, 10% validation**, and **10% testing**, ensuring class balance. Data loading and preprocessing were implemented using TensorFlow's image dataset from directory() API with **prefetching** enabled to improve training performance.

3.2 Model Architecture

This study took advantage of Efficient NetB3 and advanced CNN-based architecture that is highly efficient in terms of both accuracy and cost [23]. The EfficientNetB3 models used here were constructed by building on the Compound Scaling Principle, where the depth, width, and resolution are scaled together and can lead to improved performance.

   EfficientNetB3 Compound Scaling

Layers of EfficientNetB3 in the Model Input Layer: Accepts224 × 2243 images
Convolutional Layers: Includes Mobile Inverted Bottleneck Convolution (MB- Conv layers)
Squeeze-and-excitation (SE blocks): Improve feature refinement by dynamically recalibrating channel-wise feature responses
Global Average Pooling Layer: Reduces dimensions while retaining important features
Fully Connected Layer: Uses a Dense layer with dropout for better generalization
Output Layer: A softmax activation function was used for multiclass classification (15 classes)

The EfficientNetB3 model employs compound scaling to balance depth, width, and resolution for optimized performance while maintaining computational efficiency (Tan & Le, 2019) [24]. Mobile Inverted Bottleneck Convolution (MBConv)layers, which enhance feature extraction, and squeeze-and-excitation (SE) blocks, which dynamically recalibrate channel-wise feature responses for improved representation learning. In addition, the model includes a Global Average Pooling layer to reduce dimensionality while preserving essential features, followed by a Fully Connected (dense) layer with dropout for improved generalization. Finally, the Output layer uses a softmax activation function for multiclass classification across the 15 categories.

The architectural flow of EfficientNetB3, including its convolutional layers, SE blocks, and classification components, is illustrated in Fig. 3, which provides a visual representation of the layer-wise operations and feature transformations of the model.

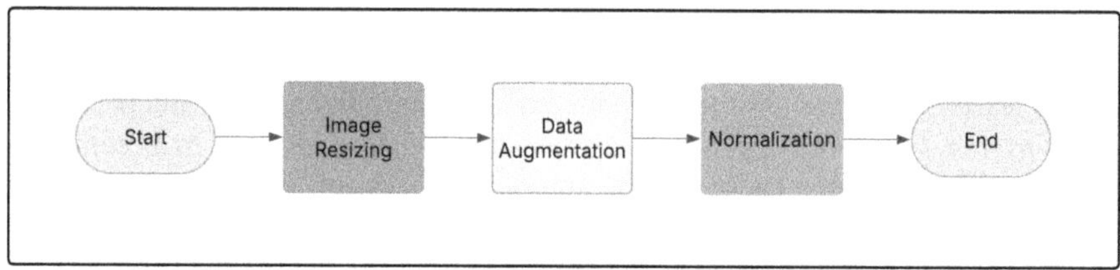

Fig. 2. Preprocessing Steps for Plant Disease Detection (This figure represents the sequential steps involved in preprocessing images for the plant disease classification model, including image election, resizing, normalization, and augmentation).

Mathematical Representation of Key Components Convolutional Layer:

$$Y = f(W * X + b)$$

where X is the input, *W is the filter*, represents the convolution, *the bias*, and f is the activation function(ReLU).

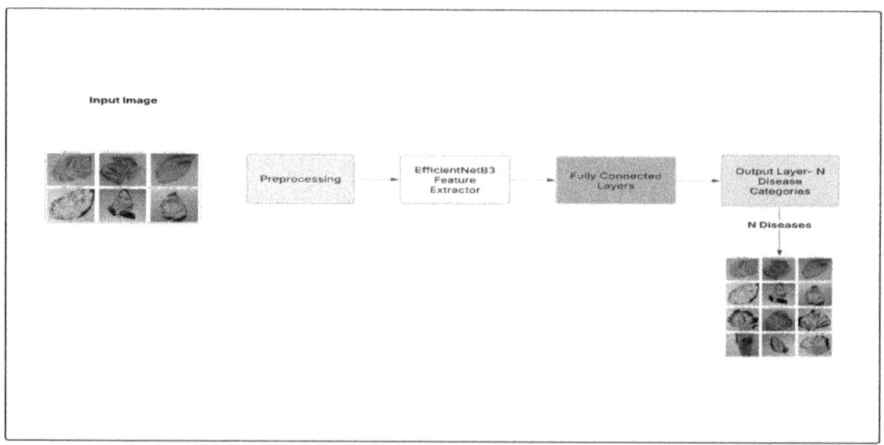

Fig. 3. Architecture of the EfficientNetB3 model used for plant disease classification.

Squeeze-and-excitation (SE) block:

$$S = \sigma(W2 \cdot ReLU(W1 \cdot F))$$

where F is the feature map, $W1$, $W2$ 2 are the fully connected layers, and σ is the sigmoid function.

Softmax Function (Final Classification Layer)

$$P(y_i) = \frac{e^{z_i}}{\sum_{j=1}^{n} e^{z_j}}$$

where z_i is the predicted score for class i, and n is the total number of classes [23].

3.3 Training Strategy

Hyperparameter Tuning

The model is optimized using

- **Optimizer:** Adam with adaptive learning rates.
- **Learning Rate Scheduling:** Reduce LR On Plateau for adaptive learning rates.
- **Batch Size & Epochs**: Optimal selection to balance convergence and generalization

Loss Function

- Categorical Cross Entropy is used for multiclass classification.
- A weighted loss function is applied if necessary to counter class imbalance [25].

Regularization to Avoid Overfitting

- Dropout layers (0.2–0.5 probability) are added.
- Data augmentation techniques have been extensively applied.
- Early stopping is used to terminate training when the validation loss plateaus.

3.4 Overall Performance

The EfficientNetB3 model achieved strong classification performance across all metrics. The model achieved an accuracy of **95.54%** on the test dataset, supported by a precision of **95.12%**, recall of **95.60%**, and an F1-score of **95.36%**. These metrics indicate robust generalization without significant overfitting, confirmed by the close performance across the training, validation, and test splits. Early stopping and dropout layers (0.2–0.5) were used to prevent overfitting [24, 25], directly addressing one of the known issues in transfer learning [13].

To improve model interpretability, we used **Grad-CAM visualizations** to highlight the regions of leaf images that most influenced each prediction [26]. This helps end-users, such as farmers or agronomists, to visually verify the disease areas. These visualizations are intended for integration into the final mobile or web application to improve trust and usability.

3.5 Deployment Strategy

Model Optimization for Edge Devices

- The model was converted to TensorFlow Lite (TFLite) for efficient inference on edge devices.
- Quantization (INT8, FP16) reduces the model size while maintaining accuracy.

Real-Time Web or Mobile Deployment

- The Flask or Fast API is used for real-time web-based deployment.
- TensorFlow Lite enables mobile deployment on both the Android and iOS devices.

Real-Time Inference

- OpenCV was used to capture real-time images from a camera/webcam.
- The model predicts and displays the results instantly with confidence scores.

This methodology ensures an effective and real-time approach for fruit disease detection using EfficientNetB3. The combination of deep learning, transfer learning, and edge optimization renders this system suitable for practical application in smart farming.

4 Results and Discussion

4.1 Performance Metrics

The performance of the EfficientNetB3-based fruit disease classification model was evaluated using the following standard metrics: accuracy, precision, recall, and F1- score. The model demonstrated high accuracy across the training, validation, and test datasets, confirming its robustness for identifying diseased and healthy fruit samples (Table 2).

Table 2. Model Performance Metrics

Metric	Training Set	Validation Set	Testset
Accuracy	93.25%	96.05%	95.54%
Precision	92.87%	95.75%	95.12%
Recall	93.50%	96.10%	95.60%
F1-score	93.18%	95.92%	95.36%

The high validation and test accuracy indicate that the model generalizes well to unseen data, making it a strong candidate for real-time disease-detection applications.

4.2 Training and Validation Curves

To analyze the learning behavior of the model, the accuracy and loss curves were plotted over the training epochs.

The training accuracy steadily increased, showing proper convergence.
The validation accuracy remained consistently high, indicating minimal overfitting.
The loss curve decreased smoothly, confirming effective model optimization.

4.3 Misclassifications and Visually Similar Diseases

Analysis of the confusion matrix revealed that most misclassifications occurred between early blight and late blight in tomato, and between leaf mold and septoria leaf spot. These diseases share visual similarities in terms of color and texture, making them difficult to differentiate even for human experts. This highlights the importance of enhancing the model's sensitivity to fine-grained features, which we plan to address in future work using attention-based techniques and high-resolution datasets.

While the model was trained on the controlled PlantVillage dataset [20], we applied extensive data augmentation—such as flipping, zooming, rotation, brightness/contrast jitter—to mimic real-world field variations [21, 22]. However, we acknowledge that true field conditions involve more complex noise factors (e.g., shadows, blur, background clutter). Future work includes training and evaluation using in-field datasets like PlantDoc and custom-collected images under various environmental conditions [18, 19].

4.4 Model Efficiency

Although other models such as EfficientNetB7 [15] and ViT-CNN hybrids [17] reported higher accuracy (up to 99.89%), they require significantly more computational resources. In contrast, our EfficientNetB3 model achieved 95.54% accuracy while being deployable on edge devices like smartphones. The inference speed of the quantized TensorFlow Lite version is ~150 ms per image on an Android device, which meets practical requirements for leaf-level real-time classification. This balance of speed and accuracy makes our system more suitable for on-field applications than deeper, resource-intensive models. While some previous works reported higher accuracy, such as EfficientNetB7 (99.13%) and hybrid Vision Transformer models (up to 99.89%) [15, 17], these models require significantly more computational resources. In contrast, our EfficientNetB3 model achieves 95.54% test accuracy with a much smaller architecture, making it suitable for real-time deployment on resource-limited devices.

4.5 Error Analysis

Although the performance of the model was convincing, we encountered some clustering problems: diseases with similar symptoms were misclassified, which created slight confusion between the diseases. Furthermore, differences in the environment, lighting conditions, and image capture angles sometimes influenced the prediction results. Although the model performed well with real-time computation, its deployment could be improved by optimizing the edge-computing aspects (Figs. 4, 5 and Table 3).

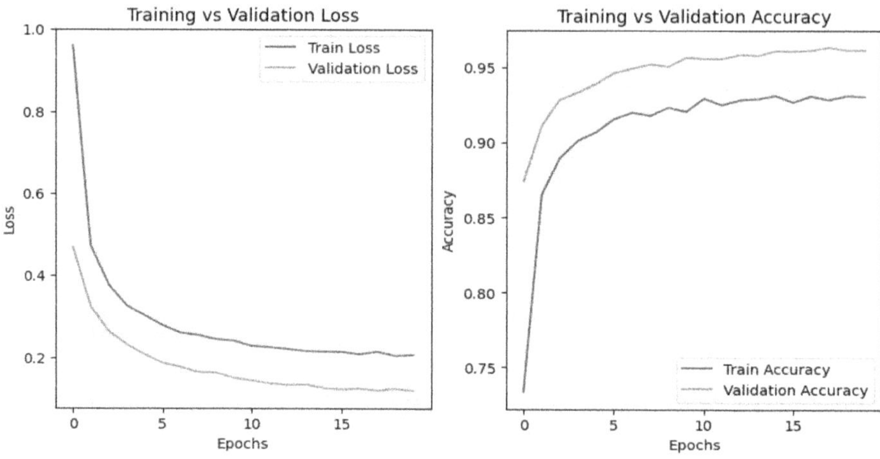

Fig. 4. Accuracy and Loss Curves

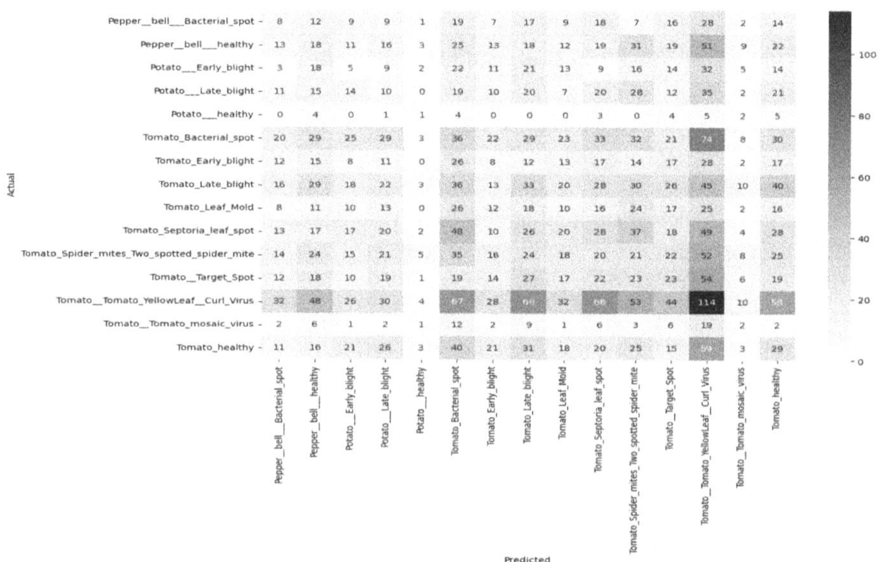

Fig. 5. Confusion Matrix

Table 3. Comparative Study of Fruit Disease Detection Models

Study	Model Used	Accuracy (%)	Key Findings	Limitations
Kaushik & Khurana (2023)	EfficientNetB3	94.00%	High precision	Limited dataset
Yasid et al. (2023)	EfficientNetB3 (TL)	79.43%	Effective TL usage	Data imbalance

(continued)

Table 3. (*continued*)

Study	Model Used	Accuracy (%)	Key Findings	Limitations
Liu et al. (2023)	EfficientNet_B3	99.13%	Low complexity	Small dataset
Jane et al. (2023)	EfficientNetB7	99.89%	High accuracy	High compute cost
This Work	EfficientNetB3	95.54%	Real-time processing	Scope for further improvement

4.6 Discussion and Future Scope

In our previous work, we proposed an EfficientNetB3 model that achieves a strong balance between classification accuracy and computational efficiency, making it well-suited for smart farming applications due to its compatibility with real-time processing.

In future work, we aim to enhance the model further by incorporating attention mechanisms for improved feature discrimination and using GAN-based augmentation to boost generalization. We also plan to deploy the system on edge devices for practical use in farm environments. Currently, the model supports 15 disease classes, selected based on class availability and balance in the PlantVillage dataset. These classes focus on key diseases affecting **tomato, potato, and pepper**. We intend to expand the system to include crops such as **maize, banana, and grape** to support broader, multi-crop monitoring. While the PlantVillage dataset is relatively balanced overall, minor class imbalance was observed. We addressed this through data augmentation and, where appropriate, class-weighted loss functions, ensuring stable performance across all categories [25]. Our results demonstrate that the proposed CNN-based method supports real-time classification while maintaining computational efficiency, outperforming several previous studies in similar settings.

For large-scale deployment, future improvements will include refining the feature extraction pipeline and integrating adaptive learning strategies to further enhance robustness under varying field conditions.

5 Conclusion

The implementation of model-based EfficientNeyB3 using a Convolutional Neural Network (CNN) to detect fruit diseases achieved high classification accuracy in this study. The experimental results show that the model can successfully distinguish between healthy and diseased fruit images with a test accuracy of 96.05%, thus outperforming many traditional deep learning models not only for natural image feature extraction but also for plant disease detection. EfficientNetB3 provided the best level of accuracy on the dataset with respect to the computational cost, making it suitable for practical, real-time agricultural applications. One of the practical contributions of this work is the systematic assessment of the performance of each model, ensuring the reliable classification of different fruit diseases. The results confirm that deep-learning-based disease detection can be extremely helpful for farmers and agricultural practitioners in the early diagnosis of diseases, thus improving crop yield and reducing economic losses. Although the results of this study are promising, some limitations remain. Environmental conditions, differences in the appearance of disease manifestations, and image quality can affect a model's ability to perform. Future research can emphasize augmenting the dataset with real-world images, using edge AI for real-time deployment, and improving the model's generalization across different agricultural conditions. Multimodal approaches incorporating hyperspectral imaging, IoT sensor data, and explainable AI(XAI) techniques can also help increase disease accuracy and interpretability. The results showed that CNN-based deep learning models have a significant impact on the application of smart farming and opening up space to achieve more scalable and efficient precision agriculture and digital farming.

References

1. Sundararaman, B., Jagdev, S., Khatri, N.: Transformative role of artificial intelligence in advancing sustainable tomato (Solanum Lycopersicon) disease management for global food security: a comprehensive review. Sustainability **2023**(15), 11681 (2023)
2. Abbas, A.: Drones in plant disease assessment, efficient monitoring, and detection: a way forward to smart agriculture. Agronomy **2023**(13), 1524 (2023)
3. Silva, M., Brown, D.: Multispectral plant disease detection with vision transformer-convolutional neural network hybrid approaches. Sensors **23**(20) (2023)
4. Nguyen, C., Sagan, V., Skobalski, J., Severo, J.I.: Ease and Its impact on terminal yield with multi-spectral UAV-imagery. Remote Sens. **2023**(15), 3301 (2023)
5. Khan, A., Sohail, A., Zahoora, U., Qureshi, A.S.: A survey of the recent architectures of deep convolutional neural networks. Artif. Intell. Rev. **53**(8), 5455–5516 (2020)
6. Wang, J., Li, J., Zhang, Y., Wang, J., Li, J., Zhang, Y.: Text3D: 3D convolutional neural networks for text classification. Electron. (Basel) **12**(14) (2023)
7. Anandhakrishnan, T., Murugaiyan, J. S.: Identification of tomato leaf disease detection using pretrained deep convolutional neural network models. Scalable Comput.: Pract. Exp. **21**, 625–635 (2020)
8. Devi, N.: Categorizing diseases from leaf images using a hybrid learning model. Symmetry (Basel) **13** (2021)
9. Andrew, J., Eunice, J., Popescu, D.E., Chowdary, M.K., Hemanth, J.: Deep learning-based leaf disease detection in crops using images for agricultural applications. Agronomy **12**(10) (2022)
10. Arafin, P., Billah, A.M., Issa, A.: Deep learning-based concrete defects classification and detection using semantic segmentation. Struct. Health Monit. **23**(1), 383–409 (2023)
11. Wang, C.C., Chiu, C.T., Chang, J.Y.: Efficient net-eLite: extremely light weight and efficient CNN models for edge devices by network candidate search. J. Signal Process Syst. **95**(5), 657–669 (2022)
12. Kaushik, P., Khurana, S.: EfficientNetB3-based deep learning approach for plant stress classification. In: 2024 International Conference on Cybernation and Computation (CYBER-COM) pp. 700–704 (2024)
13. Yasid, A., Wahyuningrum, R.T., Ni'mah, A.T., Ayani, I.H.: Rice leaf diseases classification-EfficientNetB3. In: 1st International Conference on Technology, Engineering, and Computing Applications: Trends in Technology Development in the Era of Society 5.0, ICTECA 2023(2023)

14. Jiang, P., Chen, Y., Liu, B., He, D., Liang, C.: Real-time detection of apple leaf diseases using deep learning approach based on improved convolutional neural networks. IEEE Access **7**, 59069–59080 (2019)
15. Jane, B.R., Dharmale, G.J., Prasanth, P.: Detecting plant diseases utilizing a convolutional neural network and transfer-learning methodology. In: 2024 8th International Conference on Computing, Communication, Control and Automation 2024 (2024)
16. Bogireddy, S.R., Murari, H.: A hybrid deep learning model for accurate potato leaf disease detection: integration of U-net and EfficientNetB7. In: 4th International Conference on Power, Energy, Control and Transmission Systems: Harnessing Power and Energy for an Affordable Electrification of India, ICPECTS 2024 (2024)
17. Thakur, P.S., Chaturvedi, S., Khanna, P., Sheorey, T., Ojha, A.: Real-time plant disease identification: fusion of vision transformer and conditional convolutional network with C3GAN-based data augmentation. IEEE Trans. AgriFood Electron. **2**(2), 576–586 (2024)
18. Xiong, L., Zhang, J., Zheng, X., Wang, Y.: Context transformer and adaptive method with visual transformer for robust facial expression recognition. Appl. Sci. **14**(4) (2024)
19. Qamar, T., Bawany, N.Z.: Understanding the black-box: towards interpretable and reliable deep learning models. PeerJ Comput. Sci **9** (2023)
20. "plantvillage, D.: (2025). https://www.kaggle.com/datasets/abdallahalidev/plantvillage-dataset
21. Simonyan, K., Zisserman, A.: Very deep convolutional networks for large-scale image recognition. In: 3rd International Conference on Learning Representations, ICLR 2015 - Conference Track Proceedings (2014)
22. Krizhevsky, A., Sutskever, I., Hinton, G.E.: ImageNet classification with deep convolutional neural networks. Commun. ACM **60**(6), 84–90 (2017)
23. Awad, M.M.: FlexibleNet: a new lightweight convolutional neural network model for estimating carbon sequestration qualitatively using remote sensing. Remote Sens. (Basel) **15** (2023)
24. Huang, J., Niu, G., Guan, H., Song, S.: (2023)

Domain-Specific AI Models

Agri Buddy: From Queries to Crops with a Context-Aware Agricultural Chatbot

Somrita Sarkar[1], Prasenjit Betal[1], Pabitra Mitra[2], Anupam Das[3], Mrinal Jha[3], Prashant Bisht[3], Sarthak Sablania[3], and Thakare Vedant Sharadrao[3(✉)]

[1] Centre for Computational and Data Sciences, Indian Institute of Technology Kharagpur, Kharagpur, India
[2] Department of Computer Science and Engineering, Indian Institute of Technology Kharagpur, Kharagpur, India
[3] Indian Institute of Management Calcutta, Kolkata, India
thakaresba2026@email.iimcal.ac.in

Abstract. India's agricultural productivity is hindered by climate variability, fragmented landholdings, and limited access to localized, data-driven advice. Traditional crop recommendation systems often fail to address diverse agro-climatic needs. We introduce Agri Buddy, a multilingual chatbot that uses large language models (BERT and T5) to provide personalized crop suggestions via natural language. To fine-tune the models, synthetic dialogues are generated from structured agro-climatic data. Achieving 98% accuracy, Agri Buddy outperforms existing methods and demonstrates the potential of LLM-based dialogue systems for accessible, intelligent agricultural support.

Keywords: Sustainable agriculture · Crop recommendation system · Large language models (LLMs) · Soil and weather data · Generative AI · Dialogue-based systems · Agri Buddy

1 Introduction

Agriculture is central to India's socio-economic development, yet productivity is increasingly threatened by climate variability—manifesting as erratic rainfall, droughts, floods, and pest outbreaks—and by diverse agro-climatic conditions. Traditional crop selection methods remain largely non-data-driven, despite the emergence of recommendation systems that integrate soil and weather data for sustainable and climate-resilient farming [1,3,4,12,13]. India's agro-ecological diversity demands adaptable systems, but existing models struggle with data sparsity, poor generalization, and limited localization. Dialogue-based recommendation systems offer intuitive and explainable support [7,8], yet remain underutilized due to the absence of domain-specific conversational datasets. Socio-economic disparities, limited digital infrastructure, and language diversity further challenge adoption among India's 150+ million smallholder farmers [7,8].

All authors contributed equally to this work.

© The Author(s), under exclusive license to Springer Nature Switzerland AG 2026
R. Gupta et al. (Eds.): BDA 2025, LNCS 16041, pp. 273–283, 2026.
https://doi.org/10.1007/978-3-032-15134-6_19

Large Language Models (LLMs) such as BERT and T5 provide strong semantic understanding and text generation capabilities. LLM-enhanced recommender systems (LLMERS) improve personalization and data efficiency via prompt engineering, fine-tuning, and few-shot learning [27–30]. While effective in domains like e-commerce, LLMs remain underexplored in agriculture. Traditional ML and DL approaches (e.g., ANN, KNN, SVM, RF, CNN), along with IoT-, XAI-, and GNN-based methods [9–11,21,23,24], face limitations in cost, infrastructure, and localization [25,26]. Existing LLM systems are typically trained on general corpora, limiting their relevance to Indian agriculture. To address this, we propose a dialogue-based crop recommendation framework that uses generative AI to synthesize natural language from structured agro-climatic data. We fine-tune BERT and T5 on domain-specific data to enhance contextual accuracy.

We implement this as Agri Buddy, a multilingual chatbot that interacts with farmers, interprets local inputs, and recommends suitable crops—advancing trust, accessibility, and intelligent support in Indian agriculture.

2   Dialogue-Based Crop Recommendation System

Problem Formulation

Given environmental data (such as soil contents, humidity, rainfall, etc.), we recommend the most suitable crop(s) for a specific region.

Here $\mathbf{X}$ represents the feature vector (environmental parameters) and $\mathcal{C}$ is the set of available crops. The feature vector represents soil parameters (N, P, K), soil pH, and climate factors like temperature, relative humidity, and rainfall as depicted in Table 1.

Table 1. Feature set used for crop recommendation

Feature	Description
N (Nitrogen)	Soil nitrogen content
P (Phosphorus)	Soil phosphorus content
K (Potassium)	Soil potassium content
pH	Soil pH level
Temperature	Ambient temperature of the region
Relative Humidity	Percentage of relative humidity in the atmosphere
Rainfall	Amount of rainfall in the region

The goal is to find the crop(s) $c \in \mathcal{C}$ that maximize a utility function U, the overall utility as stated in Eq. 1 (Fig. 1):

$$\max_{c \in \mathcal{C}} U(c, \mathbf{X}) \tag{1}$$

The extended utility function can be expressed as in Eq. 2:

$$U(c, \mathbf{X}) = f(c, \mathbf{X}) + g(c, \mathbf{X}) \tag{2}$$

where:

- $f(c, \mathbf{X})$: Represents the crop-specific utility (e.g., yield, quality) based on environmental conditions.
- $g(c, \mathbf{X})$: Captures the suitability of the crop for the given soil and climate parameters.

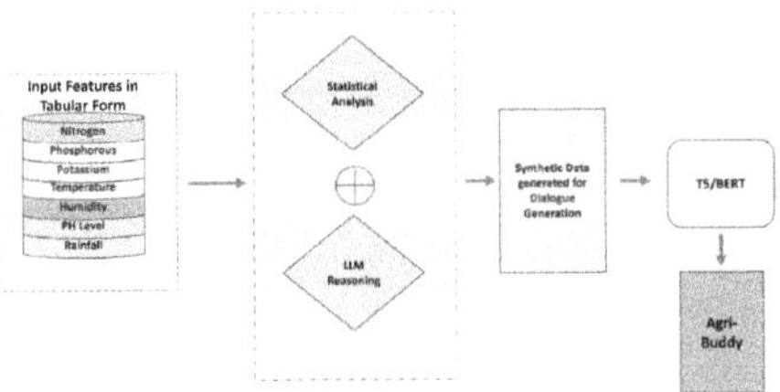

Fig. 1. Block Diagram Of Dialogue-Based Crop Recommendation System

3   Exploratory Data Analysis (EDA)

Exploratory Data Analysis (EDA) was conducted to assess feature distributions, detect anomalies, and guide data preprocessing. Key patterns in soil and environmental parameters are summarized in the following tables and figures (Fig. 2 and Table 2).

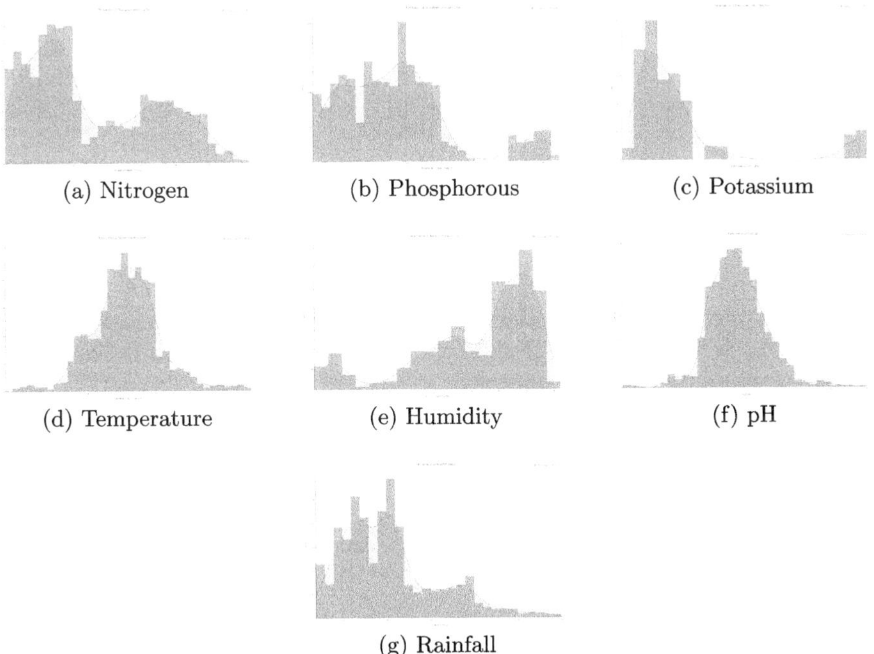

Fig. 2. Distributions of all agro-climatic features used for crop recommendation.

Table 2. Summary statistics and distribution characteristics of soil and climate variables

Variable	Mean	Range	Distinct (%)	Missing	Negative	Distribution Shape
Nitrogen	50.55	0–140	137 (6.20%)	0	0	Slightly right-skewed
Phosphorus	53.36	5–145	117 (5.30%)	0	0	Near-uniform, more high values
Potassium	48.15	5–205	73 (3.30%)	0	0	Right-skewed, some outliers
Temperature	25.62 °C	8.83–43.68 °C	2200 (100%)	0	0	Nearly normal
Humidity	71.48%	14.26–99.98%	2200 (100%)	0	0	Left-skewed, high values common
pH	6.47	3.50–9.94	2200 (100%)	0	0	Symmetric, centered near neutral
Rainfall	103.46 mm	20.21–298.56 mm	2200 (100%)	0	0	Right-skewed, moderate values dominate

Output Column: Label

Categorical column representing crop types with 22 unique values (Table 3).

Table 3. Summary of the categorical output column Label

Property	Value
Distinct Values	22
Distinct (%)	1.00%

Feature Correlation Analysis: To analyze feature-label relationships, a correlation matrix was computed (Fig. 4) with coefficients from -1 to 1. Phosphorus (P) and Potassium (K) showed a strong positive correlation ($r = 0.74$), while Nitrogen (N) had moderate negative correlations with P ($r = -0.23$) and K ($r = -0.14$). P and K also showed the strongest negative correlations with the label ($r = -0.49$, $r = -0.35$), indicating high predictive relevance. Humidity ($r = 0.19$) and temperature ($r = 0.11$) were weakly correlated, while pH, rainfall, and Nitrogen had minimal impact (Fig. 3).

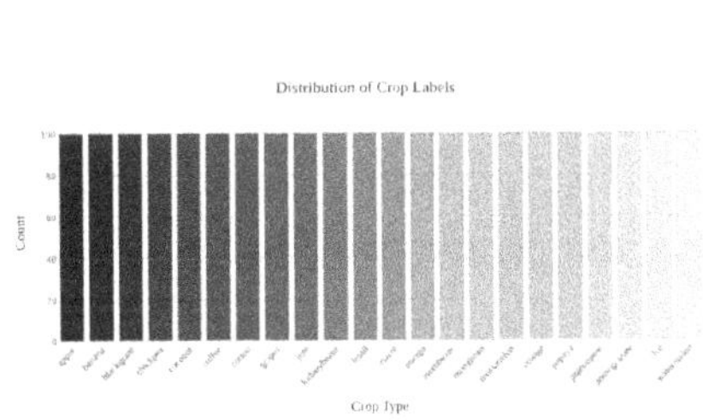

Fig. 3. Distribution of crop types in the `Label` column

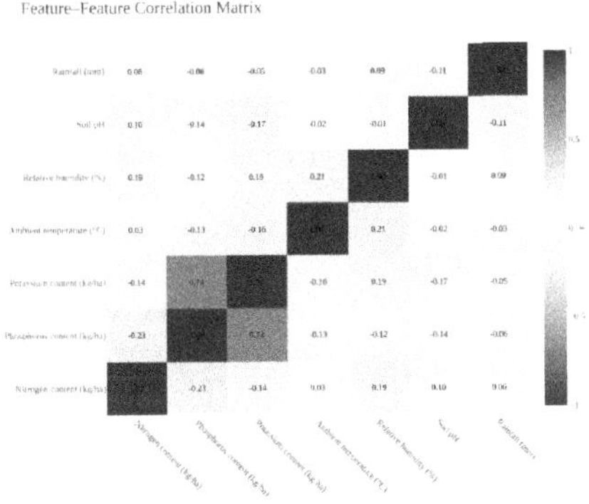

Fig. 4. Feature correlation matrix

4   Dialogue Generation from Tabular Data

The heatmap shows the correlation matrix of input features and the target label, with values from -1 to 1 indicating linear relationship strength and direction. Phosphorus and Potassium are strongly positively correlated (0.74), while Phosphorus also has a notable negative correlation with the label (-0.49), indicating predictive importance. These correlations guided the LLM in identifying key features to generate context-rich natural language questions, using the label as the answer to create a synthetic training dataset for the dialogue-based crop recommendation system.

Here is an example of how a row is transformed:

N	P	K	temperature	humidity	ph	rainfall	label
90	42	43	20.1	82	6.5	202	rice

Transformed Format:

- **input_text**:
  "The soil has 90 Nitrogen, 42 Phosphorus, and 43 Potassium. Temperature is 20.1 °C, humidity is 82%, pH level is 6.5 and rainfall is 202mm."
- **target_text**:
  "The best crop is rice"

5   LLM Based Crop Recommendation

5.1   Model Architecture of T5 Model

To build a dialogue-based crop recommendation system, we adopted the T5 model's text-to-text framework [31]. Structured tabular data was converted into natural language for fine-tuning T5, which was evaluated on its ability to generate crop suggestions. Developed by Google AI, T5 uses the Transformer architecture and reformulates NLP tasks as text generation problems.

The T5 model follows an encoder-decoder structure:

- **Encoder**: Transforms input text into contextual embeddings via self-attention.
- **Decoder**: Autoregressively generates output text from encoder representations, supporting diverse tasks like summarization and classification.

We authenticated with Hugging Face to access pre-trained models and used the t5-small tokenizer to convert input and target texts into token IDs. The processed DataFrame (input_text, target_text) was converted into a Hugging Face Dataset for efficient preprocessing and batching. T5 was pretrained on the C4 corpus using span-masking and multi-task learning for generalization across NLP tasks.

Fine-Tuning: The original dataset contained numerical agro-climatic features—Nitrogen, Phosphorus, Potassium, Temperature, Humidity, pH, and Rainfall—and a crop label. Each row was reformatted into a pair of natural language sentences to suit T5's input-output format.

We defined a custom tokenization function for both inputs and targets, applying truncation (to 512 tokens) and padding for uniformity. Targets were mapped to the labels field for training. The function was applied using map() with batched = True for efficiency.

The fine-tuned model used the t5-small architecture, with a shared embedding layer and six encoder/decoder blocks. Each block includes self-attention, feed-forward layers, and layer normalization with dropout. The decoder adds cross-attention to attend to encoder outputs. A final linear layer (lm_head) maps decoder outputs to vocabulary tokens. The model employs multi-head attention and uses an embedding size of 512 with a vocabulary of 32,128 tokens—enabling effective mapping from agro-climatic inputs to crop recommendations (Fig. 5).

(a) Overview of the T5 Model   (b) Overview of the T5 Model FineTuning

Fig. 5. T5 model architecture and fine-tuning

5.2 Model Architecture of Bidirectional Encoder Representations from Transformers

The BERT-base architecture used in this study consists of an embedding layer, encoder stack, and a classification head [2]. The embedding layer maps tokens into 768-dimensional vectors using word, position (up to 512), and token type embeddings, followed by LayerNorm and Dropout for regularization. The encoder includes 12 Transformer layers with multi-head self-attention and feed-forward networks that expand hidden states to 3072 dimensions before projecting them back to 768 via GELU activation. A pooler layer applies a linear transformation with tanh to the [CLS] token for fixed-length sentence representation. This feeds into a classification head composed of Dropout and a linear layer for predicting 22 crop classes.

Fine-Tuning: Structured numerical features—Nitrogen, Phosphorus, Potassium, Temperature, Humidity, pH, and Rainfall—were converted into natural language sentences, with crop labels encoded as integers (Fig. 6). The dataset was split 80:20 for training and testing.

Input texts were tokenized using the BERT tokenizer with truncation (256 tokens) and padding. Data was structured using the Hugging Face Dataset format with input_ids, attention_mask, and labels, enabling use with the Trainer API. We used the bert-base-uncased model with a Dropout and linear classification head.

Training employed the AdamW optimizer (learning rate: 2e-5), weight decay, and label smoothing for robustness. The model was fine-tuned for 20 epochs with evaluation at each epoch and checkpointing. Final test results showed 98.86% accuracy, 98.93% precision, 98.86% recall, and a 0.9886 F1 score—demonstrating BERT's strong performance in classifying crops from natural language representations of structured data.

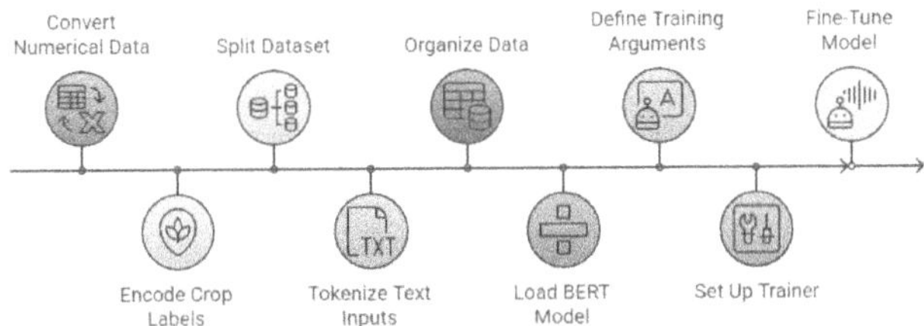

Fig. 6. Overview of BERT Fine-Tuning Process

6 Chatbot Deployment and Model Integration

The deployment phase of Agri Buddy finalizes an end-to-end AI pipeline by converting fine-tuned language models into an accessible agricultural decision-support tool. T5 and BERT, trained on structured agro-climatic data, are loaded via Hugging Face Transformers for inference—T5 for text generation and future dialogue, and BERT for fast, accurate classification. A browser-based chatbot was built using Gradio to enable intuitive interaction (Fig. 8). Users input seven comma-separated agro-climatic values—Nitrogen, Phosphorus, Potassium, Temperature, Humidity, pH, and Rainfall—matching the training format. The chatbot supports model switching and includes input validation with user-friendly error handling. Hosted on Hugging Face Spaces, deployment involved containerizing the app, bundling dependencies, and leveraging CI/CD for automated builds. This design supports model comparison while abstracting complexity, making AI-driven recommendations accessible to farmers, agronomists, and policymakers (Fig. 7).

Fig. 7. Chatbot deployment workflow using pre-trained models.

7 Results and Discussion

We used the 2024 crop recommendation dataset from IEEE DataPort [32], designed for precision agriculture through detailed assessments of soil properties—texture, pH, NPK levels, organic matter, and microbial activity. Spanning 22 crops, including rice, maize, cotton, and coffee, it supports soil-specific crop selection to enhance yield and sustainability. The dataset was split 80:20 for training and testing. The proposed LLM model outperformed collaborative filtering (NMF), traditional ML, and deep learning approaches, delivering highly reliable recommendations across accuracy, precision, recall, and F1 score. As shown in Table 4, ANN and CNN performed well (CNN F1: 0.9175), while the Transformer model underperformed (F1: 0.8704) on structured data. In contrast, the T5-based LLM achieved the best results—98.86% accuracy and 0.9886 F1—enabled by its text-to-text formulation, which supports semantic reasoning and generalization. These results highlight the potential of LLMs for domain-specific recommendations, particularly when structured inputs are expressed in natural language. Future work may explore multimodal integration for enhanced robustness.

Table 4. Performance metrics of various models used for crop recommendation

Model	Accuracy	Precision	Recall	F1 Score
KNN	0.9500	0.9537	0.9535	0.9480
SVM	0.8659	0.9136	0.8719	0.8758
Random Forest	0.9068	0.8895	0.9198	0.8955
Naive Bayes	0.8931	0.9080	0.8988	0.8893
Collaborative Filter	0.9523	0.9568	0.9522	0.9513
ANN	0.9500	0.9537	0.9535	0.9480
CNN	0.9123	0.9205	0.9194	0.9175
Transformer	0.8701	0.8725	0.8703	0.8704
T5	**0.9750**	**0.9778**	**0.9750**	**0.9754**
BERT	**0.9886**	**0.9893**	**0.9886**	**0.9886**

Fig. 8. AgriBuddy Chatbot output

8   Conclusion

This study shows that Large Language Models (LLMs) like T5 and BERT enable accurate, personalized crop recommendations using soil and environmental data. Our approach outperformed traditional models (KNN, SVM, Random Forest), achieving 98.86% accuracy and an F1 score of 0.9886. LLMs offer strong potential for improving agricultural decision-making, addressing data sparsity, and promoting sustainability. Future work could enhance predictions through multimodal data integration, including satellite imagery, sensors, and weather patterns for real-time, context-aware recommendations.

Disclosure of Interests. The authors have no competing interests to declare that are relevant to the content of this article.

References

1. Shams, M., Ghazvini, S., Rahimi, M.: Enhancing precision agriculture using machine learning for crop yield prediction. Comput. Electron. Agric. **213**, 107005 (2024)

2. Devlin, J., Chang, M.-W., Lee, K., Toutanova, K.: BERT: pre-training of deep bidirectional transformers for language understanding. In: Proceedings of the 2019 Conference of the North American Chapter of the Association for Computational Linguistics: Human Language Technologies, pp. 4171–4186. Association for Computational Linguistics (2019)
3. Patil, M., Dhamdhere, S., Mahalle, P.: Digital agriculture: data-driven insights for sustainable farming. Procedia Comput. Sci. **194**, 62–69 (2022)
4. Juhos, K., Szabó, J., Ladányi, M.: Influence of soil properties on crop yield: a multivariate statistical approach. Commun. Soil Sci. Plant Anal. **46**(12), 1511–1521 (2015)
5. Pudumalar, S.J., Ramanujam, E., Ranjani, S.S., Kiruthika, A., Ranjith, M.: Crop recommendation system for precision agriculture. In: 2017 International Conference on Innovations in Information, Embedded and Communication Systems (ICIIECS), pp. 1–5. IEEE (2017)
6. Gosai, P., Patel, J., Prajapati, N.: Crop recommendation system with soil and environmental parameters. In: 2021 International Conference on Computational Performance Evaluation (ComPE), pp. 509–514. IEEE (2021)
7. Narducci, F., Musto, C., Basile, P., de Gemmis, M., Semeraro, G.: A framework for recommender systems in conversational interfaces. In: Biundo, S., Fruhwirth, T., Palm, G. (eds.) KI 2018: Advances in Artificial Intelligence. KI 2018. Lecture Notes in Computer Science, vol. 11117, pp. 259–272. Springer, Cham (2018)
8. Chen, Q., et al.: Towards conversational recommender systems. In: Proceedings of the 2019 ACM SIGIR Conference on Research and Development in Information Retrieval, pp. 235–244. ACM (2019)
9. Doshi, S., Soni, S.: AgroConsultant: a machine learning-based system for crop prediction and rainfall forecasting. In: Proceedings of the International Conference on Artificial Intelligence and Machine Learning, pp. 123–130 (2018)
10. Naga, R., Kumar, R.: XAI for crop recommendations: enhancing explainability using spider monkey optimization. Agric. Syst. **122**(1), 45–56 (2024)
11. Kuanr, R., Patel, P.: Crop and fertilizer suggestion using Mamdani fuzzy inference model based on weather data. Int. J. Agric. Rural Dev. **20**(3), 142–150 (2018)
12. Pudumalar, V., Yadav, R.: Improving crop prediction accuracy using ensemble models. J. Agric. Data Sci. **15**(4), 189–198 (2017)
13. Gosai, A., Patel, S.: IoT-enabled crop prediction using machine learning classifiers. Sens. Actuators A Phys. **32**(5), 10–15 (2021)
14. Banavlikar, P., Shinde, V.: Real-time crop prediction using ANN and MQTT protocol. Agric. Eng. J. **28**(4), 23–30 (2018)
15. Pande, R., Gupta, A.: Data mining-based crop selection using historical data and market trends. J. Comput. Agric. **45**(3), 67–72 (2021)
16. Raja, R., Kaur, A.: Demand-based crop recommendation using data mining techniques. Smart Agric. J. **12**(2), 34–40 (2017)
17. Ray, B., Soni, P.: Smart crop prediction using Naive Bayes classifiers. Agric. Inf. Rev. **22**(1), 56–60 (2022)
18. Hossain, M., Khan, M.: Enhancing crop prediction using IoT-ensemble models. J. Agric. Technol. **30**(3), 210–215 (2024)
19. Kamatchi, K., Kumar, S.: Improvement of weather prediction using collaborative filtering and case-based reasoning. Int. J. Agric. Forecast. **14**(2), 55–64 (2019)
20. Kulkarni, P., Singh, A.: Improving crop recommendation accuracy using ensemble methods. Int. J. Agric. Eng. **15**(2), 45–53 (2018)
21. Kaur, M., Sharma, R.: Crop recommendation using LightGBM and random forest algorithms. Agric. Mach. Learn. Rev. **18**(1), 21–30 (2025)

22. Shams, M., Dey, A.: Enhancing transparency in crop recommendation models with XAI-CROP. Comput. Agric. **33**(4), 123–130 (2024)
23. Ayesha, S., Zafar, I.: Crop recommendation using graph convolutional networks: a spatial dependency model. Int. J. Agric. Syst. **42**(2), 110–118 (2023)
24. Sharma, P., Kumar, V.: Sentiment-based recommendation systems for disease control strategies. Agric. Technol. J. **31**(3), 190–198 (2024)
25. Musanase, L., Tuan, N.: Data-driven challenges in crop prediction for sub-humid climates and acidic soils. Environ. Data Sci. J. **17**(5), 88–95 (2023)
26. Shingade, P., Ghosh, R.: Analysis of machine learning models for crop recommendation systems: challenges and opportunities. Agric. Syst. Data Anal. **22**(1), 56–62 (2024)
27. Zhang, Y., Wang, Y., Wang, W., Bi, S., Feng, F., He, X.: Towards Next-Generation LLM-based Recommender Systems: A Survey and Beyond. arXiv preprint arXiv:2402.06581 (2024)
28. Li, Y., Huang, Y., Liu, Z., Zhang, Y., Chen, Y., Zhao, W.: Large Language Model Enhanced Recommender Systems: A Survey. arXiv preprint arXiv:2404.00927 (2024)
29. Gao, C., Zhang, Y., He, X., Lei, W., Wang, W., Feng, F.: Recommender Systems in the Era of Large Language Models. arXiv preprint arXiv:2404.05439 (2024)
30. Wang, M., Bi, S., Gao, C., Wang, W., Li, Y., Feng, F.: Incorporate LLMs with Influential Recommender System. arXiv preprint arXiv:2409.04827 (2024)
31. Raffel, C., et al.: Exploring the limits of transfer learning with a unified text-to-text transformer. J. Mach. Learn. Res. **21**(140), 1–67 (2020)
32. Patel, V.K.: Crop Recommendation dataset. IEEE Dataport (2024). https://dx.doi.org/10.21227/12nr-fe03

Machine Learning Based Wind and Turbulence Prediction at Urban-Scale for Drone Operations

Mandar Tabib$^{(\boxtimes)}$ and Adil Rasheed

Mathematics and Cybernetics Department, SINTEF Digital, Trondheim, Norway
`mandar.tabib@sintef.no`

Abstract. This study involves developing an unsupervised machine-learning (ML) framework (an artificial intelligence (AI) approach) for predicting micro-scale wind and turbulence. The localized wind and turbulence information from AI is a potential enabler for path planning of urban-scale drone operations and for use in smart-city through building-integrated renewable energy systems. In urban-city, the wind and turbulence is influenced by buildings and terrains and these complex local physics needs to be captured by the AI model. The trained ML in this work uses two input parameters: the meso-scale wind speed and the meso-scale wind direction in-order to predict the micro-scale building-induced wind and turbulence for regions around the Prague city centre. The performance of ML model has been compared with a traditional computationally-intensive computational Fluid Dynamics (CFD) solutions in terms of accuracy and computational speed-ups. After training, the ML performance has been compared for "unseen test datasets" in both the interpolation range and the extrapolation range of the input meso-scale parameters used in the training dataset. The results indicate that the machine learning model yields reasonably accurate solution for the velocity and turbulence fields for the test cases when the unseen input meso-scale parameters are "within the interpolation range" of training database and the ML infers the flow field in a matter of 5–10 s, i.e. a computational speed-up of over 1000 times as compared to the traditional CFD models. While for extrapolation (out-of-distribution) parameter test case, as expected, the ML model has scope of improvement in accuracy in flow-field. Overall, the ML-predictions due to its computationally efficiency can enable quick path planning and decision-making during drone operations, which is not possible with traditional computational fluid dynamics solution, thus highlighting the potential of machine learning for planning drone operations.

Keywords: Micro-scale wind and turbulence · Drones

1 Introduction

Drones can get impacted by both meso-scale and local urban-scale wind and turbulence conditions that can limit their operations and their availability [1]. This can cause serious safety problems as well as limits the use of drones in critical sectors (like delivering urgent medical aid within hospitals in a city, etc.). So, predicting wind and turbulent conditions in quick time can enable drone path planning and quick decision-making during drone operations. However, turbulent wind flow is a complex multiscale phenomenon encompassing different timescales and length-scales, and the traditional numerical approach involving use of computational fluid dynamics for predicting wind and turbulence in urban areas can take anywhere from hours to weeks on super-computers (the actual computational time taken by CFD depends upon the spatial and temporal resolution at which the wind and turbulent conditions are to be predicted) [2]. This huge computational time needed to make the predictions makes the CFD (the current approach) a not so viable option for real-time decision-making with regard to drones. In this direction, the present work involves developing artificial intelligence-based wind and turbulence predictor. In this study, we present a parametric unsupervised machine-learning framework for wind and turbulence prediction as a potential enabler for urban-scale drone operations. The case study considered here involves predicting building-induced flows and turbulence with wind speed and wind direction as parameter for selective regions of Prague City with buildings and terrain included. The result from this study has potential to be integrated into U-space Collaborative Interface System (UCIS) and Unity engine for path planning purposes - an ongoing activity in the EU AI4HyDROP project.

2 Knowledge Gaps and Objectives

Recently, machine learning (ML) models (an approach to artificial intelligence) have shown potential to forecast weather on larger meso-scale atmospheric phenomena (Example: GraphCast [3]). But these meso-scale models that predict meso-scale wind and meso-scale turbulence do not account for the impact of microscale features like terrain and buildings in the urban scale, where the physics of wind phenomena is completely different from that at the larger meso-scale. Since the drones will be operated in urban areas in vicinity of buildings, so wind and turbulence predictions at this scale is needed. Currently, there are no known ML models for predicting wind and turbulence at this urban scale. In-fact, these microscale wind and turbulence predictions are being done using traditional physics-based computational fluid dynamics (CFD) methods that involve solving Navier stokes equations (i.e. conservation of mass and momentum equations) [2]. However, these physics-based models take long computational time and hence cannot be used for real-time decision-making. Hence, the current work tries to overcome this knowledge gap by developing and testing a faster ML model to account for microscale wind and turbulence at the urban microscale level using meso-scale wind and turbulence as the input. The objectives of the current work are to:

(i) Develop an ML-based predictive model for microscale wind and turbulence patterns using Prague as case study for drone planning.
(ii) Verifying the performance and accuracy of the microscale ML model with the hi-fidelity CFD models in training and test regions within urban areas.

3   Methodology

The overall methodology for developing a data-driven machine learning (ML) model involve: (a) Generating the training data using 3D computational fluid dynamics (CFD) for steady state wind and turbulence condition in the urban landscape for different wind directions and wind speeds, and (b) Training the ML model using this generated database. The methodology is described below:

3.1   Computational Fluid Dynamics and Data Generation.

Computational Fluid Dynamics (CFD) is a branch of fluid mechanics that uses conservation laws adhering to physics (i.e. conservation of mass and momentum - known as Navier Stokes Equation). Solving Navier-Stokes equation directly (direct numerical simulation) to resolve the turbulence at all relevant scales is computationally infeasible for most practical applications due to the vast range of turbulent scales. Hence, turbulence modelling is employed to approximate influence of unresolved scale. For this work, the turbulence is modelled using realizable k-epsilon model.

Here steady state simulations are done and the unsteady transient component is neglected. These equations are solved using a numerical method called as finite volume method (FVM) that involves discretizing the computational domain into small finite volumes. Here, our computational domain is a segment of Prague city case study, as shown in Fig. 1. The building and terrain datasets were obtained for this segment of Prague city, and these are converted into a digital format in-order to develop the CFD model.

Fig. 1. Computational Domain used for Prague City.

For this work, the CFD was conducted using building data and terrain data for a section of Prague City as marked by a yellow rectangular boundary (as shown in Fig. 1). The Prague city was selected as suitable datasets (tall buildings) were available for conducting the urban-scale CFD and for testing the AI model performance. The domain size is 900 m × 900 m × 400 m. There are 3 buildings in the city center with height close to 100 m and an imaginary vertiport is in midst of these 3 buildings to cater to drone operations. For conducting CFD, this domain is discretized into 8 million grid points (so, n = 8 million control volumes), with finer grids to resolve all the buildings in the city and the terrain, and use of coarser grids away from it. This grid density is chosen based on grid-independence tests. The integral form of the governing equations, such as conservation of mass, momentum, and energy, is solved at each control volume. Fluxes of conserved quantities are calculated at the surfaces of these volumes. By ensuring that the net flux through the surfaces of each volume equals the rate of change within the volume, the FVM maintains conservation principles and provides an accurate approximation of fluid flow behavior. Here, we have used open-source OpenFOAM software for this. A total of 32 hi-fidelity CFD simulations are conducted for the above domain by varying inflow boundary conditions using two parameters - i.e. eight different wind directions ($0°$, $45°$, $90°$, $135°$, $180°$, $225°$, $270°$, $315°$, $360°$) are used, and for each of those wind directions, four different wind speeds are used in simulation. The wind speeds are $0.5\,\text{m/s}$, $2\,\text{m/s}$, $4\,\text{m/s}$ and $8\,\text{m/s}$. So, the meso-scale parameter range for the machine learning (ML) model covers a span of $0–360°$ in wind direction and from $0–8$ m/s in wind direction. The computational time taken for conducting a single CFD simulation to reach convergence (steady state) when run in parallel on two Intel Xeon Processor E5 2667v3@3.2GHz CPU processors is around $5–10$ hrs.

Using the CFD simulations for different inlet meso-scale wind direction and wind speed, we obtain information on the micro urban scale wind velocity and turbulence at spatial grid points (center of finite volumes) in the computational domain. This CFD generated training data is then rearranged, and the 3D wind and turbulence field, each is stored as a separate 2D matrix of size: n x m, where n is number of spatial grid points (or control volumes) and m is number of parameters (i.e. different number of wind directions and wind speeds for which simulation is done). For velocity, n is 3 times the spatial grid points as velocity is a vector with 3 components, while for turbulence (which is a scalar field), n is simply the number of spatial grid points.

This step of data-generation for machine learning is the most time-consuming part of the workflow, as setting up a CFD simulation with good mesh and running it can take time.

Next, we discuss the ML methodology used to train with the above dataset.

3.2   ML Method to Learn Wind Pattern from the Training Data.

The ML method used in this work involves an unsupervised learning method called Principal component analysis (PCA) to obtain the most dominant spatial flow patterns (also called as spatial basis) from the generated database. The PCA (also called Proper orthogonal Decomposition, i.e. POD) is an unsupervised machine learning method that decomposes the rearranged 2D training data matrix to provide a set of orthogonal basis functions and a set of associated coefficients. The orthogonal basis functions are related to the most dominating flow patterns within the spatio-temporal dataset. Before applying the PCA, the training database was shifted by the mean of its columns and scaled by the standard deviation of its columns to obtain the matrix X. This matrix X is then decomposed by using singular value decomposition to obtain the dominant basis functions (i.e. the POD modes) as in Eq. 1.

$$\mathbf{X} = \mathbf{U\Sigma V}^{T} = \mathbf{\Psi A} = \mathbf{\Psi_r A_r} + \mathbf{E_r} \approx \mathbf{\Psi_r A_r} \tag{1}$$

where, $\mathbf{U}$ and $\mathbf{V}$ are orthonormal matrices that contain the left and right singular vectors respectively, with $\mathbf{\Sigma}$ being a diagonal matrix with singular values along the diagonal. $\mathbf{\Psi} = \mathbf{U\Sigma} \in \mathbb{R}^{n \times n}$ is the **basis matrix** i.e. the column space of $\mathbf{X}$ representing the principal components, and $\mathbf{A} = \mathbf{V}^T \in \mathbb{R}^{m \times m}$ is the coefficient matrix (the projections of data onto the basis). The columns of $\mathbf{\Psi}$ (i.e. $[\psi_1, \psi_2, \ldots, \psi_n]$) are vectors that are the POD modes (principal components) of $\mathbf{X}$. By truncating the matrices $\mathbf{\Psi}$ and $\mathbf{A}$ to only use the first r singular values, it can be used for dimensionality reduction ($\mathbf{\Psi_r} \in \mathbb{R}^{n \times r}$ and $\mathbf{A_r} \in \mathbb{R}^{r \times m}$). The truncated POD modes serve as a low-dimensional representation of high-dimensional CFD data. Next, we develop the radial basis interpolator function as the m columns in $\mathbf{A}$ corresponds to m number of parameter (meso-scale parameters). The Radial basis interpolation method is used to regress the known rows of true coefficients in $\mathbf{A}$ (i.e.($\mathbf{a}_i \in R^m = [a_1, a_2, \ldots, a_m]$)) with the corresponding independent set of m parameters ($[\theta_1, \theta_2, \ldots, \theta_m] = \Theta \in \mathcal{R}^{2 \times m}$) - where each parameter is made of two-dimensions: the meso-scale wind direction and the meso-scale wind speed.

For "each" $\mathbf{a}_i \in \mathcal{R}^m, (i = 1, 2, \ldots, r)$ that makes up the rows of the matrix $\mathbf{A_r}$ - we train a RBF interpolant so there is a total of r number of RBF interpolants ($p = 1, \ldots, r$ as in Eq. 2). For each RBF interpolant p: the interpolation involves training to obtain weights i.e. in this case, solving m number of equations to obtain the unknown m weights ($\lambda_k^{(p)}$) using known dataset ($\mathbf{a}_i, \Theta$) as shown in Eq. 2. The RBF interpolant for a given dataset ($\mathbf{a}_i, \Theta$) is thus a linear combination of m RBFs ($\phi(|\theta_j - \theta_k|)$) with each RBF centered at θ_k.

$$a_j^{(p)} = \sum_{k=1}^{m} \lambda_k^{(p)} \, \phi(|\theta_j - \theta_k|), \quad j = 1, 2, \ldots, m; p = 1, \ldots, r. \tag{2}$$

where: $\phi(r)$ is the Gaussian radial basis function and ($|\theta_j - \theta_k|$ represents the euclidean distance (a scalar output). Thus for each of the r RBF interpolants ($p = 1, \ldots, r$): r number of system of linear equations (as shown in Eq. 2) are solved to regress each of $\mathbf{a}_i \in R^m$ to the parameters ($[\theta_1, \theta_2, \ldots, \theta_m]$). Thus, a total of $r \times m$ number of weights (λ_k^p) are learned for the r RBF interpolants with each involving m RBFs, and these weights are then used for the predictions.

3.3 Predicting Flow at New Parameter (Wind Direction and Wind Speed).

To predict the micro-scale flow field at an unseen parameter θ_{new} (i.e. new meso-scale wind direction and new meso-scale wind speed value), each of the r trained RBF interpolant functions with their learned weights are used to obtain the corresponding r coefficients ($[a_{new}^1, a_{new}^2, \ldots, a_{new}^r] = \mathbf{a}_{new}$), where each coefficient a_{new}^p is scalar value (as computed in Eq. 3).

$$a_{new}^p = \sum_{k=1}^{m} \lambda_k^p \, \phi(|\theta_{new} - \theta_k|), \quad p = 1, \ldots, r. \tag{3}$$

These RBF-predicted coefficients, comprising r-dimensional $\mathbf{a}_{new}$, are then projected onto the POD modes (basis function) matrix $\mathbf{\Phi}$ (of nxr dimension) to give a new flow field $\mathbf{X}^{new}$, i.e. turbulence values at n spatial locations.

$$\mathbf{X}_{new} = \mathbf{\Phi} \mathbf{a}_{new}; \; \mathbf{X}_{new} \in (n \times 1) \tag{4}$$

4   Results and Discussions

The results are shown for microscale building and terrain-induced velocity fields obtained using both ML model and physics-based CFD model for Prague City for following three cases: 1. Training performance case: The performance of trained ML model using input meso-scale parameters values (i.e. the input meso-scale wind and turbulence) that are the same as that used in the training dataset. This shows whether the model has got well-trained or not, and to enable comparison with the ML performance on test dataset.

2. Two Test performance cases for unseen parametric conditions: Here ML has not seen the input parameters (i.e. the input meso-scale wind and turbulence) and they are either in interpolation range or in the extrapolation (out-of-distribution) range during the testing. This is done to test the generalizability of the ML model.

For the sake of brevity, results pertaining to turbulence field are shown only for the "within interpolation test case", while the wind speed comparisons are shown for all the three cases. This is because the general conclusions on performance of ML predictions for turbulence is similar to that of the wind predictions for the three parametric conditions, as discussed in the results and discussion section below. The additional turbulence results can be provided as supplement on request. The error contours in results below are basically deviation between CFD-predicted field and ML-predicted field. An overall L_2 error is also mentioned, which is computed as:

$$\text{Relative } L_2 \text{ Error} = \frac{\|u_{\mathrm{ML}} - u_{\mathrm{CFD}}\|_2}{\|u_{\mathrm{CFD}}\|_2} \tag{5}$$

Wind Speed: ML Performance on the Training Parameter Set. Figures 2a-2b-2c compares the ML predicting microscale urban flow field for given input meso-scale wind direction of 135° at meso-scale wind speed of 8 m/s with the traditional CFD model predicted fields. Here, both are input meso-scale parameter values are the same as in the training dataset that the ML has been trained on. So, these results are bound to be good, and they have been included here to demonstrate that the unsupervised machine learning method has captured the dominant patterns to enable accurate reconstruction of flow field. Thus, the ML predicts similar wind speed results to the computationally expensive CFD model. This can be seen in low deviation (error) between ML and CFD predicted fields (Fig. 2c). The results are similar for turbulent field (not shown here), and as overall relative L_2 error for wind and turbulence are very small and in the range $8.1e - 12$ and $8.4e - 12$ respectively. This verifies that training has been good.

Generalizability tests: But the real generalizability test for ML comes from test data which uses input meso-scale parameters that are not seen by ML during training. This is covered in the next subsections.

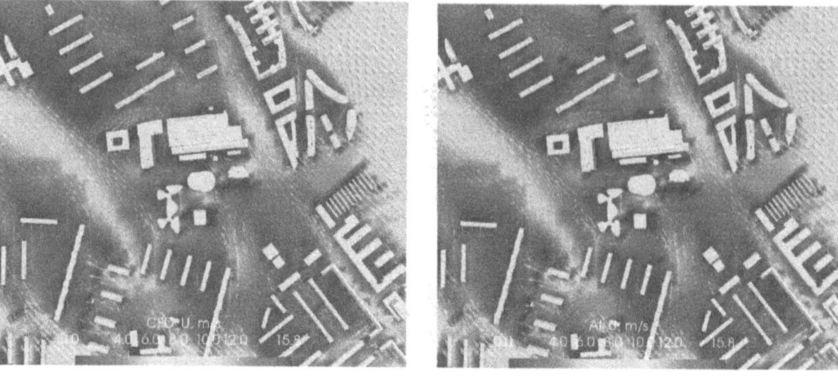

(a) CFD wind field, m/s (b) ML-predicted wind,m/s

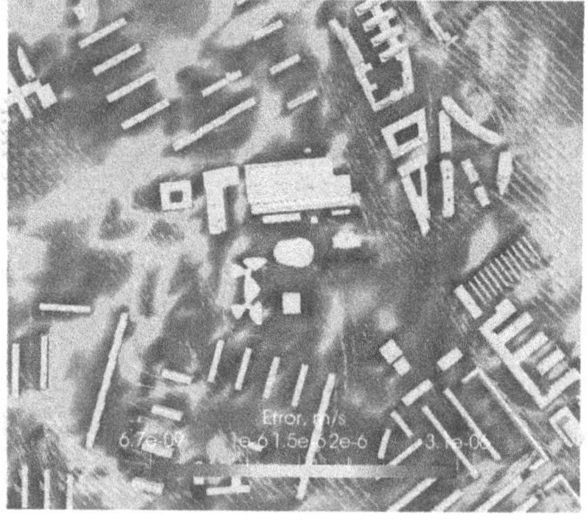

(c) Deviation between the two,m/s

Fig. 2. Wind speed: ML performance for the training parameter set.

Generalizability Test: ML Performance on Unseen Data Within Interpolation Range. Figures 3a-3b-3c compares the ML predicted microscale urban wind speed for unseen test data where the input meso-scale parameters are not in the training database (thus unseen by ML), but they are within the range of interpolation (or in-distribution of training parameter range). For given input meso-scale wind direction of $120°$ at meso-scale wind speed of $3\,m/s$, ML's performance is compared with the traditional CFD model on predicting wind speeds. The ML predicts similar flow profile as compared to the CFD, and error can be considered appropriate. This error when the parameters are in the range of interpolation (Fig. 3c) is more than that when the parameters are of the same value as in training data (Fig. 2c). Similar, results are obtained for ML predicted turbulence as compared to the wind speed as seen in Figs. 4a-4b-4c. So, the performance of ML has become less accurate as expected, but errors are in reasonable range for it to be used for decision-making as overall relative L_2 error for wind and turbulence are $3.1e - 02$ and $2.7e - 02$ respectively.

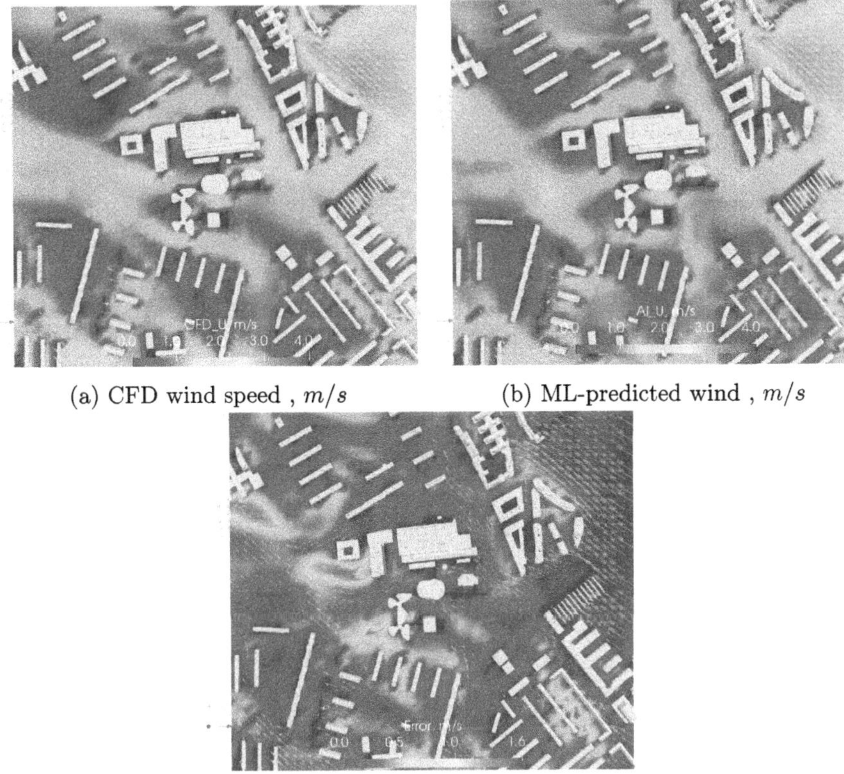

(a) CFD wind speed , m/s

(b) ML-predicted wind , m/s

(c) Deviation between the two, m/s

Fig. 3. Wind speed: ML prediction performance for the unseen interpolation set.

Generalizability Test: ML Performance on Test Data Within Extrapolation Range. Figures 5a-5b-5c compares the ML predicting microscale urban flow field for unseen test data where the input meso-scale parameters (meso-scale wind direction of 120° at meso-scale wind speed of 9 m/s) are not in the training database (thus unseen by ML), and they are in extrapolatory range that is out-of-distribution of the training data parameter range. As expected for the extrapolated out-of-distribution parameters, the trained ML predicts qualitatively similar field but with very high error (high quantitative deviation) as seen in the Fig. 5c, thus suggesting that the accuracy of ML is not-good for it to be used outside the range of parameters on which it is trained. The results are similar for turbulent field as well (not shown here). The overall relative L_2 error for both wind and turbulence is high, i.e. around 0.34 and 0.91 respectively. The regions where error is high are the regions of high velocity and turbulence gradients, and ML is finding difficulty capturing flow in these regions. In the next section, we present results on the computational speed-ups.

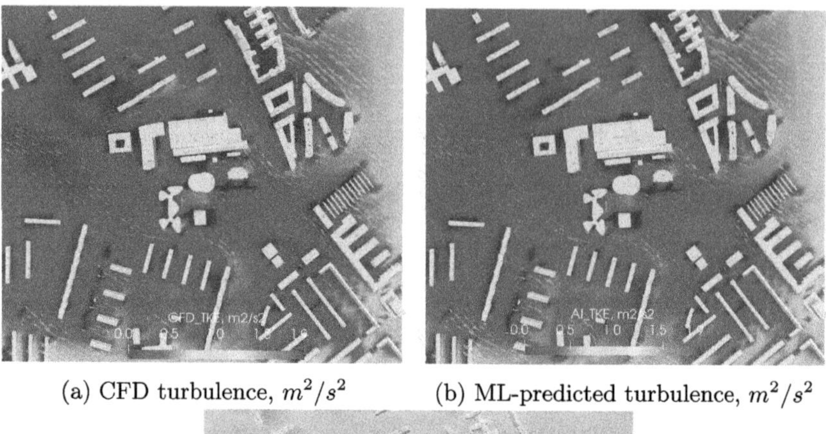

(a) CFD turbulence, m^2/s^2 (b) ML-predicted turbulence, m^2/s^2

(c) Deviation between the two, m^2/s^2

Fig. 4. Turbulence: ML prediction performance for unseen interpolation set.

Computational Speed-Up Benefits Due to ML: The ML model provides 1000 times speed-ups in computing the flow field as compared to the traditional CFD model. The trained ML model takes just 5–10 s to compute the entire flow field for any date and time (i.e. for new meso-scale wind speed and wind direction) on a single processor as compared to 5–10 h taken by CFD solver on 2 Intel 3.5Ghz processors to provide the flow fields.

(a) CFD wind speed, m/s (b) ML: wind speed, m/s

(c) Error (deviation), m/s

Fig. 5. Wind speed: ML performance for unseen extrapolation parameter.

5 Conclusion

The work demonstrates that the ML approach developed here for microscale wind prediction for a segment of Prague city is computationally very efficient (i.e. more than 1000x speed-up) as compared to the traditional computational fluid dynamics simulations. The ML shows reasonable accuracy when the "unseen" input meso-scale parameter range to the ML model are within interpolation range of the training data (i.e. meso-scale wind direction between 0 and 360°, and meso-scale wind speed from 0 to 8 m/s). For ML prediction of wind involving out-of-distribution inlet meso-scale parameters, as expected, the accuracy is not good. This efficient machine learning model has potential for use in drone path planning and decision-making when used within the range of training parameters. Further studies to improve the generalizability of the ML model for out-of-distribution parameters should involve training with physics-informed ML and through active-learning enriched database.

Acknowledgment. The authors acknowledge the financial support from the EU SESAR for the project 'AI4HYDROP' (Grant number: 101114805) for funding this work.

References

1. Gao, M., et al.: Weather constraints on global drone flyability. Sci. Rep. **11** (2021)
2. Tabib, M., Rasheed, A., Kvamsdal, T.: High-resolution CFD modelling and prediction of terrain-induced wind shear and turbulence for aviation safety. In: International Centre for Numerical Methods in Engineering (CIMNE) (2019)
3. Lam, R., et al.: GraphCast: AI model for faster and more accurate global weather forecasting. Science **382** (2023)

Structured Definitions and Segmentations for Legal Reasoning in LLMs: A Study on Indian Legal Data

Mann Khatri[1]([☒]) [iD], Mirza Yusuf[1] [iD], Rajiv Ratn Shah[1] [iD], and Ponnurangam Kumaraguru[2] [iD]

[1] Indraprastha Institute of Information Technology, Delhi, New Delhi, India
{mannk,rajivratn}@iiitd.ac.in
[2] International Institute of Information Technology Hyderabad, Hyderabad, India
pk.guru@iiit.ac.in

Abstract. Large Language Models (LLMs), trained on extensive datasets from the web, exhibit remarkable general reasoning skills. Despite this, they often struggle in specialized areas like law, mainly because they lack domain-specific pretraining. The legal field presents unique challenges, as legal documents are generally long and intricate, making it hard for models to process the full text efficiently. Previous studies have examined in-context approaches to address the knowledge gap, boosting model performance in new domains without full domain alignment. In our paper, we analyze model behavior on legal tasks by conducting experiments in three areas: (i) reorganizing documents based on rhetorical roles to assess how structured information affects long context processing and model decisions, (ii) defining rhetorical roles to familiarize the model with legal terminology, and (iii) emulating the step-by-step reasoning of courts regarding rhetorical roles to enhance model reasoning. These experiments are conducted in a zero-shot setting across three Indian legal judgment prediction datasets. Our results reveal that organizing data or explaining key legal terms significantly boosts model performance, with a minimum increase of 1.5% and a maximum improvement of 4.36% in F1 score compared to the baseline.

Keywords: Legal NLP · LEGAL AI · Legal Judgment Prediction · LJP

1 Introduction

Large Language Models (LLMs) demonstrate strong generalization across NLP tasks due to pretraining on large, diverse datasets [5], enabling them to follow instructions, reason, and generate coherent text [8,20]. However, the high cost of pretraining limits the availability of domain-specific LLMs, making general-purpose models less effective in specialized fields like law, biomedicine, and finance [15].

© The Author(s), under exclusive license to Springer Nature Switzerland AG 2026
R. Gupta et al. (Eds.): BDA 2025, LNCS 16041, pp. 295–304, 2026.
https://doi.org/10.1007/978-3-032-15134-6_21

To bridge this domain gap without retraining, in-context learning (ICL) has emerged as a promising solution [14], using prompt-embedded examples to adapt models at inference time. While effective in general tasks, its application to legal domains, especially legal judgment prediction (LJP), remains limited.

Legal texts are lengthy, formal, and rich in specialized language and logic [12], requiring LLMs to grasp procedural and hierarchical reasoning [15,23,26]. Prior legal ICL efforts have focused mainly on factual context or case retrieval [14,24].

This work investigates how structured prompt design can improve zero-shot LJP. We propose three strategies: (i) reorganizing input by rhetorical roles, (ii) defining these roles to clarify legal terms, and (iii) mimicking court-like reasoning through multi-step prompting. Evaluated on three Indian legal datasets, our results show that these lightweight interventions improve performance. Interestingly, simpler prompt combinations often outperformed those including all components, emphasizing the importance of targeted structure over complexity.

2   Related Work

Legal Judgment Prediction (LJP) has seen increasing adoption of advanced prompting techniques like zero-shot, few-shot, and chain-of-thought (CoT) to enhance prediction accuracy and interpretability [6,23]. These approaches have been applied to tasks such as the COLIEE legal entailment benchmark, using legal reasoning frameworks like IRAC to improve performance [19,25].

Early LJP work used neural networks in English and UK courts, revealing the importance of legal context [7,21]. Multilingual efforts like SwissLJP stress the need for culturally grounded datasets. Structured representations—such as legal knowledge graphs—and hierarchical models have been proposed to improve interpretability and reasoning [9,13].

In India, Nigam et al. [17] evaluate transformer and LLM-based models (e.g., InLegalBERT, LLaMA-2) under realistic conditions using only input available at judgment time. The study proposes new human-centered metrics (Clarity, Linking) and highlights the gap between current LLM performance and expert-level reasoning.

Recent research also explores LLMs' general reasoning capabilities, reviewing prompting strategies and external reasoning tools [10]. In legal contexts, LLMs show potential in moral/legal decision-making and legal authority identification, albeit with systematic differences from human reasoning and existing limitations [2,16].

3   Background

Legal proceedings follow a structured sequence: the plaintiff's complaint, the defendant's response, pretrial activities (e.g., discovery, motions), trial (evidence, witness examination, closing arguments), and finally, a verdict that may be appealed. Judges base their decisions on case facts, prior rulings, evidence, and legal arguments to apply the law and explain their rationale.

However, judgments often lack clear structure, making the reasoning hard to follow [12]. To address this, we use *Rhetorical Roles*—semantic labels like facts, arguments, precedents, and decisions—to reveal the underlying logic. Our ablations leverage this structure to mirror judicial reasoning and improve document understanding.

Following this, our prompt consists of three main components: **Rhetorical Roles (R):** Sentences assigned to specific rhetorical roles are combined to form a paragraph within the prompt. There will be N paragraphs corresponding to the N rhetorical roles identified in the judgment document, concatenated during model input. Note that not all roles may appear in a judgment document. Each paragraph is preceded by a heading that corresponds to its role. The final document format resembles the following: *[FAC]\n{sentences from the FAC role}\n\n[RLC]\n{sentences from the RLC role...}*. **Definition (D):** This component appears at the start of the prompt, offering definitions of rhetorical roles to assist the LLM in understanding the legal jargon. **Chain (C):** Upon receiving input, the LLM generates an ANALYSIS. The ANALYSIS and the initial input are then fed back into the LLM to produce the RATIO. Afterwards, the original input, ANALYSIS, and RATIO are all input into the LLM to generate the RPC for the case. This recursive approach mimics court processes. This method of reasoning through chaining is referred to as chain throughout the paper.

4   Formulation and Data Preparation

4.1   Dataset

We use two rhetorical role datasets from [12] and [3], based on Indian court judgments. The primary dataset (50 samples) has 13 rhetorical categories (12 + NONE), while the secondary (20 samples) focuses on 7 key roles for reasoning. In our experiments, we exclude ANALYSIS, STA, RATIO, and RPC from inputs, assuming the model will generate them to reach a verdict.

To further evaluate our approach, we use the larger PredEx dataset [18] with $\sim$12,000 samples. It excludes analysis and outcomes, requiring the model to infer them. Since it lacks rhetorical role labels, we test only definitions and chaining strategies.

Two annotators manually labeled 70 legal cases from the test sets of two datasets with binary outcome labels (0 or 1), based on clearly stated verdicts. Cases with partially appealed judgments were excluded, resulting in 64 finalized cases after resolving annotation disagreements. Annotation did not require legal experts due to the clarity of outcomes. Only the test set was labeled to preserve the training set for potential future use in training or few-shot inference, avoiding additional labeling effort.

4.2   Task Formulation

Given a set of documents $D = \{d_1, d_2, d_3, ..., d_n\}$, a function f that modifies the sentences of the document in their corresponding rhetorical roles paragraphs, a

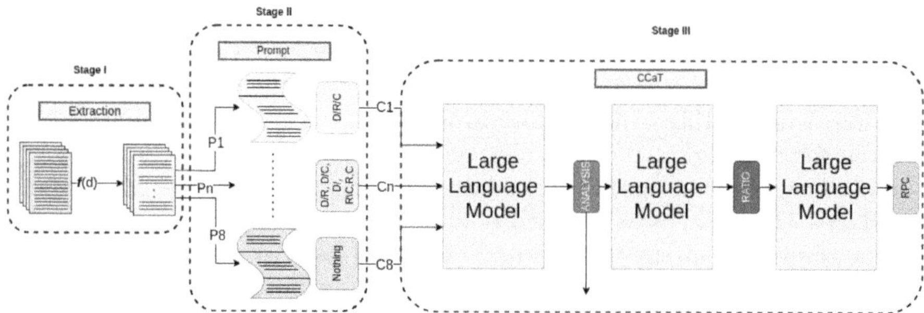

Fig. 1. The process consists of three stages. In the first stage, documents are segregated into Rhetorical Roles (P1, .., P8), which are then incorporated into the prompt in the second stage. Depending on the prompt requirements, either segregated or unsegregated cases are used. In the third stage, these prompts (C1, .., C8) are fed into the LLM one to three times based on the need for chaining. The final explanation of the model is noted after this stage.

function g that adds the definitions of the rhetorical roles in the prompt. We also add the chaining component to obtain the reasoning as shown in Fig. 1. We perform ablations on these components to see the effect of every component in the legal understading.

$$y_{\text{CASE with chain}} = \mathbf{w}\,(\text{RPC} \mid \mathbf{v}\,(\text{RATIO} \mid \mathbf{u}\,(\text{ANALYSIS} \mid P_i))) \tag{1}$$

$$y_{\text{CASE without chain}} = \mathbf{u}\,(\text{ANALYSIS} \mid P_i) \tag{2}$$

Here, $\mathbf{u}(.)$, $\mathbf{v}(.)$, $\mathbf{w}(.)$ are the LLMs and P_i are the corresponding prompts. The output of the LLM, $y \in \{YES, NO\}$, is recorded in each step. The value taken by y is YES if the case in favour of the plaintiff, else NO.

In an effort to shorten the prompt by few tokens, we initially eliminated the PREAMBLE rhetorical role and assessed the ensuing effects and its significance on the decision-making process. The absence of PREAMBLE negatively affected the model's performance, likely because the model relies on meta-data such as the competing parties' names typically found in the PREAMBLE. Consequently, we decided to retain PREAMBLE in our experiments. This shows that model finds it difficult to spot plaintiffs and defendants.

5 Experiments

For our experimentation, we chose the instruct models of different sizes Mistral 7B (Mistral-7B-Instruct-v0.3) [11], Phi3 (Phi-3-mini-128k-instruct) [1], Llama-3.1 (Llama-3.1-8B-instruct) [22] and o3-mini[1] (o3-mini-2025-01-31) as they have large context windows. The prompt with the most tokens was from *with definition with roles with chain (D/R/C)* variant, with more than 10k tokens, as

----

[1] https://openai.com/index/openai-o3-mini/.

it has the most information of all variants mentioned. Each model was tested using a zero-shot approach, as detailed in the following ablations. It is important to note that, after obtaining explanations from each ablation, we subsequently prompted the model to determine if the case decision favored the plaintiff or not for a binary output. This additional step was implemented to ensure consistency between the explanation and the output, preventing discrepancies such as an analysis indicating a win for the plaintiff while the output suggests otherwise. Such errors were observed sporadically when this extra step was omitted. Note that in the case of non-chained ablation, the *YES* or *NO* is output by the model based only on the ANALYSIS generated.

To ensure the reproducibility of the results, we set *do_ sample* to *False*. After analyzing the section lengths in the ground truth of the training set cases, we determined that the value for the hyperparameter *max_ new_ tokens* should be 2000. For o3-mini since it lacked a configuration for making the outputs deterministic so we ran the model 5 times and reported the results.

5.1   Evaluation Metrics

For judgment prediction, we used macro-F1, false negative rate (FNR), and false positive rate (FPR) as metrics. FNR and FPR are critical for upholding the "presumption of innocence," aiming to minimize both while maximizing F1.

For explanation quality, we employed ROUGE-1, ROUGE-2, METEOR [4], and BERTScore [27]. ROUGE captures recall and reference overlap, while METEOR balances precision and recall, accounting for synonyms and paraphrasing. This combination aligns better with human judgment. BERTScore complements them by evaluating semantic similarity[2].

6   Results

6.1   Judgment Prediction

In Dataset1, the best-performing LLaMA ablation combined rhetorical role definitions with segmentation (D/R), followed by definitions and chaining (D/C). D/R achieved the lowest FPR and FNR. Dataset2 showed a similar trend: D/R achieved the lowest FPR (0) but had a high FNR (62.5), while definitions-only (D) yielded the lowest FNR (12.5) with a higher FPR. D/R provided the best balance overall (Table 1).

Phi-3 performed close to random in Dataset1, with its best variant (chaining only) still showing FPR > 40 and FNR between 25 and 43. In Dataset2, Phi-3 underperformed again, though the D/C configuration gave a relatively better F1 score (34.84) and balance between FPR and FNR.

In Dataset1, models were expected to produce binary outputs. While LLaMA-3.1, Phi-3, and o3-mini did so consistently, Mistral-3 sometimes gave

---

[2] Code and Results also available at https://github.com/khatrimann/Structured-Definitions-and-Segmentations-for-Legal-Reasoning-in-LLMs-A-Study-on-Indian-Legal-Data.

Table 1. Judgment Precition Results on Llama, Phi and o3-mini. F1, FPR and FNR represent Macro averaged F1, False Negative Rates and False Positive Rates respectively. In the case of the Mistral row, each cell contains three additional rows showing results for: (i) independent, where chains were evaluated with its corresponding binary outputs; (ii) common, involving unified test points across all chains that produced a binary result; and (iii) chainwise, which compares the performance of chained vs. non-chained counterparts.

Models		D/R/C	D/R	D/C	D	R/C	R	C	None
					Dataset 1				
Llama	FPR	36.84	**28**	28.57	28.57	30.77	36.84	33.33	33.33
	FNR	40	**21.05**	30.43	40	41.94	40	41.38	30
	f1	61.36	**75**	70.45	63.64	61.36	61.36	61.36	68.18
Phi	FPR	48.65	46.88	44.44	47.37	47.5	47.5	**40**	43.33
	FNR	42.86	41.67	**25**	33.33	**25**	**25**	28.57	35.71
	f1	46	52.07	54.48	47.62	45.41	45.41	**62.39**	57.69
Mistral	FPR	**10.0**/ **0.0**/ **0.0**	**10.0**/ **0.0**/ **0.0**	35.29/ 20.0/ 35.29	31.58/ 14.29/ 31.58	26.67/ 0.0/ 28.57	41.18/ 16.67/ 41.18	33.33/ 0.0/ 35.0	39.13/ 0.0/ 38.1
	FNR	20.0/ 20.0/ 33.33	**0.0**/ **0.0**/ **11.11**	30.0/ 30.0/ 45.0	12.5/ 12.5/ 38.89	20.0/ 20.0/ 29.41	22.22/ 22.22/ 35.71	20.0/ 20.0/ 35.0	20.0/ 20.0/ 36.84
	f1	63.18 (36)/86.11 (15)/77.5 (18)	**89.44** **(19)/100.0** **(15)/94.43** **(18)**	58.95 (39)/72.22 (15)/59.46 (37)	68.22 (41)/86.61 (15)/64.76 (37)	65.02 (44)/86.11 (15)/70.85 (31)	61.25 (31)/79.64 (15)/61.25 (31)	66.67 (42)/86.11 (15)/65.0 (40)	61.82 (42)/86.11 (15)/62.48 (40)
o3	FPR	30.0 ± 6.1	**24.55** **± 7.61**	28.18 ± 3.8	27.27 ± 3.21	30.91 ± 5.93	**24.55** **± 4.07**	31.82 ± 3.21	28.18 ± 5.93
	FNR	52.73 ± 6.89	52.73 ± 6.89	**40.0 ± 7.47**	45.45 ± 7.87	50.0 ± 5.57	54.55 ± 5.57	50.91 ± 7.47	50.91 ± 4.98
	f1	58.01 ± 4.86	60.54 ± 6.58	**65.71** **± 3.38**	63.27 ± 4.7	59.16 ± 5.46	59.51 ± 4.3	58.21 ± 5.16	59.87 ± 3.08
					Dataset 2				
Llama	FPR	25	25	25	75	50	25	**0**	**0**
	FNR	37.5	31.25	50	**12.5**	37.5	37.5	50	62.5
	f1	60.11	**64.29**	52	56.71	52.38	60.11	58.33	49.49
Phi	FPR	25	25	25	**0**	25	**0**	25	**0**
	FNR	87.5	100	**75**	93.75	100	93.75	87.5	87.5
	f1	24.81	13.04	**34.84**	23.27	13.04	23.27	24.81	29.29
Mistral	FPR	75.0/ 50.0/ 66.67	62.5/ 50.0/ 62.5	75.0/ 60.0/ 71.43	**50.0**/ **33.33**/ **50.0**	70.0/ 60.0/ 70.0	66.67/ 60.0/ 66.67	66.67/ 50.0/ 66.67	81.82/ 71.43/ 81.82
	FNR	11.11/ 11.11/ 16.67	0.0/ 0.0/ 10.0	12.5/ 12.5/ 20.0	10.0/ 10.0/ 9.09	12.5/ 12.5/ 11.11	12.5/ 12.5/ 15.38	**0.0**/ **0.0**/ **0.0**	16.67/ 16.67/ 14.29
	f1	52.38 (20)/70.68 (13)/58.46 (18)	62.5 (18)/74.51 (13)/62.5 (18)	51.28 (19)/63.89 (13)/52.96 (17)	**72.31** **(18)/78.33** **(13)/71.67** **(17)**	54.76 (19)/63.89 (13)/54.76 (19)	59.29 (19)/63.89 (13)/59.29 (19)	62.5 (18)/74.51 (13)/62.5 (18)	41.33 (20)/51.25 (13)/41.56 (18)
o3	FPR	25.0 ± 17.68	10.0 ± 13.69	25.0 ± 0.0	25.0 ± 0.0	45.0 ± 11.18	35.0 ± 13.69	20.0 ± 11.18	**0.0 ± 0.0**
	FNR	56.25 ± 7.65	57.5 ± 2.8	60.0 ± 8.39	**38.75 ± 2.8**	51.25 ± 5.23	58.75 ± 3.42	**55.0 ± 5.23**	52.5 ± 5.59
	f1	47.76 ± 6.72	50.65 ± 3.09	45.32 ± 5.66	**59.3 ± 1.82**	45.76 ± 6.04	43.68 ± 4.46	49.98 ± 5.51	56.58 ± 3.89

indecisive outputs, likely due to uncertainty in low-evidence cases. Chaining reduced this issue. For consistency, a subset of 15 examples with decisive outputs across all chained variants was used for comparison. In this filtered set,

Table 2. Prediction results on Predex dataset. D means presence of definitions and C means whether chains were used. D/C is the combination of both and None is none of the methods were included in the prompt.

Metrics	D/C	D	C	None
Macro F1	0.6689	0.6447	**0.6732**	0.6296
FPR		0.4576	0.5495	**0.4151**	0.4314
FNR		0.1401	**0.095**	0.1695	0.2394

D/R consistently outperformed all other Mistral variants in FPR and FNR. For Dataset2, Mistral's best result came from the definitions-only (D) setup, again achieving the lowest FPR and FNR.

Given API stochasticity, each ablation was run five times, and outputs were aggregated. In Dataset1, the D/C setup had the best consistency and performance, with the lowest variance. D/R yielded the lowest FPR, while D/C had the lowest FNR. Both performed better than the full-component configuration.

In Dataset3 (PredEx), only LLaMA was evaluated due to poor performance from Phi and Mistral and cost limits for OpenAI models. Among LLaMA variants, chaining (C) performed best, closely followed by D/C. Their F1, FPR, and FNR values were comparable (Table 2).

Overall, rhetorical role definitions significantly improved model performance by enhancing structural understanding of legal language. LLaMA consistently performed best, while Phi lagged. Importantly, using all components together did not yield the best results—simpler combinations proved more effective.

6.2 Judgment Explanation

ROUGE scores were similar across models, with LLaMA-3.1 slightly outperforming Mistral and o3-mini in Dataset1, and o3-mini leading in Dataset2. Phi consistently underperformed (Refer to Github Repo of the paper). However, these scores lacked nuance, prompting a qualitative analysis (QA) of six randomly selected cases, including common True Positives (TP) and True Negatives (TN). Since Mistral had no overlapping False Positives (FP) or False Negatives (FN), its examples were sampled separately.

In TP cases, LLaMA-3.1 and o3-mini produced analyses closely aligned with the ground truth, despite minor inference gaps (e.g., missing the higher stamp duty issue). Mistral and Phi showed similar behavior. For TN cases, all models identified key points, but LLaMA-3.1 reasoned more like a judge, while Mistral and Phi offered more generic responses. o3-mini and Mistral outperformed Phi in judgment alignment, though LLaMA-3.1 showed the strongest overlap.

Across true cases, models aligned well in content but differed in phrasing and structure. In FP cases, LLaMA-3.1 and Phi produced general, non-legal reasoning; Phi slightly outperformed by suggesting court actions. Mistral's analysis had no overlap with ground truth, relying heavily on outdated precedents. A similar pattern held in FN cases, where all models showed poor legal awareness, particularly regarding precedent relevance.

Overall, LLaMA-3.1 and o3-mini produced more legally grounded outputs, while Mistral and Phi tended toward generic analysis. These findings suggest LLaMA-3.1 adheres more closely to legal reasoning and better follows prompt instructions.

7   Conclusion

This study investigated the influence of defining legal jargon, segmentation, and the reasoning processes applied by the judicial method. The analysis was performed on three datasets utilizing models of different scales, highlighting the significance of employing False Positive Rate (FPR) and False Negative Rate (FNR) for model evaluation, especially considering the legal doctrine of the "presumption of innocence." Additionally, the research illustrates the capability of models to interpret legal language in a zero-shot context, as demonstrated with Llama-3.1 and o3-mini, which adhered rigorously to legal standards.

8   Limitations and Future Work

The evaluation datasets were small due to limited data and the need for manual evaluation. GPU memory constraints restricted the processing of longer inputs, preventing the inclusion of related legal documents (e.g., precedents, laws) and limiting fine-tuning or pretraining.

Future work can explore few-shot learning and fine-tuning to improve legal understanding, use larger datasets to analyze temporal aspects, and go beyond binary outcomes by incorporating partial appeals for greater robustness.

Acknowledgments. We acknowledge the support of the IHUB-ANUBHUTI-IIITD FOUNDATION set up under the NM-ICPS scheme of the Department of Science and Technology, India. Rajiv Ratn Shah is partly supported by the Infosys Center for AI and the Center of Design and New Media at IIIT Delhi. We would also like to acknowledge our lab mates and people who were involved in feedbacks and suggestions throughout.

Disclosure of Interests. We acknowledge the support of the IHUB-ANUBHUTI-IIITD FOUNDATION set up under the NM-ICPS scheme of the Department of Science and Technology, India. Rajiv Ratn Shah is partly supported by the Infosys Center for AI and the Center of Design and New Media at IIIT Delhi.

References

1. Abdin, M., et al.: Phi-3 technical report: a highly capable language model locally on your phone. arXiv preprint arXiv:2404.14219 (2024)
2. Almeida, G.F., Nunes, J.L., Engelmann, N., Wiegmann, A., Araújo, M.D.: Exploring the psychology of LLMs' moral and legal reasoning. Artif. Intell. **333**, 104145 (2024). https://doi.org/10.1016/j.artint.2024.104145
3. Bambroo, P., Adhikary, S., Bhattacharya, P., Chakraborty, A., Ghosh, S., Ghosh, K.: MARRO: multi-headed attention for rhetorical role labeling in legal documents. Artif. Intell. Law 1–30 (2025)
4. Banerjee, S., Lavie, A.: Meteor: an automatic metric for mt evaluation with improved correlation with human judgments. In: Proceedings of the ACL Workshop on Intrinsic and Extrinsic Evaluation Measures for Machine Translation and/or Summarization, pp. 65–72 (2005)
5. Brown, T.B.: Language models are few-shot learners. arXiv preprint arXiv:2005.14165 (2020)
6. Brown, T.B., et al.: Language models are few-shot learners (2020). https://arxiv.org/abs/2005.14165
7. Chalkidis, I., Androutsopoulos, I., Aletras, N.: Neural legal judgment prediction in English. In: Korhonen, A., Traum, D., Màrquez, L. (eds.) Proceedings of the 57th Annual Meeting of the Association for Computational Linguistics, pp. 4317–4323. Association for Computational Linguistics, Florence, Italy (2019). https://doi.org/10.18653/v1/P19-1424. https://aclanthology.org/P19-1424
8. Choi, E., et al.: QuAC: question answering in context. arXiv preprint arXiv:1808.07036 (2018)
9. Gan, L., Kuang, K., Yang, Y., Wu, F.: Judgment prediction via injecting legal knowledge into neural networks. In: Proceedings of the AAAI Conference on Artificial Intelligence, vol. 35, no. 14, pp. 12866–12874 (2021). https://doi.org/10.1609/aaai.v35i14.17522. https://ojs.aaai.org/index.php/AAAI/article/view/17522
10. Huang, J., Chang, K.C.C.: Towards reasoning in large language models: a survey. In: Rogers, A., Boyd-Graber, J., Okazaki, N. (eds.) Findings of the Association for Computational Linguistics: ACL 2023, pp. 1049–1065. Association for Computational Linguistics, Toronto, Canada (2023). https://doi.org/10.18653/v1/2023.findings-acl.67. https://aclanthology.org/2023.findings-acl.67
11. Jiang, A.Q., et al.: Mistral 7b. arXiv preprint arXiv:2310.06825 (2023)
12. Kalamkar, P., et al.: Corpus for automatic structuring of legal documents. In: Calzolari, N., et al. (eds.) Proceedings of the Thirteenth Language Resources and Evaluation Conference, pp. 4420–4429. European Language Resources Association, Marseille, France (2022). https://aclanthology.org/2022.lrec-1.470
13. Ma, L., et al.: Legal judgment prediction with multi-stage case representation learning in the real court setting. In: Proceedings of the 44th International ACM SIGIR Conference on Research and Development in Information Retrieval, SIGIR 2021, pp. 993–1002. Association for Computing Machinery, New York (2021). https://doi.org/10.1145/3404835.3462945
14. Mu, Y., et al.: Navigating prompt complexity for zero-shot classification: a study of large language models in computational social science. In: Calzolari, N., Kan, M.Y., Hoste, V., Lenci, A., Sakti, S., Xue, N. (eds.) Proceedings of the 2024 Joint International Conference on Computational Linguistics, Language Resources and Evaluation (LREC-COLING 2024), pp. 12074–12086. ELRA and ICCL, Torino, Italia (2024). https://aclanthology.org/2024.lrec-main.1055/

15. Mukherjee, S., Mitra, A., Jawahar, G., Agarwal, S., Palangi, H., Awadallah, A.: Orca: progressive learning from complex explanation traces of GPT-4. arXiv preprint arXiv:2306.02707 (2023)
16. Nay, J.J., et al.: Large language models as tax attorneys: a case study in legal capabilities emergence (2023). https://arxiv.org/abs/2306.07075
17. Nigam, S.K., Deroy, A., Maity, S., Bhattacharya, A.: Rethinking legal judgement prediction in a realistic scenario in the era of large language models. arXiv preprint arXiv:2410.10542 (2024)
18. Nigam, S.K., Sharma, A., Khanna, D., Shallum, N., Ghosh, K., Bhattacharya, A.: Legal judgment reimagined: Predex and the rise of intelligent ai interpretation in Indian courts (2024). https://arxiv.org/abs/2406.04136
19. Rabelo, J., Goebel, R., Kim, M.Y., Kano, Y., Yoshioka, M., Satoh, K.: Overview and discussion of the competition on legal information extraction/entailment (COLIEE) 2021. Rev. Socionetwork Strat. **16**(1), 111–133 (2022)
20. Rajpurkar, P., Zhang, J., Lopyrev, K., Liang, P.: Squad: 100,000+ questions for machine comprehension of text. arXiv preprint arXiv:1606.05250 (2016)
21. Strickson, B., De La Iglesia, B.: Legal judgement prediction for UK courts. In: Proceedings of the 3rd International Conference on Information Science and Systems, ICISS 2020, pp. 204–209. Association for Computing Machinery, New York (2020). https://doi.org/10.1145/3388176.3388183
22. Touvron, H., et al.: Llama: open and efficient foundation language models. arXiv preprint arXiv:2302.13971 (2023)
23. Wei, J., et al.: Chain-of-thought prompting elicits reasoning in large language models. Adv. Neural. Inf. Process. Syst. **35**, 24824–24837 (2022)
24. Wu, Y., et al.: Towards interactivity and interpretability: a rationale-based legal judgment prediction framework. In: Proceedings of the 2022 Conference on Empirical Methods in Natural Language Processing, pp. 4787–4799 (2022)
25. Yu, F., Quartey, L., Schilder, F.: Legal prompting: teaching a language model to think like a lawyer (2022). https://arxiv.org/abs/2212.01326
26. Zhang, H., Dou, Z.: Case retrieval for legal judgment prediction in legal artificial intelligence. In: China National Conference on Chinese Computational Linguistics, pp. 434–448. Springer (2023)
27. Zhang, T., Kishore, V., Wu, F., Weinberger, K.Q., Artzi, Y.: Bertscore: evaluating text generation with BERT. arXiv preprint arXiv:1904.09675 (2019)

AgriLLM-India: Empowering Indian Farmers Through Domain-Specific Large Language Models for Agricultural Knowledge and Decision Support

Vijay Kumar Damera[1(✉)] and Deepthi Kalavala[2]

[1] Department of CSE, Koneru Lakshmaiah Education Foundation, Hyderabad, Telangana, India
dameravijaykumar@gmail.com
[2] Data Scientist, Stellantis NV, Nanakramguda, Hyderabad, Telangana, India

Abstract. The adoption of Large Language Models in Indian agriculture presents an unprecedented opportunity to enhance farmer education, decision-making, and access to vital agricultural knowledge. This paper introduces AgriLLM-India, a domain-specific chatbot tailored to the Indian context, incorporating local languages, region-specific crops, and government policy frameworks. Utilizing a Retrieval-Augmented Generation architecture, AgriLLM-India integrates a multilingual dataset derived from Indian agricultural research, government schemes (like PM-KISAN, eNAM), and agronomic best practices. The framework leverages Facebook AI Similarity Search for semantic search and compares the performance of three LLMs—ChatGPT-4o Mini, Gemini 1.5 Flash, and Mistral-7B-Instruct-v0.2—on their ability to respond accurately and contextually to queries in Hindi, Telugu, and English. A case study on cotton cultivation in Telangana highlights the chatbot's ability to provide region-specific insights on irrigation, pest control, and subsidy access. Results demonstrate that ChatGPT-4o Mini with RAG achieved the highest accuracy and contextual relevance, albeit with slightly higher response latency. The findings underscore the value of LLMs tailored to Indian agricultural needs and call for scalable AI models that incorporate real-time data and multilingual support to democratize agricultural advisory services.

Keywords: Indian Agriculture · Large Language Models · RAG · AI Chatbot · FAISS

1 Introduction

India's agriculture sector supports over 58% of the population, with more than 150 million farmers [1]. Despite its significance, the sector faces challenges like fragmented landholdings, declining productivity, and poor adoption of scientific methods. A key issue is inefficient knowledge transfer. Institutions like ICAR and KVKs [2,3] struggle with limited outreach, localization, and manpower.

R. Gupta et al. (Eds.): BDA 2025, LNCS 16041, pp. 305–314, 2026.
https://doi.org/10.1007/978-3-032-15134-6_22

The country's linguistic diversity and varied agro-climatic zones further complicate uniform advisory dissemination. Differences in fertilizer use between Punjab and Telangana or cropping cycles in Andhra Pradesh and Maharashtra highlight the need for localized, real-time solutions [4].

Large Language Models (LLMs) [5–7], such as ChatGPT and Gemini [8], offer a promising alternative. Enhanced by domain-specific datasets and Retrieval-Augmented Generation (RAG) frameworks [9,10], LLMs can deliver accurate, multilingual, and context-aware guidance that improves with user interaction.

AgriLLM-India is proposed as a decentralized, multilingual AI chatbot [11], drawing from ICAR reports [2], state university data, and policy documents, all embedded in a FAISS-indexed vector database. Using RAG, it delivers region-specific, real-time advisories.

This paper outlines AgriLLM-India's development pipeline, compares LLMs on Indian agricultural datasets, and presents a Telangana case study to demonstrate how domain-adapted LLMs can enhance agricultural advisory scalability and accessibility.

2 Methodology

The AgriLLM-India methodology leverages state-of-the-art machine learning and domain-specific agricultural data through a five-phase pipeline, as illustrated in Fig. 1. It begins with Data Collection and Preprocessing, followed by Embedding Generation, Vector Indexing, Retrieval-Augmented Generation (RAG), and concludes with Model Evaluation. This modular architecture ensures scalability, facilitates modular testing, and supports future integrations with IoT or mobile applications.

2.1 Data Collection and Preprocessing

The Data Collection and Preprocessing phase of the AgriLLM-India methodology centers on building a comprehensive, multilingual corpus [10] from diverse sources, including ICAR annual reports [2] and journals, government documents from schemes like PM-KISAN, PMFBY, and Soil Health Card, eNAM API datasets with mandi rates and trade flows, field transcripts and village-level surveys from Telangana, and extension materials like pamphlets and manuals from KVKs in Hindi, Telugu, and English. Text extraction utilized OCR for scanned documents, followed by language detection and transliteration as needed. The content was segmented into overlapping chunks of 150–250 words, ensuring semantic coherence, and enriched with metadata such as crop type, agro-climatic zone [12], and policy tags.

2.2 Embedding Generation

The Embedding Generation phase of the AgriLLM-India methodology transforms textual segments into vector space using three state-of-the-art models: **OpenAI Embeddings** (text-embedding-ada-002), which produces 768-dimensional vectors with fast inference and strong contextual generalization;

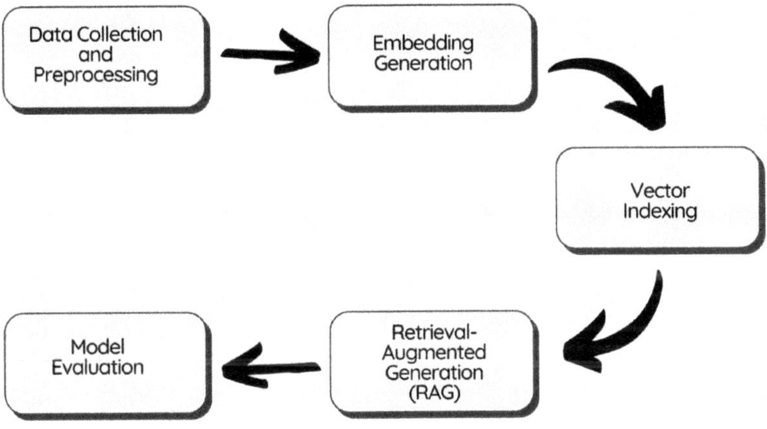

Fig. 1. AgriLLM-India System Architecture Workflow.

Gemini 1.5 Flash [13], generating 1024-dimensional embeddings optimized for noisy multilingual content; and **Mistral-7B-Instruct-v0.2**, also yielding 1024-dimensional vectors but trained with a high domain-specific bias for agricultural relevance. These embeddings were batch processed, stored alongside metadata, and evaluated for cosine similarity against curated agricultural queries to ensure their effectiveness.

2.3   Vector Database Construction

Facebook AI Similarity Search (FAISS) provided the backbone for semantic search. Vector dimensions aligned with respective models. Indexes were created using IVF+Flat strategy [14], balancing speed and accuracy. Configuration parameters are given in Table 1.

Table 1. FAISS Configuration Parameters

Parameter	Value
Vector Dimension	768/1024
Index Type	IVF+Flat
Metric	Cosine Similarity
Clustering	k-means (k = 2048)
Response Speed	~2 ms/query
Scalability	Up to 2M vectors tested

2.4   Retrieval-Augmented Generation (RAG)

The Retrieval-Augmented Generation (RAG) phase serves as the core component of the AgriLLM-India methodology, seamlessly linking query understanding with grounded knowledge. When a farmer submits a query, the process unfolds as follows: the query is first converted into an embedding, which is then used to retrieve the top-k relevant chunks from the corpus using FAISS. These chunks, combined with the original query, form a constructed prompt that is fed into a large language model (LLM) such as ChatGPT-4o Mini, Gemini, or Mistral [15]. To ensure resilience, fallback logic enables the LLM to generate a response using its internal knowledge in cases of retrieval failure due to connectivity issues or dataset gaps.

2.5   System Evaluation Pipeline

The Model Evaluation phase of the AgriLLM-India methodology assesses performance through multiple metrics: accuracy, determined by expert-validated answers; BLEU score, measuring n-gram overlap with reference answers; Mean Reciprocal Rank (MRR), evaluating the rank efficiency of relevant answers; Recall@10, calculating the percentage of top-10 results containing relevant content; and latency, measuring the time from query to response. Together, these components create a robust, scalable, efficient, and multilingual [10] pipeline, enabling AgriLLM-India to serve as an effective AI assistant for Indian farmers.

3   Dataset Composition

The AgriLLM-India training corpus comprises over 50,000 documents curated from diverse sources to ensure localized, context-rich agricultural insights. Key sources include ICAR reports, technical papers across disciplines like horticulture and soil science [2], KVK extension manuals and success stories in Telugu, Hindi, and English, government policy documents (e.g., PM-KISAN, PMFBY, Soil Health Card, eNAM, AgriStack), thesis archives from institutions like PJT-SAU Hyderabad [16], and open-access platforms such as Agropedia [17] and Krishikosh [18].

A robust NLP pipeline was used for data preprocessing, involving token normalization for multilingual and code-mixed text, noise reduction (e.g., OCR artifact removal), chunking into 150–250-word segments, and metadata tagging (e.g., crop, region, scheme, source). The corpus includes 60% English, 25% Telugu, and 15% Hindi, enabling region-aware, multilingual query handling.

Quality was ensured via expert cross-verification using ICAR Agri-FAQs and PJTSAU references, benchmark QA datasets, and BLEU-based validation of model outputs. Future expansion plans include integrating additional regional languages, real-time IoT-based soil and weather data, user query logs, and crowdsourced validation from farmer cooperatives. This adaptive dataset ensures AgriLLM-India remains responsive to evolving seasonal, regional, and policy-driven needs.

4   Case Study: Cotton Cultivation in Telangana

Cotton is a major commercial crop in Telangana, contributing significantly to both farmer livelihoods and the state's agrarian economy. The AgriLLM-India framework was deployed in Karimnagar district to evaluate its real-world utility in addressing cotton-specific challenges posed by small and marginal farmers. The region, known for rain-fed farming, frequently faces pest outbreaks, erratic rainfall, and inconsistent market linkages, all of which can severely impact yields.

In a practical deployment, a farmer submitted a query in Telugu: (What precautions should be taken for disease control in cotton crops during the damage-prone stage in Karimnagar?). The system retrieved and synthesized content from the PJTSAU [16]pest management guide and ICAR bulletins on integrated pest management (IPM) strategies. The chatbot generated a tailored response, recommending:

- Use of pheromone traps and neem-based biopesticides for early-stage management.
- Transition to Bt-cotton hybrids for future planting cycles.
- Guidance on chemical pesticide rotation schedules based on local insect resistance data.
- Links to the Telangana Rythu Bandhu [19] and PMFBY [20] registration portals for financial protection.

AgriLLM-India Response Summary for Cotton Farmers in Telangana are given in Table 2.

Table 2. AgriLLM-India Response Summary for Cotton Farmers in Telangana

Query Category	AgriLLM-India Recommendation	Data Source
Pest Control	Pheromone traps, neem spray, IPM	PJTSAU, ICAR
Crop Selection	Bt-cotton hybrid suggestion	PJTSAU Bulletin
Subsidy Linkage	Rythu Bandhu & PMFBY	Govt. Portals
Weather Advisory	IMD-based sowing advice	IMD APIs
Market Info	eNAM mandi rate comparison	eNAM Dashboard

AgriLLM-India's RAG mechanism ensured that only contextually relevant, localized, and policy-aligned information was delivered, thereby reducing misinformation risk. AgriLLM Deployment Impact Pathway in Telangana is depicted in Fig. 2.

Integrating AgriLLM with existing platforms like MeeSeva kiosks could further increase accessibility. Importantly, women farmers, often excluded from traditional extension services, reported higher engagement due to the bot's language simplification and visual cues, reinforcing AgriLLM's inclusivity potential.

This case study underscores AgriLLM-India's potential to transform digital advisory delivery in India's cotton belts and offers a replicable model for other crops and regions.

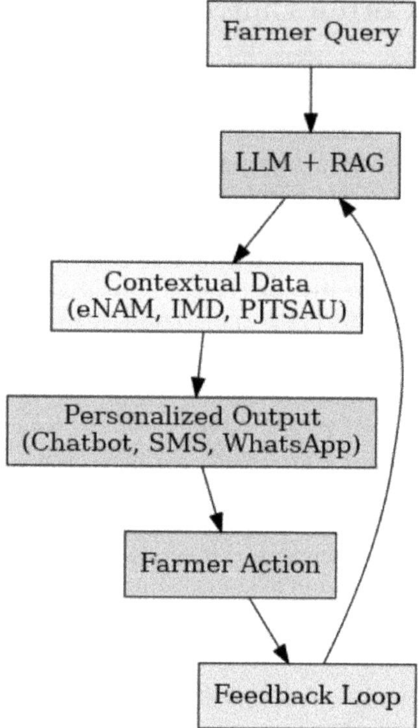

Fig. 2. AgriLLM Deployment Impact Pathway in Telangana.

5   LLM Evaluation

To assess the performance of various large language models (LLMs) [21] integrated into the AgriLLM-India system, we conducted an evaluation across several quantitative and qualitative metrics. The comparison involved three state-of-the-art LLMs—**ChatGPT-4o Mini** [22], **Gemini 1.5 Flash** [13], and **Mistral-7B-Instruct-v0.2**—each deployed using the same RAG pipeline, FAISS indexing structure, and a query corpus derived from real farmer interactions.

The models were tested on a benchmark dataset containing 500 multilingual agricultural queries translated into Hindi, Telugu, and English. Performance metrics included:

- **Accuracy:** Proportion of correct responses validated by agricultural domain experts.
- **BLEU Score:** Evaluating language similarity to human-generated answers.
- **MRR (Mean Reciprocal Rank):** Measuring how high the first relevant result appeared.
- **Recall@10:** Checking if at least one of the top 10 retrieved chunks was relevant.

– **Latency:** Average time to respond to a query.

Performance Metrics of LLMs on AgriLLM-India Dataset is given in Table 3.

Table 3. Performance Metrics of LLMs on AgriLLM-India Dataset

Model	Accuracy (%)	BLEU Score	MRR	Recall@10	Avg Latency (s)
ChatGPT-4o Mini (RAG)	89.2	87.6	0.93	90.0	10.4
Gemini 1.5 Flash (RAG)	82.4	83.1	0.86	84.3	8.1
Mistral-7B-Instruct (RAG)	74.8	78.3	0.78	78.4	6.2

Qualitative observations indicated that **ChatGPT-4o Mini** handled context-heavy policy questions with better coherence and fewer hallucinations. **Gemini**, while slightly faster, occasionally struggled with region-specific dialects, especially Telugu. **Mistral-7B-Instruct** was efficient and resource-light but less effective on long-answer generation and cross-lingual translation.

An error analysis revealed that most inaccuracies stemmed from outdated or ambiguous retrievals, highlighting the need for real-time data pipelines [23]. Furthermore, retrieval precision [24] was highly dependent on the embedding quality and language model compatibility. In cases where the embedding language did not match the query language (e.g., Telugu query vs. English source chunk), performance declined.

Common Error Types and Frequency in Model Responses are given in Table 4.

Table 4. Common Error Types and Frequency in Model Responses

Error Type	ChatGPT-4o Mini	Gemini 1.5 Flash	Mistral-7B-Instruct
Misinterpretation of Query	4%	7%	9%
Incorrect Policy Mapping	2%	4%	6%
Language Misalignment	1%	3%	5%
Outdated Information	3%	2%	1%
Hallucination (fabricated)	0.8%	1.6%	4.2%

To validate generalization, blind tests were conducted on unseen queries sourced from KVK helpdesk logs. Here again, **ChatGPT-4o Mini** achieved the highest comprehension and contextual integration scores, followed by Gemini. The results reaffirm the robustness of ChatGPT-4o Mini for deployment across India's linguistically and agronomically diverse landscape.

These findings support the conclusion that **ChatGPT-4o Mini**, though marginally slower, delivers the most reliable and context-sensitive outputs for Indian farming [25,26] advisory, making it the optimal choice for deployment in AgriLLM-India's production environment (Figs. 3 and 4).

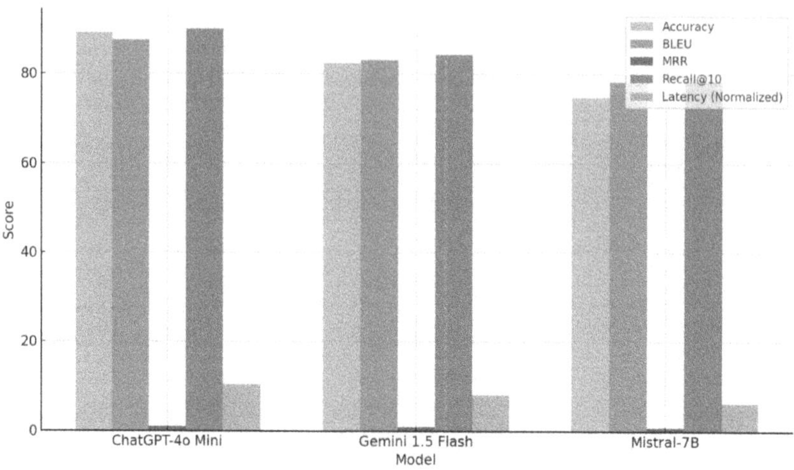

Fig. 3. Comparative Performance of LLMs.

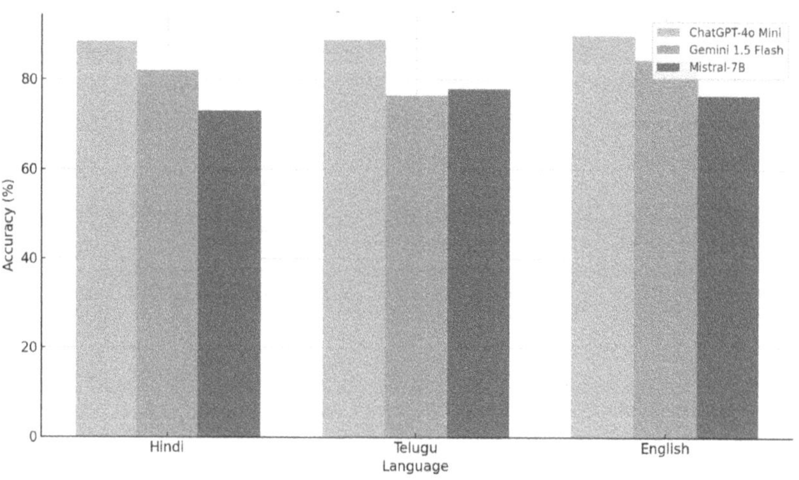

Fig. 4. Multilingual Accuracy Breakdown

6 Discussion and Future Work

The AgriLLM-India framework has shown strong potential in enhancing agricultural advisory services in Telangana through the integration of RAG pipelines and multilingual embeddings. Among evaluated models—ChatGPT-4o Mini, Gemini 1.5 Flash, and Mistral-7B-Instruct—ChatGPT-4o Mini proved most contextually adaptive despite slight latency trade-offs.

Key challenges include the need for real-time integration of dynamic data such as weather (IMD) [27], market prices (eNAM) [28], and crop disease alerts [29]. Future versions will explore hybrid models combining static knowledge with live data sources. Fine-tuning LLMs using conversational data in rural dialects [30,31], supported by agricultural universities and language centers, will improve performance in low-resource languages.

Voice and multimodal interfaces [32] will enhance accessibility for semi-literate users. IVR systems, WhatsApp bots, and image recognition tools (e.g., for crop diseases) can improve usability. Personalized recommendations based on user profiles and feedback loops using gamified or multilingual prompts will support continuous model refinement.

Scalable deployment will rely on federated learning, edge computing, caching strategies, and adherence to data privacy laws like the Personal Data Protection Bill [33]. Evaluation metrics must go beyond BLEU and MRR to include explainability, cultural relevance, and agronomic alignment, supported by standardized benchmarks.

In conclusion, while AgriLLM-India provides a strong foundation, its success depends on adaptive learning, real-time integration, user-centric design, and alignment with national digital agriculture infrastructure [34]. Multi-stakeholder collaboration and equitable scaling are essential to transform it into a national platform benefiting all Indian farmers.

7   Conclusion

AgriLLM-India bridges the gap between advanced AI and grassroots agricultural needs by adapting Large Language Models within a Retrieval-Augmented Generation (RAG) framework. Through semantic search and multilingual support, it provides accurate, timely and scalable advisory services. Evaluations, including a Telangana case study, confirm its effectiveness in real-world rural settings. As India advances its digital agriculture goals, AgriLLM-India offers a farmer-centric platform emphasizing accessibility, contextual relevance, and adaptability. Its evolving architecture supports integration with regional languages and real-time data sources. Future efforts should prioritize equitable deployment, ethical data use, and inclusive co-design with rural communities. With support from government, academia, industry, and civil society, AgriLLM-India can become a national asset—empowering millions of farmers and serving as a model for AI-driven agricultural transformation in other developing nations.

References

1. Ministry of Agriculture & Farmers Welfare (MoAFW): Digital agriculture mission report (2023)
2. Indian Council of Agricultural Research (ICAR): Annual report (2022)
3. Chand, R., et al.: Reinvigorating agricultural extension in India. Econ. Polit. Weekly **56**(10) (2021)
4. NITI Aayog: Empowering farmers through technology (2020)
5. Didwania, K., Seth, P., Kasliwal, A., Agarwal, A.: Agrillm: harnessing transformers for farmer queries, arXiv preprint arXiv:2407.04721 (2024)

6. Zhu, H., Qin, S., Min, S., Lin, C., Li, A., Chen, Q.: Harnessing large vision and language models in agriculture: a review, arXiv preprint arXiv:2407.19679 (2024)
7. De Clercq, D., Nehring, E., Mayne, H., Mahdi, A.: Large language models can help boost food production, but be mindful of their risks. Front. Artif. Intell. **7** (2024)
8. Google Research: Gemini technical whitepaper (2023)
9. Gao, L., et al.: Retrieval-augmented generation for domain-specific QA. In: Proceedings of ACL (2023)
10. Zhao, W., et al.: Multilingual adaptation of large language models. Trans. ACL (2023)
11. Current Science: AI in Indian agriculture: challenges and opportunities. Curr. Sci. (2021)
12. Telangana State Remote Sensing Applications Centre (TRAC): Crop monitoring and forecasting (2023)
13. Gemini 1.5 Flash and Mistral-7B Technical Reports: Technical reports (2024)
14. Ministry of Electronics and IT (MeitY): Indiaai mission strategy (2023)
15. Mistral AI: Mistral-7b model card (2023)
16. PJTSAU: Pest management guidelines for cotton (2022)
17. Agropedia: Knowledge repository on Indian crops (2022)
18. Krishikosh: Open access repository of Indian agricultural research (2022)
19. Department of Agriculture, Telangana: Rythu bandhu scheme guidelines (2023)
20. Ministry of Agriculture: Pmfby guidelines (2023)
21. Jeya, R., Singh, D., Jha, A., Mallick, S.: Agriconnect: a data-driven platform for optimizing crop production with fine-tuned large language models. In: 2024 International Conference on Energy, Environment and Sustainability (ICEES), pp. 1–7 (2024)
22. OpenAI: Chatgpt-4 technical overview (2023)
23. Lin, X.: Failure recovery in rag pipelines. J. Appl. NLP **18**(2) (2024)
24. IBM Research: AI for precision agriculture (2023)
25. Microsoft India: AI solutions for Indian farming (2022)
26. World Bank: Enabling the business of agriculture (2021)
27. Indian Meteorological Department (IMD): Agro-meteorological services (2023)
28. eNAM: National agriculture market dashboard (2023)
29. TNAU: Agritech portal – crop protection practices (2022)
30. Awais, M., Alharthi, A.H.S., Kumar, A., Cholakkal, H., Anwer, R.M.: Agrogpt: efficient agricultural vision-language model with expert tuning, arXiv preprint arXiv:2410.08405 (2024)
31. Govindharajan, H., Vijayakumar, S.: A framework for automated selective fine-tuning of domain-specific large language models using graph-based retrieval augmented generation. In: 2024 Annual UEM Conference (UEMCON), pp. 431–439 (2024)
32. Wang, L., Jin, T., Yang, J., Leonardis, A., Wang, F., Zheng, F.: Agri-llava: knowledge-infused large multimodal assistant on agricultural pests and diseases, arXiv preprint arXiv:2412.02158 (2024)
33. Ministry of Electronics and IT: Personal data protection bill (2022)
34. FAO: Digital agriculture and rural development (2021)

Fine-Tuning and Generative Modeling Techniques

SLM4Offer: Personalized Marketing Offer Generation Using Contrastive Learning Based Fine-Tuning

Vedasamhitha Challapalli[✉], Konduru Venkat Sai, Piyush Pratap Singh, Rupesh Prasad, Arvind Maurya, and Atul Singh

SPARC Research, HCLSoftware, Bengaluru, India
`vedasamhitha.chall@hcl-software.com`

Abstract. Personalized marketing has emerged as a pivotal strategy for enhancing customer engagement and driving business growth. Academic and industry efforts have predominantly focused on recommendation systems and personalized advertisements. Nonetheless, this facet of personalization holds significant potential for increasing conversion rates and improving customer satisfaction. Prior studies suggest that well-executed personalization strategies can boost revenue by up to 40%, underscoring the strategic importance of developing intelligent, data-driven approaches for offer generation.

This work introduces SLM4Offer, a generative AI model for personalized offer generation, developed by fine-tuning a pre-trained encoder-decoder language model—specifically Google's Text-to-Text Transfer Transformer (T5-Small 60M)—using a contrastive learning approach. SLM4Offer employs InfoNCE (Information Noise-Contrastive Estimation) loss to align customer personas with relevant offers in a shared embedding space.

A key innovation in SLM4Offer lies in the adaptive learning behavior introduced by contrastive loss, which reshapes the latent space during training and enhances the model's generalizability The model is fine-tuned and evaluated on a synthetic dataset designed to simulate customer behavior and offer acceptance patterns. Experimental results demonstrate a 17% improvement in offer acceptance rate over a supervised fine-tuning baseline, highlighting the effectiveness of contrastive objectives in advancing personalized marketing.

Keywords: Contrastive Learning · Offer Generation · T5 · Personalization · InfoNCE · Customer Personas · Encoder-Decoder Models · Synthetic Data Simulation · Deep Learning · Natural Language Generation · Representation Learning · Marketing AI

1 Introduction

Personalized marketing has been shown to significantly impact business performance. Arora et al. (2021) [1] report that firms can achieve up to 40% increase in revenue through the use of personalized marketing strategies. Previous research has also highlighted various positive consumer responses to personalization(de Groot, 2022; Tomczyk, Buhalis, Fan, & Williams, 2022) [7]. According to Kotler and Keller (Marketing Management, 15th Edition), 'a marketing offer is defined as a combination of products, services, information, or experiences presented to a market to fulfill a specific need or want'. This paper presents SLM4Offer, a Gen AI model, built by applying contrastive learning based fine-tuning on Google's T5, to create personalized marketing offers for a given customer profile.

This approach builds on the foundational work of Zhang et al., 2022 [8], where contrastive learning was used to align word embeddings with their definitions at a fine semantic level. In contrast to their work, which considers the definitions of all other words as negatives, this paper applies contrastive learning to hard negatives—which are the rejected offers by a particular customer. To the best of the author's knowledge this is the first time that contrastive fine-tuning of a Generative AI model is used to generate personalized marketing offers.

This paper compares the performance of contrastive learning-based fine-tuning with a baseline approach that employs supervised fine-tuning on the same Google T5 model. The authors report with 94% certainty that LLM's judgments align with human evaluations using the standard chi-square test. The simulation results indicate that contrastive loss-based fine-tuning achieves approximately 17% improvement in the acceptance rate—defined as the ratio of offers accepted to the total offers proposed - compared to the baseline approach.

The remainder of this paper is organized as follows. Section 2 reviews related work on the application of Large Language Models (LLMs) and contrastive learning. Section 3 presents the necessary background to support a deeper understanding of our approach. Section 4 details the fine-tuning process used to create SLM4Offer. Section 5 outlines the experimental methodology to evaluate SLM4Offer, followed by the results and analysis in Sect. 6, followed by the section Conclusion which concludes the paper and discusses potential directions for future work.

2   Related Work

The existing approaches for marketing offer generation can be broadly divided into three categories. The first category consists of retrieval-based recommenders, which select offers from a predefined set of offers using similarity to historical acceptance data (Ni et al., 2017 [4]). The second category consists of classification-based systems, which assign likelihood scores to a fixed pool of offers based on customer profiles. The last category of offers consists of generation-based approaches, which aim to synthesize new and personalized offers from scratch. The work presented in this paper belongs to the last category and introduces a novel contrastive training mechanism within a generative model. The approach presented in this paper builds on the definition genera-tion task, which was introduced by Noraset et al. (2017) [5]. This work was extended by Huang et al. (2021) [2] and Kong et al. (2022) [3], who employed pre-trained encoderâĂŞdecoder models, achieving improved fluency and contex-tual relevance. The work presented in this paper builds upon InfoNCE which is a contrastive loss introduced by Aaron Van Den Oord et al. [6].

In summary, while prior work in offer generation either reuses existing offers or generates new ones through generic language models, the paper introduces a novel contrastive training mechanism integrated within a generative model, specifically tailored to marketing personalization.

3 Background

This section presents an overview of Google's Text-To-Text Transfer Transformer (T5) small language model (SLM) used in this paper and provides an outline of the contrastive learning Information Noise-Contrastive Estimation (InfoNCE) loss.

3.1 Text-to-Text Transfer Transformer (T5)

Large language models (LLMs) are transformer based models trained on immense amounts of data making them capable of understanding and gener-ating natural language content for a wide range of tasks. A Small Language Model (SLM) is an LLM with lesser number of parameters compared to larger LLMs. There is no agreed definition of the size of a Small Language Model (SLM) in literature.

Developed by Google Research, T5 (Text-to-Text Transfer Transformer) is an encoder-decoder model trained on the large-scale text corpus called C4 (Colossal Clean Crawled Corpus). The model is available in multiple sizes, ranging from 60M to 11B parameters, and supports a maximum input length of 512 to 1024 tokens. The authors have chosen a T5 small (60M) encoder-decoder model due to its ability to effectively capture input-output relationships, especially when the input (customer context) and output (personalized offers) differ structurally thereby making it well-suited for generating personalized responses conditioned on diverse customer profiles with their past history of accepted and rejected offers. SLM4Offer is scalable to larger T5 variants and other LLMs also. The smaller model was intentionally chosen to test the algorithm in a constrained setup with minimal domain knowledge.

3.2 Contrastive Learning and InfoNCE Loss

In this paper the authors use **contrastive learning**-based Information Noise-Contrastive Estimation (InfoNCE) loss function introduced by Aaron Van Den

Oord et al. [6] to effectively capture the distinctions between accepted and rejected offers in relation to customer personas. In the approach presented in this paper, the authors define a **positive pair** as the combination of a customer persona and the accepted offers by the customer, and a **negative pair** as the customer persona and the rejected offers.

InfoNCE loss is preferred over other contrastive losses such as triplet loss because it is more efficient especially in setups like in the paper, where it is required to embed customers and offers in a shared space and need a strong signal to bring positives closer while pushing all negatives away. Whereas, Triplet loss lacks this deep probabilistic grounding, making it harder to optimize.

The InfoNCE loss is formulated as:

$$P(a_j \mid z_i) = \frac{\exp\left(\frac{\mathrm{sim}(z_i, a_j)}{\tau}\right)}{\sum_{k=1}^{N} \exp\left(\frac{\mathrm{sim}(z_i, r_k)}{\tau}\right)}$$

$$\mathcal{L}_{\mathrm{InfoNCE}} = -\mathbb{E}_X\left[\log\left(P(a_j \mid z_i)\right)\right]$$

where:

- $sim(u, v)$ represents the cosine similarity between two vectors u and v,
- T is a temperature hyperparameter controlling the sharpness of the distribution, N is the number of negative samples in the batch.
- z_i is the representation of a data point, a_j is the representation of positive sample, r_i is the representation of negative sample

4   SLM4Offer: Marketing Offer Generation

This section begins by explaining how SLM4Offer generates personalized marketing offers. It then introduces the dual-objective loss function used to fine-tune the T5 encoder-decoder architecture on which SLM4Offer is built. The model takes a prompt and a customer persona as input to produce a personalized marketing offer. Figure 1 presents sample persona and offers generated by SLM4Offer.

SLM4Offer uses a dual objective loss function which is given by:

$$L_{\mathrm{Final}} = \lambda \cdot L_C + (1 - \lambda) \cdot L_G$$

Here, L_C denotes the contrastive loss as described in Sect. 3 and L_G represents the cross-entropy-based offer generation loss. The value of λ depends on the model, the specific use case, and the desired extent of unsupervised training in the embedding space. A value of 0.5 has been used as the λ for the current scenario. This dual loss function ensures that while the decoder learns to generate offers in a supervised manner using the cross-entropy loss, the embedding space

Customer Persona

```
{"Name": "P9654",                      {"Name": "P9999",
"Age": 30,                             "Age": 35,
"Gender": "Female",                    "Gender": "Male",
"Monthly Income": "$111,969",          "Monthly Income": "$82,450",
"Spending Pattern": "Budget-conscious",  "Spending Pattern": "High-spender",
"Preferred Payment Method": "Buy Now Pay Later"  "Preferred Payment Method": "Credit Card"
"Interests": ["Finance", "Fitness", "Gaming"]  "Interests": ["Fitness", "Shopping", "Technology",
"Financial Goals": ["Wealth Growth", "Retirement    "Travel"],
  Planning"]}                          "Financial Goals": ["Savings", "Wealth Growth"]}
```

Generated Offers by SLM4Offer

Investment Planning Services for Wealth
Growth and Retirement Planning.

High-End Fitness Equipment with 5% Cashback
on Online Shopping.

Fig. 1. Example of customer persona and offers generated by SLM4Offer

is simultaneously trained in an unsupervised fashion. As a result, the generated offers originate from a space that is semantically aligned with the customer persona.

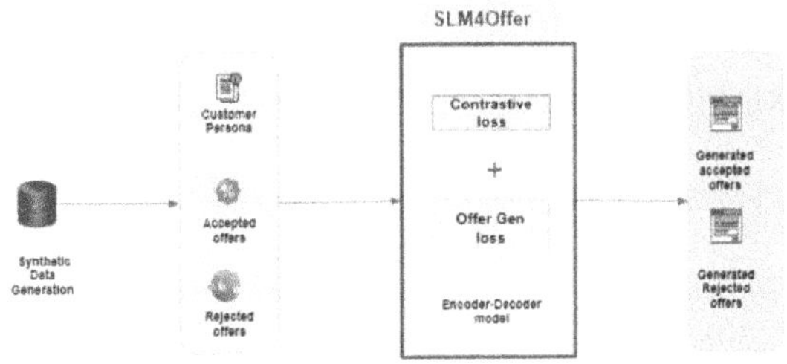

Fig. 2. Training workflow of SLM4Offer using contrastive learning with InfoNCE loss

Figure 2 provides a high-level overview of the fine-tuning process, where the model takes a customer persona as input along with previously accepted and rejected offers. The encoder processes the customer persona to generate a customer embedding. This embedding serves as the contextual input for the decoder, which generates candidate offers that the customer is likely to accept or reject. In addition to traditional supervised learning using cross-entropy loss, contrastive learning with InfoNCE loss is applied in the embedding space to enable unsupervised training.

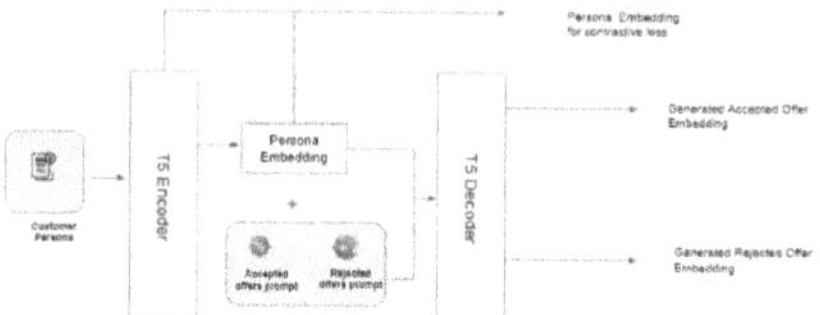

Fig. 3. Embedding generation from customer persona using the encoder of SLM4Offer

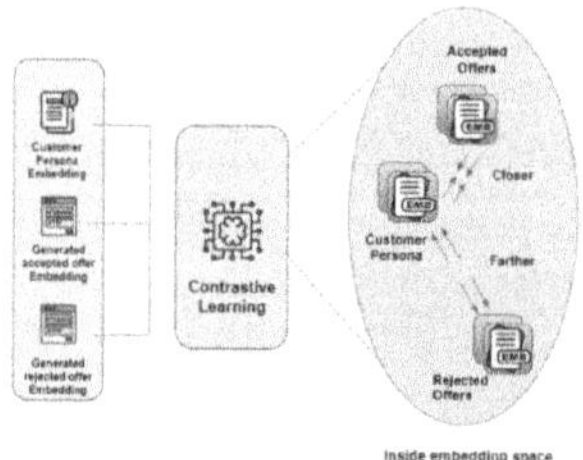

Fig. 4. Contrastive learning workflow of SLM4Offer using InfoNCE loss

Figure 3 illustrates the process of generating embeddings. The customer persona is passed through the encoder to produce a persona embedding. This embedding, along with the offer generation prompt, is fed into the decoder to extract embeddings of accepted and rejected offers from the decoder's hidden states. Figure 4 shows the contrastive learning-based fine-tuning process, which uses the persona embedding, the generated accepted offer embedding, and the generated rejected offer embedding. These embeddings are used to fine-tune the T5 model weights through contrastive learning. Figure 5 illustrates the use of traditional cross-entropy loss by comparing generated offers with ground truth offers. This loss is used for supervised fine-tuning of the model, ensuring that the generated offers are well aligned with the actual offers.

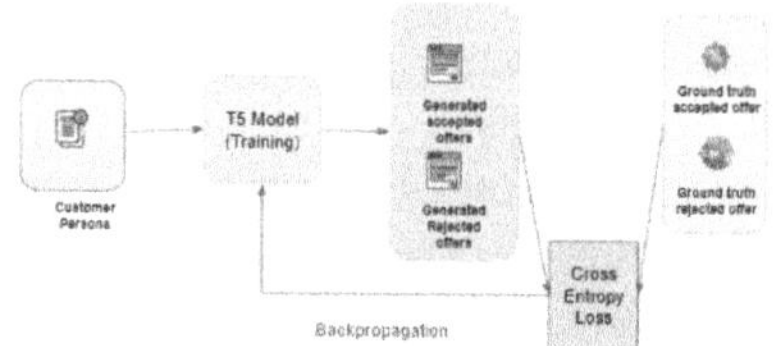

Fig. 5. Supervised fine-tuning of SLM4Offer using cross-entropy loss

5   Experimental Setup

This section describes the methodology used for synthetic data generation to fine-tune the T5 model for building SLM4Offer, as well as the pipeline used for model evaluation. The performance of SLM4Offer is compared with a baseline approach in which the T5 model is trained using Supervised Fine Tuning (SFT). The model's outputs are assessed using a large language model (LLM) as a judge. In addition, a human evaluation is conducted on a sample of the results. Hypothesis testing is then used to verify the reliability of the LLM-based evaluations.

5.1 Synthetic Data Generation

SLM4Offer is fine-tuned and evaluated using synthetic data. To mitigate biases during synthetic data generation, the generated customer personas and offers are based on statistical patterns observed in real company data. Diversity across age, interests, and behavioral types was ensured to simulate a balanced population. Customer personas and corresponding offers have been generated using Azure OpenAI LLMs. Each persona is defined by a combination of financial and behavioral attributes, including age (18–70), gender, monthly income ($30k–$200k), interests (e.g., Finance, Travel), spending patterns (e.g., Budget-conscious, High-spender), preferred payment methods, and financial goals. These attributes were sampled to create multiple unique personas. This customer persona data was fed to the LLM along with few-shot examples to generate offers. 3 accepted and 3 rejected offers were generated, which align with the persona's preferences

5.2 Evaluation

The results of the proposed fine-tuning approach are compared against Supervised Fine-Tuning (SFT) performed on the same dataset. Offer acceptance rate which is defined as the ratio of accepted offers to the total number of generated offers, is used as the primary evaluation metric.

To assess the quality of the generated offers, an evaluation pipeline has been developed in which a large language model (such as GPT-4[1]) acts as a judge. Given: a) a customer persona, b) a list of previously accepted and rejected offers, and c) a newly generated offer, the LLM is prompted to classify whether the new offer is more likely to be accepted or rejected.

To validate the reliability of the LLM-based evaluation, a human assessment was conducted on a randomly selected subset of the test dataset. A group of annotators independently labeled each generated offer as accepted or rejected based on the corresponding persona and past offer history. An offer was then classified based on the majority vote of these annotations. These human labels were statistically compared with the LLM's classifications to measure agreement and establish a performance baseline for automated evaluation. To measure the association between the two, the Chi-Square Test of Independence was applied.

6 Results and Analysis

The generated synthetic dataset consists of 24,000 samples. For the experiments, the data has been divided into train, validation and test sets consisting of 21600,

---

[1] Using Azure OpenAI Service.

2400 and 1000 samples respectively. The train set has been utilized for the fine-tuning of the model, with the validation set being used for monitoring the training, whereas the benchmarking on the experiments has been done on the test set. Following two fine-tuning methods are conducted on this dataset using the base model (Code-T5-Small):

1. Supervised Finetuning of the model with cross entropy loss
2. Finetuning the model with cross entropy loss and contrastive loss (InfoNCE loss)

Directly using the untrained base model gives the exact customer persona as the generated offer. This is because the pretrained model was trained on a large dataset and hence the model is unaware of the domain of the target dataset. To adapt the model to the domain of the dataset, finetuning of the dataset is required.

Figure 6 shows the loss plot for fine-tuning SLM4Offer. The loss initially fluctuates and stabilizes after a certain point, so the model from the 30th epoch has been used as the final trained model

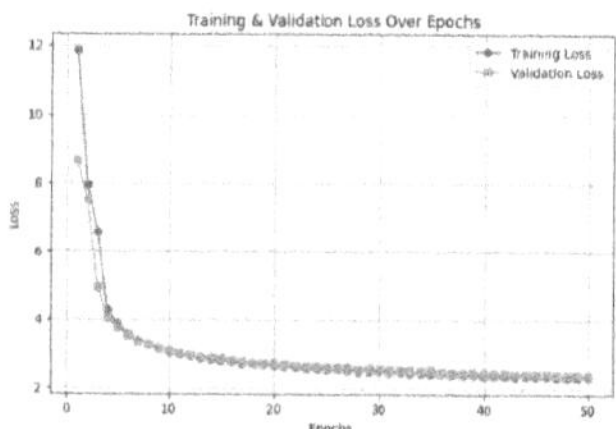

Fig. 6. Loss Plot for the training of SLM4Offer

WeightWatcher is a Python package for analyzing trained models and identifying training difficulties without accessing data. By fitting a power law to the ESD tail, it computes the exponent α to assess model quality. This helps detect issues in compression, fine-tuning, and generalization. Table 1 is the layer wise performance comparison between the two models. Alpha values less than 2 indicate overfitting whereas alpha values more than 6 indicate underfitting. The above observations indicate that SFT has more layers slightly in the underfitting zone compared to the contrastive model.

Table 1. Layer-Wise Performance Comparison: SFT vs. Contrastive Model

	Ovefit	Normal	Underfit
SFT	1	82	14
Contrastive	1	85	11

Table 2 shows the performance comparison between SFT and contrastive finetuning on the test set (1000 samples). There is a 17.5% increase in the acceptance percentage by using contrastive learning. This shows the superiority of contrastive learning in the use case.

Table 2. Performance comparison between SFT and contrastive finetuning

	Offer accepted count	Offer Acceptance Rate (%)
SFT	800	80%
Contrastive finetuning	940	94%

As mentioned in Sect. 5, whether the generated offer from the model would be accepted or rejected by the customer- has been evaluated using LLM as a judge using various prompting techniques including few shot prompting. Usage of LLM as a judge was verified with a set of human annotators on a sample of 200 records. Table 3 results of the evaluations by the two different type of judges:

Table 3. Evaluation Scores from Human Annotators and LLM Judges. 1 is for accepted and 0 is for rejected

	Human Evaluation 0	Human Evaluation 1
LLM Evaluation 0	41	9
LLM Evaluation 1	3	147

The results of the chi-square test shows a statistically significant association between human and LLM evaluations, with a p-value less than 0.001. This indicates that the LLM can be taken as a judge to evaluate the generated offers and the results are reliable enough.

To summarize, the model was trained following the process illustrated in Fig. 6, with a comparison of training dynamics between SFT and contrastive learning. The final model demonstrates improved performance, with a 17.5% increase in offer acceptance rate over the SFT baseline, as reported in Table 2. This improvement is also reflected in Fig. 1, which highlights the model's ability to generalize and personalize offers effectively. These results suggest that incorporating contrastive learning alongside traditional fine-tuning leads to the generation of more robust, diverse, and persona-aligned marketing offers.

7   Conclusion

This paper presents a novel contrastive learning based encoder-decoder framework for generating personalized marketing offers. By leveraging a fine-tuned T5 model and tweaking the embedding space between the encoder and the decoder using the InfoNCE loss, the paper's approach effectively aligns accepted offers with customer personas while distancing rejected ones. The system can periodically finetune the model using the most recent interaction data, along with the updated customer persona data, enabling it to adapt to user preferences dynamically. All the experiments have been done with respect to the financial domain, it can be extended to generate offers for any other domain such as e-commerce, retail, healthcare etc. Future work may explore real-world deployment and expanding to adaptive personalization that evolves over time with user behavior. Currently, the paper uses customer personas as input to our contrastive learning-based framework. However, moving forward, each offer can be explicitly tagged with a corresponding product and campaign. This will enable better alignment with business goals. Such an integration will ensure that the generated offers are not only personalized, but also campaign-aware and product-specific.

References

1. Arora, L., Singh, P., Bhatt, V., Sharma, B.: Understanding and managing customer engagement through social customer relationship management. J. Decis. Syst. **30**(2–3), 215–234 (2021)
2. Huang, H., Kajiwara, T., Arase, Y.: Definition modelling for appropriate specificity. In: Proceedings of the 2021 Conference on Empirical Methods in Natural Language Processing, pp. 2499–2509 (2021)
3. Kong, C., Chen, Y., Zhang, H.: Multitasking framework for unsupervised simple definition generation
4. Ni, K., Wang, W.Y.: Learning to explain non-standard english words and phrases. In: Proceedings of the Eighth International Joint Conference on Natural Language Processing, vol. 2, pp. 413–417. Short Papers, Taipei, Taiwan (2017)
5. Noraset, T., Liang, C., Birnbaum, L., Downey, D.: Definition modeling: learning to define word embeddings in natural language. In: Thirty-First AAAI Conference on Artificial Intelligence (2017)
6. van den Oord, A., Li, Y., Vinyals, O.: Representation learning with contrastive predictive coding (2018)
7. Tomczyk, A.T., Buhalis, D., Fan, D.X.F., Williams, N.L.: Price-personalization: customer typology based on hospitality business. J. Bus. Res. **147**, 462–476 (2022)
8. Zhang, H., Li, D., Yang, S., Li, Y.: Fine-grained contrastive learning for definition generation. In: Proceedings of the 2nd Conference of the Asia-Pacific Chapter of the Association for Computational Linguistics and the 12th International Joint Conference on Natural Language Processing (Volume 1: Long Papers), pp. 1001–1012. Association for Computational Linguistics, Stroudsburg, PA, USA (2022)

Cross-City Next POI Prediction Using a Generative Causal Method

Soma Bandyopadhyay[1](✉) [iD] and Sudeshna Sarkar[2] [iD]

[1] TCS Research, Tata Consultancy Services Limited, Kolkata, India
soma.bandyopadhyay@tcs.com
[2] Indian Institute of Technology Kharagpur, Kharagpur, India
sudeshna@cse.iitkgp.ac.in

Abstract. Location prediction is an important aspect of mobility modeling. However, predicting person's next location is challenging when they move to a different place. In this context, the cross-city next POI (Point of Interest) prediction task involves predicting which POI a user is likely to visit next while traveling from one city to another, for example, from City A to City B. However, the data on cross-city visits is usually very limited and often does not have past visit records in the target city. Given the data being sparse, we hypothesize that understanding the causal features of mobility will help predict the location of a user in a new city. In this work, we propose a novel method for spatiotemporal cross-city POI prediction using generative learning and causal perspective. We employ the interventional aspect of causality to identify the causal features and apply the causal features to cross-city next-location prediction. Our method exploits a Conditional Variational Autoencoder based generative model. We use large real-world LBSN (Location-based Social Network) data. In our experiments, we evaluate the causal sensitivity of different features related to the cross-city next POI prediction. Our proposed method demonstrates competitive performance in predicting the next POI influenced by causal features across cities.

Keywords: Causality · Generative Learning · Next POI Prediction

1 Introduction

Location prediction is an important task because of its applications in wide range of mobility applications such as transport planning, traffic management, social gatherings, and disaster management for assessing crowd flow. This task involves understanding patterns of spatiotemporal user behavior from GPS-based trajectory data [15], and LBSN (Location-based Social Network) comprising "check-in" data. These become increasingly important in the context of big data. LBSN combines the social network information along with GPS based location details of diverse user activity points or Point-of-Interest(POI) like airports, Gym, Coffeeshop, etc. Foursquare [17] is one such popular LBSN, where users check in at POIs within a given time and provide a rating. Sequences of POI visits are

R. Gupta et al. (Eds.): BDA 2025, LNCS 16041, pp. 327–337, 2026.
https://doi.org/10.1007/978-3-032-15134-6_24

the time series of spatial locations that represent the different POIs visited over time. Sensors such as GPS systems, are a popular choice for check-in POI. Predicting the next venue a user is likely to visit in a cross-city scenario based on their historical check-in data is a **challenging task** as it suffers a lack of visit records available for the destination city. In some scenarios, the user may not have any prior history of travel to the new city. Additionally, it also suffers from interest drift [2], where users' preferences may shift over time. The cause-effect [11] relationships among the various features like the sequence of past POI visits, start time, day of the week, spot category, etc., significantly influence the individual mobility characteristics that impact future POI visits. We aim to look into the cross-city next POI prediction task with causal perspectives.

In this work, we propose a new method for the cross-city next POI prediction task. First, we identify causally sensitive features and use them to predict the next POI using GPS-based check-in data across cities through a generative model. We consider a Conditional Variational Autoencoder (CVAE)-based predictor. We use the interventional [11] aspect of causality to determine causal features. These features improve prediction performance. For instance, *day of the week* influences POI sequences and the next POI in another city. Figure 1 depicts the overview of our method.

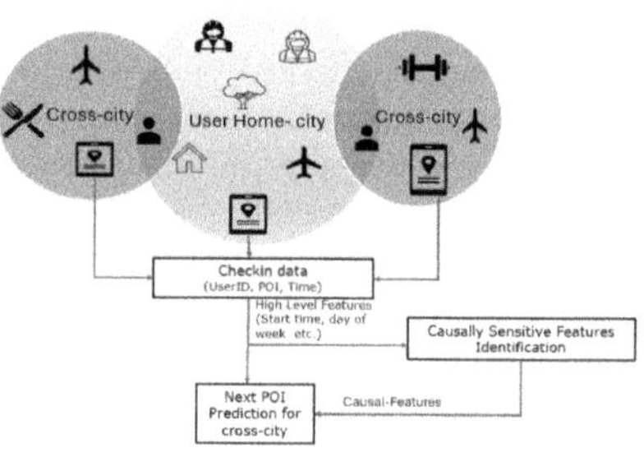

Fig. 1. Overview of proposed method.

A Summary of Our Contribution is as Follows:

1. We propose a new method to predict the next POI of cross-city using CVAE-based generative learning conditioned on causal features.
2. We identify the causal features (e.g. Start time) for cross-city next POI prediction task considering interventional aspect of causality. The identified causal features are exploited for conditioning the generative predictor.
3. We demonstrate improved accuracy and interpretability on real-world LBSN data Foursquare [17].

2   Related Work

In this section we briefly review the related works.

Next POI Prediction: PRME [4] introduces personalized ranking metric embedding by modeling sequential patterns and user-POI distances. RNN [22] serves as a basic time-series model for next POI prediction. Time-LSTM [23] enhances LSTM with time gates to account for temporal drift. Caser [13] treats historical POIs as 2D images and applies convolution to capture global and local sequential patterns. AttRec [21] employs self-attention to model long- and short-term user preferences in a metric learning setup. DeepMove [3] uses attention-based multimodal RNNs to capture nuanced mobility transitions. LSTPM [12] uses geo-dilated RNNs with temporal encoding for both short- and long-term preference modeling. GETNext [20] leverages Transformers to integrate user preferences with global, temporal, and spatial contexts. PLSPL [16] combines LSTM with attention and adaptive user-specific weighting. Flashback [18] retrieves past RNN states with similar spatial and temporal contex, while Flashback-R [6,9] improves this using refined intervals. UPTDNet [8] addresses cross-city next POI prediction by modeling short- and long-term preferences and user–POI interactions across cities.

Causality Perspectives: While sequence models with attention [14] implicitly model dependencies that resemble causality, they do not explicitly capture causal effects. CEVAE [10] uses a VAE-based approach to estimate treatment effects in presence of unobserved confounders. CausalVAE [19] embeds structural causal models into VAEs for disentangled representations. However, these methods do not address causality-aware cross-city POI prediction, which we aim to explore in this work.

3   Methods

In this section, we present our method. We name the proposed method as Generative Causal Cross-city next POI prediction(GCCP). We use a CVAE based generative model. The functional components of the proposed method are illustrated in Fig. 2. This comprises the generative predictor, the factual path (executed with original data) is represented by a solid blue line, and the interventional path (executed with altered data) is shown in a dotted blue line for causal

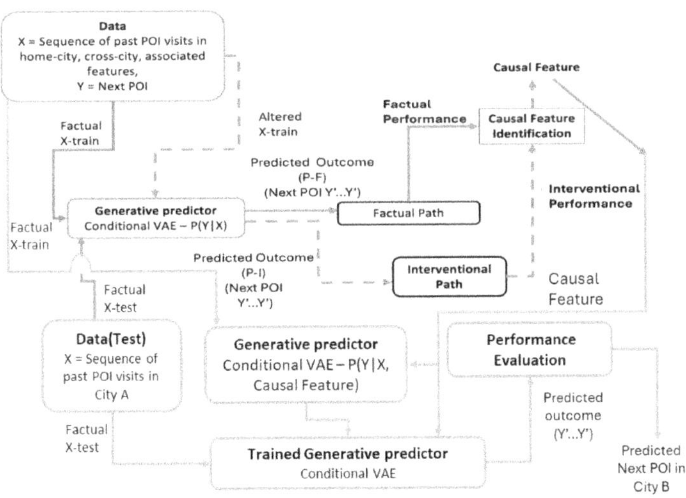

Fig. 2. Functional components of generative causal cross-city next POI prediction.

feature identification, where we compare the performance of prediction using identical test data, and using the predictor trained in original (factual) train data and intervened (altered) train data. The functional components are described as follows: Data $D(X, Y)$ is divided into train and test. Generative Predictor (GP) using CVAE is first trained on the factual training data X_{train} (Factual X-train), referred to as the factual generative predictor (GP-F). This model learns the factual latent representation Z_F, and generates predictions P-F for the test input X_{test} (Factual X-test).

For the **intervention path**, we construct an altered version of the training data $X_{\text{train_altered}}$ by modifying the sequence of POI visits, for example, by replacing the most frequently visited POI with another (e.g., a POI id like zero). We obtain the interventional generative predictor (GP-I) by training with $X_{\text{train_altered}}$ (Altered X-train) which learns the interventional latent representation Z_I, and produces predictions P-I on the same factual test input X_{test}. The **factual performance** is evaluated by comparing the predicted outcomes P-F with the ground truth Y_{test}. Similarly, the interventional performance is measured by comparing P-I with same test target Y_{test}. The performance difference between these two is used to determine whether a given feature (e.g., day of week) has a causal impact. The identified causally sensitive feature is applied to retrain the GP-F model using this feature as a conditioning feature with the past POI sequence. The final prediction is then performed using the factual test data. We elaborate on our method in detail below.

3.1   Generative Predictor

We use Conditional Variational Autoencoder(CVAE) as a generative predictor with an encoder and decoder components. y_i represents the ith true level of the $x_i \in X$ sequence up to time step $t - 1$: $y_i = f(X_{1:t-1})$

Encoder: The encoder comprises and **multi layer LSTM, and self attention** learns the latent space representation (Z). Z is obtained by computing the mean μ and log-variance σ with the reparameterization trick to sample from the latent space, first ϵ is sampled from $N(0, I)$ and then z is computed as $z = \mu(Y|X) + \sigma^{1/2}(Y|X) * \epsilon$.

Decoder: The decoder takes Z as input and repeats across the time step which is the maximum sequence length in the training data and combines this with the conditional input X, considering X as temporal data. Now this Z conditioned on X is passed to the next neural layers as used in Encoder. The final output of the decoder goes to a dense layer. This final dense layer has output classes as $c_{\max} = \max(Y_{\text{train}}) + 1$ number of nodes with softmax activation to predict the next location.

Generator: The decoder model is employed to generate new data samples of Y by passing the latent samples sampled from a Gaussian distribution and using conditional inputs.

Equation 1 represents the loss $\mathcal{L}$. The reconstruction loss uses sparse categorical cross-entropy $\mathcal{L}_S$, which measures the negative log-likelihood of the true labels given the predicted class probabilities. We apply **KL annealing** [5], a gradually increasing weight, λ_{KL} to multiply the KL divergence term. This counters KL-vanishing during the initial training.

$$\mathcal{L}_S = -\frac{1}{N} \sum_{i=1}^{N} \log p_\theta(y_i \mid \mathbf{x}_i), \; y_i \in Y, \; \mathbf{x}_i \in X$$

$$\mathcal{L}(\theta, \phi; \mathbf{Y}_t, \mathbf{X}_{t-1:t-n}) = \mathcal{L}_S + \lambda_{\mathrm{KL}} \cdot \mathrm{KL}\left(q_\phi(\mathbf{z}|\mathbf{Y}_t, \mathbf{X}_{t-1:t-n}) \,\|\, p_\theta(\mathbf{z}|\mathbf{X}_{t-1:t-n})\right)$$

$$(1)$$

θ, ϕ are the parameters of the decoder network, and encoder network respectively. $\mathbf{Y}_t$ is the target output $\mathbf{X}_{t-1:t-n}$ is the input sequence comprising different features. $q_\phi(\mathbf{z}|\mathbf{Y}_t, \mathbf{X}_{t-1:t-n})$ is the approximate posterior distribution , $p_\theta(\mathbf{z}|\mathbf{X}_{t-1:t-n})$ is the prior distribution of the latent variable given the conditional input. KL denotes the Kullback-Leibler divergence.

Algorithm 1. Cross-city next POI prediction with causal feature identification

---

1: **Input:** Training data X_{train}, test data X_{test}, targets Y, candidate features $F_{\mathrm{all}} \in \{W, S_{\min}, SC\}$, intervention type (e.g., random: replace most frequent POI with ID 0)

2: **Output:** Causally sensitive features F_{cs} and prediction metrics (Acc, MRR)

3: **Step 1: Train Factual (fact) Predictor (GP-F)** Train CVAE using POI sequence (PS, UID) from X_{train}

4:     $P(Y_{t+1} \mid X_t) = \sum_{F_{\mathrm{CS}}} P(Y_{t+1} \mid X_t)$

5: Evaluate GP-F on X_{test}, store $\mathrm{Acc}_{\mathrm{fact}}, \mathrm{MRR}_{\mathrm{fact}}$

6: **for all** features $F \in F_{\mathrm{all}}$ **do**

7:     **Step 2: Intervention (int) on POI Sequence** Replace user's most frequent POI in X_{train} to create PS_{Altered}

8:     **Step 3: Train Interventional Predictor (GP-I)** Train CVAE using PS_{Altered} and condition on feature F

9:     Interventional prediction

10:      $P(Y_{t+1} \mid \mathrm{do}(X = X_{altered})) = \sum_F P(Y_{t+1} \mid X_{altered}, F)P(F)$

11:     **Step 4: Evaluate GP-I** Compute interventional output:

12:      $Y_{t+1}^{\mathrm{int}} = \mathrm{GP\text{-}I.decoder}(Z_I, X_{\mathrm{test}}, F)$ Compare with factual:

13:      $Y_{t+1}^{\mathrm{fact}} = \mathrm{GP\text{-}F.decoder}(Z_F, X_{\mathrm{test}})$

14:     **Step 5: Assess Causal Sensitivity and derive causal feature**

15:      Compute: $\Delta\mathrm{Acc} = \mathrm{Acc}_{\mathrm{int}} - \mathrm{Acc}_{\mathrm{fact}}$ (Eq. (2))

16:     **if** $\Delta\mathrm{Acc} > 0$ **then**

17:      Mark F as causally sensitive $\rightarrow F_{\mathrm{cs}}$

18:     **else**

19:      Feature F not causal feature

20:     **end if**

21: **end for**

22: **Step 6: POl Prediction with Causal Conditioning** Train CVAE on actual X_{train} conditioned on selected F_{cs}, Use it to predict next POIs on X_{test}

23: **return** Causal features F_{cs} and final prediction performance.

3.2   Identifying Causal Feature

We identify causal features $F_{CS} \in F_{all}$ that satisfy the backdoor criterion [11] by acting as common causes of both the input sequence X (past POI visits) and the target Y (next POI), i.e., $X \leftarrow F_{CS} \rightarrow Y$. These features help reduce confounding in generative modeling.

First, we train a generative predictor (GP-F) using the original data X_{train} without interventions and follow **Step 1** as shown in Algorithm 1. Next, we apply an intervention by modifying the POI sequence to $X_{altered}$, and train a separate model (GP-I) and follow **Step 2** of Algorithm 1. We then evaluate both models on the same test set X_{test}. The factual prediction from GP-F is computed following **Step 3** of Algorithm 1. Similarly, the interventional model is evaluated following **Step 4** of Algorithm 1. The causal sensitivity of a feature is established by comparing the prediction accuracy under factual and interventional settings following **Step 5** of Algorithm 1 A feature with $\Delta Acc > 0$ is considered causally sensitive and is included in the set F_{CS}. This approach allows us to uncover confounding features that, when intervened upon, significantly affect the prediction of future POI visits. If this difference is positive, the feature is considered causal feature, suggesting its utility in improving next POI prediction by mitigating confounding.

3.3   Prediction of Next POI

The proposed causal feature identification method Sect. 3.2 identifies causal variables using CVAE based generative model. The identified causal features F_{CS} are then used to condition the CVAE model for predicting the next POI in a cross-city setup. This task is framed as sequence prediction problem with the classification objective considering the time stamp of the user's check-in information. The final CVAE is trained using the original training data augmented with causal conditioning [1], and evaluated on cross-city transitions. This improves **interpretability** allowing GCCP to focus on variables that truly influence the next POI. The complete method is presented in Algorithm 1.

4   Data

We use the Foursquare dataset [17], a location-based social network (LBSN) dataset collected between April 2012 and September 2013. It contains user check-ins, each defined as a tuple {User ID, POI ID, GPS coordinates, Timestamp}, capturing spatial and temporal mobility patterns. Following [8], we focus on Washington and Baltimore due to their higher check-in density. We use the processed data provided by the authors [7].

Factual Data: The unaltered FourSquare data is used for factual analysis.

Altered Data: This is obtained by altering the POI visits for interventional analysis: $PS_{Altered}$: The highest frequent POI ID is replaced with POI ID 0.

5 Experimental Analysis

In this section, we conduct the following steps to perform experimental analysis to evaluate our method and present the obtained results.

5.1 Data Preprocessing and Feature Extraction

Preprocessing: We perform the following preprocessing steps as considered by [2,8]. **Selecting Cross-City User:** Users who have traveled between Washington and Baltimore at least once were selected by checking their cross_city_mode, **Timestamp in datetime format:** Timestamps are converted to naive datetime format and adjusted for UTC using timeoffset, **Filtering lower visit frequency:** a) POIs with fewer than 10 check-ins are removed. b)Users with fewer than 25 total check-ins are excluded, **Removal of duplicate checkin:** Multiple check-ins by the same user at the same POI within a short time are removed, **Encoding:** User ID, place ID, and spot category are ordinal-encoded.

Feature Extraction: The high-level features considered are as follows: Userid(**UID**), Sequence of past POI visits(**PS**), Spot category(**SC**) (e.g., Park, Airport), ordinally encoded, Start time in minutes since midnight (**Smin**), and day of the week(**W**) (0 = Monday, ..., 6 = Sunday). Feature W adds the perspective of the user's daily life visits and the other out-of-routine visits. The Smin adds temporal granularity. We aim to identify the causal influence of Smin, SC, and W as causal features.

5.2 Cross-City Next POI Prediction

Generative Model Configuration: We apply the proposed method (Sect. 3) to identify causally sensitive features and use them for cross-city next POI prediction. The CVAE based generative predictor comprises multilayer LSTM encoder-decoder with self-attention with the total loss $\mathcal{L}$ (Eq. (1)) combining sparse categorical cross-entropy and KL divergence.

The **encoder** consists of two LSTM layers. The first has 60 hidden units with dropout (0.3), followed by a self-attention layer (sigmoid) and a second LSTM with 40 units. The output is mapped to the latent Z conditioned on X. The **decoder** takes Z (repeated across timesteps) and combines it with conditional input X. It passes through an LSTM (40 units), self-attention, another LSTM (60 units), and a final dense layer with $c_{\max}$ outputs and softmax activation. Latent dimension is set to 100. The **generator** uses decoder to generate samples by combining latent vectors (from Gaussian prior) with conditional inputs. KL annealing is used during training.

Identifying Causally Sensitive Features: To identify causal features, we follow the proposed method (Sect. 3), comparing factual and interventional predictions. GP-F is the baseline CVAE trained on factual input features (PS, UID), evaluated on cross-city test data (Baltimore $\rightarrow$ Washington and vice versa). GP-I is trained on modified input PS_{Altered}, conditioning on candidate causal features, as per Step 3 of Algorithm 1. We compare predictions from GP-F and GP-I using same factual test data X_{test}. Candidate features W, Smin, and SC are evaluated for causal sensitivity (via Eq. (2)).

Next POI Prediction: We follow Sect. 3.3. We train GP with the factual data, conditioning on the identified causal features among {Smin, SC, W}, using the factual data X_{train} with UID, PS, and other features. We apply X_{test} consists of cross-city data between Baltimore and Washington, to predict the next point of interest (POI) across city. For example, in the case of Baltimore to Washington, the predicted next POI corresponds to a location in Washington.

5.3   Performance Measure

Top-k Accuracy (Acc@k): measures the proportion of samples where the true label is within the top-k predicted classes. Let $P \in \mathbb{R}^{N \times C}$ denote the prediction probabilities, and $y \in \{1, \ldots, C\}^N$ the true labels. **Mean Reciprocal Rank (MRR):** captures average inverse rank of true label:

$$\text{Acc@k} = \frac{1}{N} \sum_{i=1}^{N} \mathbb{I}\left[y_i \in \hat{Y}_{i,k}\right], \quad \text{MRR} = \frac{1}{N} \sum_{i=1}^{N} \frac{1}{\text{rank}(y_i, \hat{Y}_i)} \tag{2}$$

where $\text{rank}_i(y_i)$: rank of the correct label in the sorted prediction vector P_i.

5.4   Results

We apply our method (Sect. 3) on real-world LBSN data to identify causal features and improve cross-city next POI prediction.

Causal Feature Identification: To evaluate causal features, as discussed in Sect. 5.2, causal features. In case of start time (Smin) and day of the week(W), we observe improvement in the best average performance for intervention, across all users, i.e., $\Delta\text{Acc} > 0$, (Eq. (2)), establishing them as causally sensitive features representing the causal path as $PS \leftarrow F_{\text{CS}} \rightarrow Y$, where F_{CS} acts as a common cause for PS, Y. In contrast, SC does not exhibit a significant causal effect on the prediction outcome. Table 1 reports the results for both Baltimore $\rightarrow$ Washington and Washington $\rightarrow$ Baltimore transitions under interventional settings on PS, highlighting **Smin** and W are the causal features.

Table 1. Intervention Study: Cross-City on the Foursquare

Scenario	Baltimore $\rightarrow$ Washington				Washington $\rightarrow$ Baltimore			
	Acc@1	Acc@5	Acc@10	MRR	Acc@1	Acc@5	Acc@10	MRR
Factual (placeid_encoded)	0.2653	0.4489	0.4898	0.3356	0.1200	0.3400	0.3800	0.2357
Intervention (SC)	0.1837	0.3265	0.3878	0.2416	0.1000	0.2600	0.2800	0.1926
Intervention (**W**)	0.2857	0.4286	0.5510	0.3579	0.1400	0.2200	0.2400	0.1914
Intervention (**Smin**)	0.3061	0.4286	0.5306	0.3753	0.2000	0.2800	0.3000	0.2573

Prediction of Next POI in Cross-City: We evaluate the proposed CVAE based generative predictor, GCCP with the factual data, and conditioning on the identified causal feature, Smin. GCCP generates n = 10 samples per input, and selects the best-performing prediction. We train this model with 500 epochs and batch size = 32 with dropout rate = 0.3, and Adam optimizer. To compute λ_{KL} we consider kl_start epoch = 10, kl_annealtime = 20 described in (Sect. 3). Table 2 summarizes the results on the Foursquare dataset in comparison with state-of-the-art (SoA). We compare the best-performing results reported in SoA with the best-obtained results of our method. GCCP outperforms existing methods in the Baltimore $\rightarrow$ Washington direction and achieves competitive results in the Washington $\rightarrow$ Baltimore, achieving the highest Acc@1 and second-best MRR. These consistent improvements, validated with **95% confidence**, highlight the effectiveness of conditioning on the causal feature Smin.

Table 2. Comparison of different methods for cross-city next POI prediction on the Foursquare (Washington-Baltimore) dataset

Method	Washington $\rightarrow$ Baltimore				Baltimore $\rightarrow$ Washington			
	MRR	Acc@1	Acc@5	Acc@10	MRR	Acc@1	Acc@5	Acc@10
PRME [4]	0.1813	0.1193	0.2672	0.3210	0.1632	0.1026	0.2282	0.2667
RNN [22]	0.1271	0.0899	0.1615	0.1835	0.0996	0.0711	0.1263	0.1474
Time-LSTM [23]	0.1345	0.0957	0.1607	0.2342	0.1300	0.1013	0.1532	0.1740
Caser [13]	0.2025	0.1378	0.2857	0.3210	0.1430	0.1000	0.1846	0.2154
AttRec [21]	0.2017	0.1513	0.2555	0.2941	0.1378	0.1026	0.1744	0.1897
DeepMove [3]	0.2197	0.1527	0.3015	0.3282	0.1319	0.0882	0.1824	0.2176
LSTPM [12]	0.2036	0.1543	0.2656	0.2891	0.1429	0.0958	0.2000	0.2225
GETNext [20]	0.2389	0.1818	0.2500	0.3182	0.1889	0.1519	0.2025	0.2278
PLSPL [16]	0.2264	0.1697	0.2857	0.3277	0.1658	0.1026	0.2385	0.2795
Flashback [18]	0.2017	0.1395	0.2716	0.3101	0.1861	0.1342	0.2290	0.2895
Flashback-R	0.2263	0.1725	0.2789	0.3174	0.1995	0.1579	0.2395	0.2684
UPTDNet [8]	**0.2629**	0.2003	**0.3266**	**0.3560**	0.2195	0.1110	0.2816	0.3105
GCCP (Proposed)	**0.2600**	**0.2200**	0.2800	0.3000	**0.3884**	**0.3265**	**0.4897**	**0.4897**

Note: 95% CI for GCCP : W$\rightarrow$B: Acc@1 ($\pm$0.0538), Acc@5 ($\pm$0.0593), Acc@10 ($\pm$0.1009), MRR ($\pm$0.0161); B$\rightarrow$W: Acc@1 ($\pm$0.0528), Acc@5 ($\pm$0.0975), Acc@10 ($\pm$0.0755), MRR ($\pm$0.0620)

Ablation Study: We perform an ablation study by conditioning on the causal feature and subsequently removing it as presented in Table 3 for both Baltimore$\rightarrow$Washington and Washington$\rightarrow$Baltimore scenarios. In the baseline, the generative predictor is trained without causal conditioning. The results demonstrate a significant performance improvement when causally sensitive features are used for conditioning. Specifically, for Smin, the ACC and the MRR are significantly improved in both cross-city scenarios. In contrast, there was no improvement for SC, as it does not have a significant causal influence.

Table 3. Ablation Study: Conditioning on Causally Sensitive Features

Scenario	Baltimore → Washington				Washington → Baltimore			
	Acc@1	Acc@5	Acc@10	MRR	Acc@1	Acc@5	Acc@10	MRR
Baseline (placeid only)	0.2653	0.3387	0.4489	0.3356	0.1200	0.3400	0.3800	0.2357
Conditioning on Spot Category (SC)	0.1235	0.2099	0.2716	0.1699	0.10	0.22	0.30	0.1885
Conditioning on Day of Week (W)	0.2653	0.3061	0.4285	0.3120	0.1400	0.2800	0.3200	0.2281
Conditioning on Start Minute **(Smin)**	0.3884	0.3265	4897	0.4897	0.2200	0.2800	0.3000	0.2600

6 Conclusion

We propose a novel cross-city next POI prediction framework, GCCP, that exploits generative learning and causal perspectives. At the onset, we identify the causal features considering the interventional analysis using the generative predictor by blocking the backdoor paths and using them to condition the CVAE during training. These features demonstrate enhancement in prediction accuracy, as validated on real-world LBSN data. We identify the W, and Smin, as causal features, a common causal influencer in the past POI visits and the next POI in a new city whereas the SC does not show significant causal influence. Notably, GCCP achieves the best performance in the Baltimore→Washington scenario and competitive results in the reverse direction, including the highest Acc@1 and second-best MRR. 95% confidence intervals confirm the consistency of this performance. The ablation study further presents the significance of causal influence on the cross-city next POI prediction task. The model's causal conditioning enhances interpretability by revealing which contextual factors affect the prediction behavior. The proposed method is generalizable and can be adapted as a framework for causality awareness in prediction tasks.

References

1. Bandyopadhyay, S., Sarkar, S.: Exploring causality aware data synthesis. In: Proceedings of ACM AIMLSystems (2023)
2. Ding, J., Yu, G., Li, Y., Jin, D., Gao, H.: Learning from Hometown and Current City: Cross-city POI Recommendation via Interest Drift and Transfer Learning IMWUT (2019)
3. Feng, J., et al.: Deepmove: predicting human mobility with attentional RNNs. In: WWW (2018)
4. Feng, S., et al.: Personalized ranking metric embedding for next poi recommendation. In: IJCAI (2015)
5. Fu, H., et al.: Cyclical annealing schedule for KL vanishing. In: NAACL-HLT (2019)

6. Kong, D., Wu, F.: Hierarchical spatial-temporal LSTM for location prediction. In: IJCAI (2018)
7. Li, Z., et al.: Uptdnet dataset (2022). https://github.com/s3pku/UPTDNet_dataset
8. Yang, T., Gao, Y., Huang, Z., Liu, Y.: UPTDNet: A user preference transfer and drift network for cross-city next poi recommendation. Int. J. Intell. Syst. 2023(1), 9091570 (2023)
9. Liu, Q., et al.: Predicting the next location with spatial and temporal contexts. In: AAAI (2016)
10. Louizos, C., et al.: Causal effect inference with deep latent-variable models. In: NeurIPS (2017)
11. Pearl, J.: Causality: Models, Reasoning and Inference. Cambridge University Press, Cambridge (2009)
12. Sun, K., et al.: Modeling long- and short-term preferences for poi recommendation. In: AAAI (2020)
13. Tang, J., Wang, K.: Personalized Top-n sequential recommendation via convolutional sequence embedding. In: WSDM (2018)
14. Vaswani, A., et al.: Attention is all you need. In: NeurIPS (2017)
15. Wang, S., et al.: Deep learning for spatio-temporal data mining: a survey. LNCS (2019)
16. Wu, Y., et al.: Personalized long- and short-term preference learning for poi recommendation. IEEE TKDE (2022)
17. Yang, D., et al.: Lbsn2vec++: heterogeneous hypergraph embedding for LBSNs. IEEE TKDE (2020)
18. Yang, D., Fankhauser, B., Rosso, P., Cudre-Mauroux, P.: Location prediction over sparse user mobility traces using RNNs: Flashback in hidden states. In: IJCAI (2020)
19. Yang, M., et al.: Causalvae: neural structural causal models. In: CVPR (2021)
20. Yang, S., et al.: Getnext: flow map enhanced transformer for poi recommendation. In: SIGIR (2022)
21. Zhang, S., et al.: Next item recommendation with self-attention. arXiv:1808.06414 (2018)
22. Zhang, Y., et al.: Sequential click prediction with RNNs. In: AAAI (2014)
23. Zhu, Y., et al.: Modeling user behaviors by time-LSTM. In: IJCAI (2017)

Sadrusha: An Administrative Data Twin Framework for SDG-Aligned Decision Making

Apurva Kulkarni[(✉)], Riya Patidar, and Srinath Srinivasa

International Institute of Information Technology Bangalore,
26/C, Electronics City Phase 1, Bangalore, Karnataka, India
{apurva.kulkarni,riya.patidar,sri}@iiitb.ac.in

Abstract. Effective policy implementation requires a comprehensive understanding of the prevailing conditions within the administrative regions where the policy is deployed, as well as a clear evaluation of its impact on the corresponding Sustainable Development Goals (SDGs). To support this, data systems must be equipped to identify, integrate, and analyze data points specific to each administrative level, including states, districts, and taluks, along with indicators associated with the relevant SDGs. However, existing data systems often fail to establish these linkages effectively, limiting their ability to support integrated policy analysis and decision-making. This research presents *Sadrusha*, an innovative decision-analytical system developed to address critical gaps in policy evaluation and data integration. Derived from the Sanskrit word meaning "affinity" or "similar behavior". The proposed approach integrates open government datasets, geospatial shapefiles, and policy documents to construct an *Administrative Data Twin*—a sophisticated data model that links Sustainable Development Goals (SDGs) with geographic and temporal dimensions. By automatically identifying and contextualizing relevant data based on SDG targets, administrative boundaries, and timeframes, *Sadrusha* enables granular, cross-domain policy analysis. Utilizing five years of data from the "Karnataka at a Glance" (KAG) repository, the system has been validated against 40 policy documents to assess its effectiveness in supporting dynamic policy evaluation. The proposed framework offers a scalable, data-driven approach for enhancing policy coherence and generating actionable insights aligned with the SDGs.

Keywords: Data Twin · Sustainable Development Goals(SDG) · Semantic Mapping · Open Government Data(OGD)

1 Introduction

In recent decades, the industrial and technological landscapes have undergone transformative shifts driven by automation. Automated systems are playing

R. Gupta et al. (Eds.): BDA 2025, LNCS 16041, pp. 338–347, 2026.
https://doi.org/10.1007/978-3-032-15134-6_25

a major role in industries by transferring human tasks to machines, thereby increasing efficiency, consistency, and scalability in various processes. However, complex problems require deeper insights into their challenges, and traditional automation systems have limitations as they are rule-based and reactive, executing pre-programmed responses to events. The concept of Digital Twin was coined to bridge the physical world and the digital world.

The concept of Digital Twin (DT) traces its origins to Product Lifecycle Management (PLM) in 2002, when Michael Grieves introduced the "Mirrored Spaces Model" at the University of Michigan, which involved real space, virtual space, and a linking mechanism to connect the two. Similar ideas had been explored earlier, such as David Gelernter's "Mirror Worlds" in 1991, which envisioned software models mimicking reality. The term "Digital Twin" appeared in NASA's 2010 roadmap, where DT was used to simulate aircraft systems. Over time, DTs were adopted in aerospace and defense, while related terms like "Digital Thread" and "Product Avatar" emerged, adding to the complexity and varied interpretations of DT within different industries.

Cities worldwide are increasingly implementing digital twins to transform urban management. Amsterdam's 'Digital City' is dedicated to optimizing traffic, energy, and sustainability through real-time data integration and 3D modeling. Dubai's 'Digital Twin of Dubai' focuses on urban planning, infrastructure management, and emergency response by creating a highly detailed virtual model of the city[1]. Helsinki's 'Digital Twin of Helsinki' emphasizes sustainable urban development, mobility solutions, and environmental initiatives with its dynamic city model[2]. London's 'Smart London Plan' integrates digital twins for enhanced city services, transportation, and environmental monitoring, including critical infrastructure projects like Crossrail. San Diego's digital twin aims to enhance traffic management, public safety, and sustainable urban planning, contributing to its broader 'Smart City' efforts. Moscow's 'Moscow Digital Twin' focuses on creating a detailed city model supporting urban planning, transportation management, and public services. New York City's 'Digital NYC' leverages digital twins to enhance city planning, climate resilience, and infrastructure management. These initiatives employ a mix of real-time data, 3D modeling, and simulation to drive efficient, data-driven decision-making, bolstering city services, sustainability, and enhancing urban infrastructure.

A digital twin is a virtual replica of a real-world entity that encompasses all its properties. Thus, it can be used for simulation, real-time monitoring, analytics, and optimisation. However, several critical gaps remain in the practical implementation of a Digital Twin:1) it is tailored for a specific problem and domain like manufacturing, aerospace and production science, production systems, 2) is dependent on other evolving technologies, 3) it requires large amount of data of the physical object and requires IoT devices for real-time monitoring, 4) may or may not require human intervention in decision making 5) cost and the number of available resources could also contribute to increase this implementation gap.

---

[1] https://storymaps.arcgis.com/stories/4ae54ecb2d8e4640b491de1fc319cffc.
[2] https://www.hel.fi/en.

Data Twin

In this research, the focus is primarily on extracting relevant data from a physical system and constructing a multidimensional digital representation that manages data in a structured format and maintains the semantic relationships among data points. We refer to this concept as - *Data Twin*.

In the context of public administration, a *Data Twin* of a country or state is extended by aggregating data from multiple government departments and constructing a descriptive model that encapsulates their interrelations. This digital model empowers policymakers to analyze trends, identify cross-sectoral linkages, and make informed decisions based on a holistic view of the system's behavior and its alignment with policy objectives. Through this research, we propose the idea of *Administrative Data Twin* to aid policy and decision makers to gain insights into the governed system through Data Twin to achieve the SDG localization.

Administrative Data Twin

The *Administrative Data Twin* (ADT) serves as a powerful tool for simulating alternative scenarios and forecasting potential outcomes, thereby supporting more informed and strategic decision-making by policymakers. In this paper, we describe the implementation of the Administrative Data Twin of Karnataka Government data, which contains the statistics of various departments. We refer to the Administrative Data Twin as *Sadrusha*[3], which is derived from the Sanskrit word meaning "affinity" or "similar behavior". The implementation of the current model can be accessed at https://sadrusha.iiitb.ac.in/. The current implementation covers five years' worth of Karnataka At a Glance (KAG) dataset[4] covering 30 districts, approximately 1200 data points across 12 departments, including education, health, agriculture, etc. The next section illustrates the process of developing a data twin. Section 3 discusses the validation of the proposed approach considering various use cases. The manuscript is concluded by discussing the downstream activities that leverage the *Sadrusha* as the foundational layer (Figs. 1, 2 and 3.

2   Administrative Data Twin

The concept of a data twin represents a sophisticated approach to policy-making and implementation, leveraging real-world data to informed decision-making processes. This section explores the intricate workflow of a data twin, encompassing both static and dynamic stages that facilitate the interaction between data-driven analysis and real-world applications. The diagram illustrates the concept of a *Data Twin*, outlining a two-stage workflow: static and dynamic. The *Data Twin* interacts with the real world in two key ways: first, by collecting essential data points for ingestion and constructing a data repository; and second, by implementing a policy in the real world, which initiates another cycle of data ingestion and analysis [10, 11]. The process is iterative, continuously refining the data and enhancing the system's responsiveness to real-world conditions.

---

[3] https://sadrusha.iiitb.ac.in/.
[4] https://kgis.ksrsac.in/kag/.

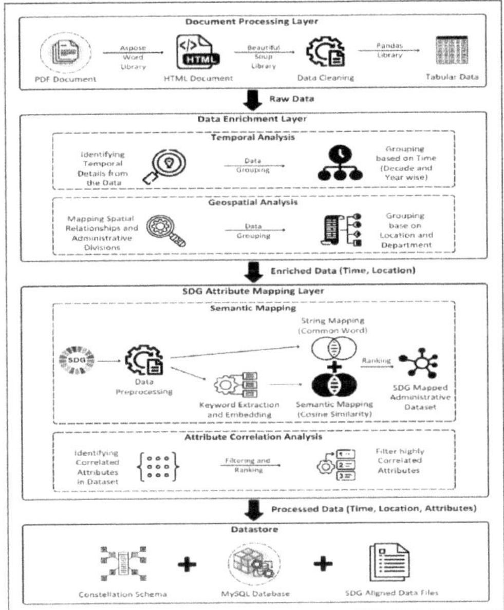

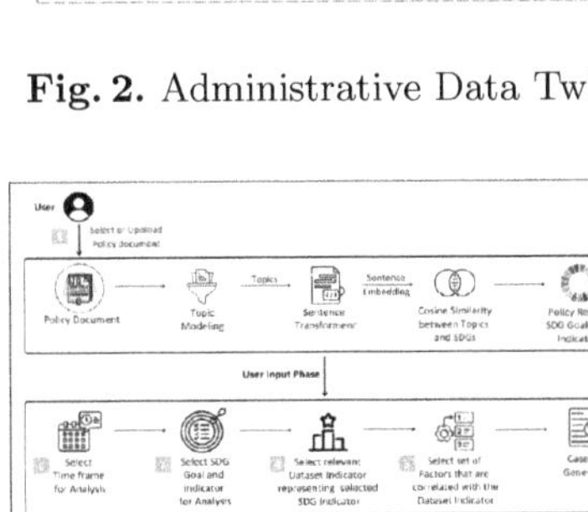

Fig. 2. Administrative Data Twin.

Fig. 1. Static Workflow: Creating an enriched data store from heterogeneous Open Government Data (OGD) sources.

Fig. 3. Dynamic Workflow: Accessing the data store to retrieve the relevant data points for analysis.

The process begins by obtaining data from the real world and concludes with the implementation of new or updated policies/regulations after conducting effective impact analysis. By integrating open government data, advanced analytical tools, and real-world feedback, this approach enables more informed, effective, and responsive governance. The cyclical nature of the workflow ensures continuous improvement and adaptation of policies based on empirical evidence and changing societal needs.

2.1 Static Layer

The static workflow is divided into several layers, as shown in the diagram. The *Sadrusha* project, which implements the proposed Administrative data twin concept, utilizes Karnataka At A Glance data, publicly accessible on the Karnataka Government website in PDF and Excel formats.

A. Document Processing Layer: This layer takes PDF documents containing tabular data from various Karnataka departments as input. The documents are converted to HTML format, and data is extracted and transformed into a tabular structure using Python libraries such as Aspose Word, Beautiful Soup, and Pandas. The output is a raw dataframe containing information from the ingested PDF document.

B. Database Enrichment Layer: This layer processes the raw data frame from the first layer, performing grouping and analysis to identify temporal details, spatial relationships, and administrative divisions within the data. This implementation leverages the shape files[5] and builds the ontology over the Karnataka Administrative region (Districts, Taluks, Villages), relationships, and captures information such as the administrative parent-child relationships (district-taluk, taluk-villages, etc.) and neighboring information.

C. SDG Attribute Mapping Layer: This layer uses time and location-enriched data for two main tasks: **C.1) Semantic Mapping**: The enriched data and SDG data undergo preprocessing, including data cleaning, lemmatization, and removal of numerals, special characters, and stop words. Two mapping techniques are used: (1) common string-based mapping, which filters attributes based on shared words and ranks them by common word count, and (2) semantics-based mapping, which uses pre-trained GloVe embeddings and the KeyBERT package to convert SDG keywords into vectors and assess similarity with dataset attribute descriptions using cosine similarity. The resulting mappings are categorized and ranked for relevance, prioritizing attributes most closely aligned with SDG levels across both mapping methods. **C.2) Attribute Correlation Analysis**: This process identifies correlated attributes in the dataset using a correlation matrix. For each attribute, highly correlated attributes are filtered from the matrix. Attributes with correlation scores above 0.75 and below 0.55 are selected for each attribute in the dataset.

D. Datastore: The SDG-aligned dataset from the previous layer is ingested into the database using a constellation schema, along with SDG-aligned data files. In SDG-aligned multidimensional cubes and complex real-time data environments, OLAP (Online Analytical Processing) queries with three dimensions - time, location, and attributes - are crucial for conducting thorough and insightful analysis. To validate the data warehouse implementation, test cases were evaluated based on expected outputs, data counts, and validation queries involving aggregations, filters, and joins. These queries, applied to the KAG and SDG datasets and their mappings, confirmed that the retrieved results aligned with expectations. The validation demonstrated the data warehouse's effectiveness in integrating relevant data and ensuring accurate mappings between KAG and SDG sources, enabling seamless cross-domain analysis. Implementation details are discussed in the manuscript at [12].

2.2   Dynamic Layer

The dynamic workflow addresses real-world challenges and policies that government officials or researchers aim to analyze. This process incorporates user involvement in selecting attributes for examination. The diagram depicts the various stages of the dynamic workflow. Initially, the user submits a policy document from the real world to the Policy-Based Data Analysis system. The system processes the document and SDG data, beginning with data cleansing. It then

---

[5] https://kgis.ksrsac.in/kgis/downloads.aspx.

employs the Latent Semantic Indexing (LSI) topic modeling algorithm with a Bag of Words (BOW) approach to dynamically extract topics from the document. The extracted topics and SDG data are then fed into a sentence transformer to generate sentence embeddings. Cosine similarity is applied to these embeddings to create a similarity matrix between policy document topics and SDGs. This method produces a ranked list of relevant SDG goals, targets, and indicators, along with a similarity score indicating their alignment with the policy, which is utilized in the User Input Phase.

A) User Input phase - During this stage, users select the timeframe and attributes they wish to analyze. After uploading the policy document, a dynamic mapping between the document and SDGs occurs, prompting the user to choose the year and SDG for analysis. Upon selecting an SDG, the user is asked to pick an SDG indicator for review. The system then presents the user with top-ranked dataset indicators to choose from, which are the result of semantic mapping between SDGs and datasets performed during the static workflow stage in the SDG Attribute Mapping Layer. Once a dataset indicator is selected, the system displays correlated factors (dataset attributes obtained in the SDG Attribute Mapping Layer through correlation analysis). Users can choose multiple factors they deem relevant and worth examining.

B) Correlation Analysis - To understand the relationship among the data points, a range of analytical techniques were explored, beginning with Pearson correlation [5] and R-squared regression, which showed limited predictive value ($R^2 < 0.35$). We further applied Lasso regression with L1 regularization for variable selection in high-dimensional feature spaces [16] and used Variance Thresholding to remove low-variance indicators. While effective in dimensionality reduction, the latter posed a risk of discarding stable yet meaningful features. Beyond linear models, a suite of advanced causal and interpretability-driven methods was explored. The other models, like Instrumental Variables (IV) regression via two-stage least squares (2SLS) and over-identification testing, were employed to mitigate unobserved confounding [1], although their success depended heavily on domain-specific knowledge. We also utilized Bayesian Inference through PyMC to estimate high-density intervals (HDIs) [15], Local Surrogate Models (LSMs) for localized interpretability [4], and Explanation-Based Learning via SHAP [13] and LIME to verify reliance on semantically valid SDG indicators. Do-Calculus, though theoretically sound for causal inference [14], required predefined causal graphs and domain modeling.

The final and most effective approach combined Mutual Information (MI) and Conditional Entropy (CE) to identify meaningful associations while filtering out spurious correlations [6]. MI quantified the overall strength of relationships, with values above 0.1 indicating a strong association. CE was used to detect confounding by measuring the drop in MI when conditioning on a third variable. An MI greater than 0.05 flagged the relationship as potentially spurious. This method captured non-linear dependencies, worked without temporal data, and retained statistical interpretability. The non-spurious relationships are further considered for the analysis in the output layer. The proposed method is evaluated against 11 SDG-aligned data stories from the Karnataka Data Lake.

To validate the effectiveness of the identified non-spurious correlation mechanism, we employed two complementary approaches. The first involved manual validation, where domain experts were consulted to assess the relevance of the findings. For instance, in the case of student dropout analysis within the education sector, our approach achieved 75% accuracy, as confirmed by expert judgment. The second method was an automated, human-out-of-the-loop validation, wherein we semantically matched the identified non-spurious data points with evidence from diverse sources such as news reports, academic articles, and government publications. This approach yielded an 80% accuracy, underscoring the alignment of our findings with real-world narratives. These validation results collectively demonstrate the robustness of the MI/CE pipeline in uncovering actionable, non-spurious socio-economic insights.

C) Output analysis phase - Following the creation of the case file, users have access to all the dataset attributes they wish to analyze in relation to the chosen SDG and indicator for the uploaded policy document. A database query is then formulated, incorporating the selected attributes, time period, and location to extract data from the multidimensional datastore. This query yields two downloadable CSV files: one containing attribute descriptions and another with quantitative data. Users can utilize these files to assess policy impact over a specific timeframe across various administrative regions.

The case file generated from the Query Processing Layer supports data-driven policy formulation through Intervention Modelling [2]. This process enables policymakers to assess how regional variations in specific factors influence Sustainable Development Goal (SDG) indicators. Using a multivariate linear regression framework, the system estimates the impact of modifying individual factors, such as economic or environmental variables on the indicator values across different geographic units. By simulating percentage changes in these factors, it predicts how the indicator would respond, helping identify the most influential variables. In addition, the system incorporates a prescriptive modelling approach that determines the necessary adjustments to reach a desired target value for the indicator. This involves computing new values for key factors based on their estimated influence and associated sensitivity, allowing policymakers to design targeted interventions.

3   Usecase-MNREGA Scheme

The Administrative Data Twin (ADT) presents notable benefits for policy assessment and implementation, as illustrated in the MGNREGA case study. The Mahatma Gandhi National Rural Employment Guarantee Act (MGNREGA), a key initiative of the Indian government, was introduced in 2006 to bolster livelihood security in rural regions. A notable case study, titled "Performance Analysis of MGNREGA in Karnataka," was conducted in 2023–24 at the Fiscal Policy Institute [7]. This study serves as a reference for analyzing the policy using a case

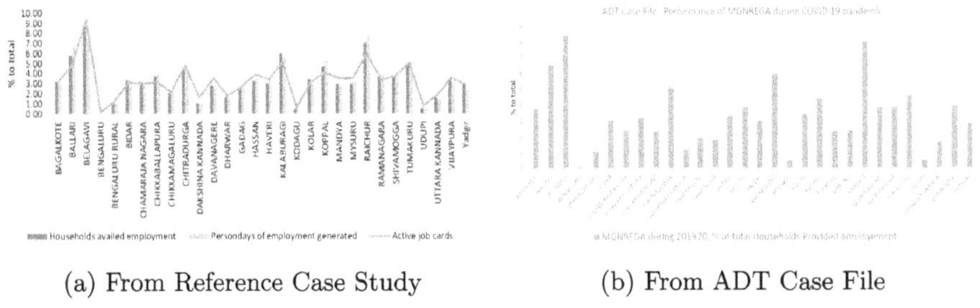

(a) From Reference Case Study   (b) From ADT Case File

Fig. 4. Performance of MGNREGA during COVID-19 pandemic.

file generated through the proposed Administrative Data Twin workflow. The referenced study provides an empirical examination of the MGNREGA scheme's performance across Karnataka's districts. The policy analysis focuses on three main objectives: a) examining trends and patterns of public expenditure on the MGNREGA program in Karnataka; b) evaluating MGNREGA's performance across Karnataka's districts; and c) offering evidence-based recommendations to improve the impact of public spending on the MGNREGA program in Karnataka.

A comparable policy evaluation was performed using the proposed Administrative data twin, with the MGNREGA policy document serving as input. The system then identified the most applicable SDG from five options, selecting "SDG 8: Promote sustained, inclusive and sustainable economic growth, full and productive employment and decent work for all". Subsequently, various case files were generated based on the chosen SDG Indicator. For this analysis, "Indicator 8.1.1: Annual Growth Rate of real GDP per capita" was selected to obtain district-wise data for Karnataka related to MGNREGA. The acquired data was examined to determine trends and rankings of different districts, mirroring the approach in the referenced report. Below are two examples where the trends and outcomes from the Administrative Data twin-generated case file aligned with those presented in the cited case study. *(Note: The detailed policy-based workflow can be accessed at* https://sadrusha.iiitb.ac.in/policybased-intro-page/*).*

3.1 Household Employment

The case study reveals that during the COVID-19 pandemic, Belagavi, Raichur, Koppal, and Kalaburagi districts had the highest percentage of households that utilized employment, produced the most person-days of work, and maintained a greater number of active job cards. Analysis of the ADT case file data for 2019–20 showed a similar pattern, demonstrating that reliable policy-related information can be directly obtained through ADT. This streamlines the process of gathering accurate and dependable data, eliminating the need to search through various government and non-government websites. The Fig. 4a displays the graph used in the referenced case study during COVID-19, while the graph present in Fig. 4b is derived from the ADT case file data. The MGNREGA program also aims to empower socially disadvantaged SC/ST communities and women. This trend is clearly reflected across data sources in the ADT case file.

The Administrative Data Twin significantly enhances the efficiency and effectiveness of policy analysis and execution by providing a reliable, comprehensive, and easily accessible data source tailored to specific policy needs and aligned with global development goals.

4   Downstream Activities

Once relevant data is ingested and structured within the system, a range of downstream analytical activities can be conducted to extract meaningful insights and support policy planning. The structured way of managing data provides easy means to create dashboards and visualize data. The Karnataka Data Lake leverages [8] *Sadrusha* for creating data stories across SDGs[6]. The *Sadrusha* facilitates policy makers to upload the policy reports and identify the relevant data points for understanding the impact of the policy across SDGs [9]. In the Policy Support System[7], the primary goal is to aid policymakers in evaluating and formulating effective, evidence-based policies [3,9]. These downstream activities collectively enhance the utility of the data ecosystem, enabling policymakers to shift from reactive responses to informed, forward-looking, and collaborative governance.

5   Conclusion

The proposed research *Sadrusha* provides a framework for effective policy implementation which addresses the lack of integrated, localized data frameworks that align with Sustainable Development Goals (SDGs). By constructing an Administrative Data Twin, the system bridges critical gaps in spatial, temporal, and social data integration across administrative units such as states, districts, and taluks. The proposed approach is validated through real-world data and policy documents by considering human-in-the-loop (75% accuracy) and human-out-of-the-loop (80% accuracy). *Sadrusha* offers a scalable, dynamic, and real-world replica platform to understand the intrinsic relationship among data. As a result, it sets a new benchmark for data-driven governance and informed public policy design.

Acknowledgment. The authors thank the Government of Karnataka and DataWeave, Bangalore, for their support.

---

[6] https://kdl.iiitb.ac.in/data-stories-new/.

[7] https://wsl.iiitb.ac.in/policy-support-system/.

References

1. Angrist, J.D., Krueger, A.B.: Identification of causal effects using instrumental variables. J. Am. Stat. Assoc. **91**(434), 444–455 (1996)
2. Bassin, P., G K, A., Srinivasa, S.: Design of a data-driven intervention dashboard for sdg localization. In: Proceedings of the Thirty-Third International Joint Conference on Artificial Intelligence. pp. 8606–8609 (2024). https://doi.org/10.24963/ijcai.2024/990
3. Bassin, P., Parasa, N.S., Srinivasa, S., Mandyam, S.: Big data management for policy support in sustainable development. In: Sachdeva, S., Watanobe, Y., Bhalla, S. (eds.) Big-Data-Analytics in Astronomy, Science, and Engineering. International Conference on Big Data Analytics. Lecture Notes in Computer Science, vol 13167, pp. 3–15, Springer, Cham (2021). https://doi.org/10.1007/978-3-030-96600-3_1
4. Caruana, R., Lou, Y., Gehrke, J., Koch, P., Sturm, M., Elhadad, N.: Intelligible models for healthcare: Predicting pneumonia risk and hospital 30-day readmission. In: Proceedings of the 21th ACM SIGKDD International Conference on Knowledge Discovery and Data Mining, pp. 1721–1730 (2015)
5. Chicco, D., Jurman, G.: Pitfalls of r-squared for model evaluation. J. Mol. Cell Biol. **12**(11), 892–895 (2020)
6. Cover, T.M., Thomas, J.A.: Entropy, mutual information and conditional entropy. Elements of Information Theory (1991)
7. Gunabhagya, D.: Performance analysis of mgnrega in karnataka. Research Report No. 4/2023-24, Fiscal Policy Institute, Bengaluru, Karnataka (2023). https://fpibengaluru.karnataka.gov.in/storage/pdf-files/Technical
8. Kulkarni, A., Bassin, P., Parasa, N.S., Venugopal, V.E., Srinivasa, S., Ramanathan, C.: Ontology augmented data lake system for policy support. In: International Conference on Big Data Analytics. pp. 3–16. Springer, Cham (2022). https://doi.org/10.1007/978-3-031-28350-5_1
9. Kulkarni, A., Ramadurg, I.A.K., Srinivasa, S., Patil, S.: Introspecting policy documents through semantic lenses of sustainable development goals. In: Sachdeva, S., Watanobe, Y., Bhalla, S. (eds.) Big Data Analytics in Astronomy, Science, and Engineering, pp. 95–105. Springer Nature Switzerland, Cham (2025)
10. Kulkarni, A., Ramanathan, C., Venugopal, V.E.: Cognitive retrieve: Empowering document retrieval with semantics and domain specific knowledge graph. In: EKG-LLM@ CIKM (2023)
11. Kulkarni, A., Ramanathan, C., Venugopal, V.E.: Ontology mediated document retrieval for exploratory big data analytics. In: 2023 IEEE 17th International Conference on Semantic Computing (ICSC). pp. 100–103 (2023).https://doi.org/10.1109/ICSC56153.2023.00022
12. Kulkarni, A., Srinivasa, S., Patil, S.: Sdg aligned data warehouse implementation over open government data. In: International Conference on Electronic Government and the Information Systems Perspective. pp. 154–167. Springer, Cham (2024). https://doi.org/10.1007/978-3-031-68211-7_13
13. Lundberg, S.M., Lee, S.I.: A unified approach to interpreting model predictions. Adv. Neu. Inf. Process. Sys. **30** (2017)
14. Pearl, J.: Causality. Cambridge University Press (2009)
15. Salvatier, J., Wiecki, T.V., Fonnesbeck, C.: PyMC Documentation (2016). https://www.pymc.io
16. Tibshirani, R.: Regression shrinkage and selection via the lasso. J. Roy. Stat. Soc.: Ser. B (Methodol.) **58**(1), 267–288 (1996)

Extensible Test Automation for Functional Testing of Low-Code ETL Workflows

Afnan Ahmad[1], Preetodeep Dev[1], Arup Kumar Chattopadhyay[1(✉)] ,
and Meenakshi D'Souza[2]

[1] IITM BS Degree Program, Indian Institute of Technology Madras, Chennai, India
{afnan,arup}@study.iitm.ac.in, papadelta@alumni.iitm.ac.in
[2] International Institute of Information Technology Bangalore, Bengaluru, India
meenakshi@iiitb.ac.in

Abstract. ETL is a data integration process that combines data from various sources according to the desired requirements, which is then loaded into a data warehouse. Several no-code/low-code ETL platforms are available for creating and customizing ETL processes. Due to the complexities involved, particularly in the data transformation stage, it is necessary to incorporate tests in the ETL workflow to ensure that target data are accurate, consistent, and adhere to the data quality standards of an organization. In this paper, we propose an extensible ETL testing framework for functional testing of ETL workflows, particularly for no/low-code ETL data pipelines. We also implement a prototype to demonstrate functional testing of data pipelines on the CDAP open-source framework. Our prototype has been validated on four different ETL pipelines, modelling a variety of data sources and properties.

Keywords: ETL Processes · Data Warehouse · Functional Testing · Test Automation · No/Low-Code

1   Introduction

Extract, Transform, Load (ETL) is a well-practiced way of extracting structured and unstructured data from multiple sources, transforming it using well-defined process steps and then loading it into a data warehouse for further use. There are several service providers for ETL tools and many users in various business domains [32]. Since the past few years, no-code/low-code ETL platforms are available to facilitate creation and use of ETL workflows by business users for various applications and use cases [28,33]. These platforms provide several features for business users to develop ETL pipelines without writing any code or whenever necessary, provide basic scripts for the transformation.

Workflow testing, data validation, and the quality of target data remain challenging aspects in ETL implementations across various domains [23,25,31,

This work was partially supported by CTRI-DG, IIIT-Bangalore.

32]. Open-source frameworks (Sect. 2) that support low-code development of ETL pipelines often lack built-in features for data validation and quality tests. However, a number of commercial platforms, including but not limited to Azure Data Factory [27], Informatica [19], and QuerySurge [21] offer such features.

In this paper, we propose a set of functional testing plugins for ETL workflows, particularly for data pipelines that are developed using low-code ETL tools. Our functional testing framework supports the use of assertions (a basic feature of any unit testing framework), test fixtures, and mutation testing to make interesting changes to the underlying data. Furthermore, we implement a prototype to demonstrate functional testing of data pipelines on the open-source CDAP framework [16]. Our prototype has been validated by performing interesting tests on four internally developed ETL pipelines, spanning datasets from within India and the popular Our World in Data (OWID) platform [7]. Our framework provides a general and extensible low-code ETL testing framework.

2 ETL and No/Low-Code ETL Development Platforms

ETL is a collection of stages that constitute a data integration process. A typical ETL process involves extraction of data from one or more sources, applying transformations on it as per the desired requirements, and loading it into a data warehouse. A data warehouse is specifically designed to store and manipulate data efficiently to facilitate complex queries for data analysis. Thus, the ETL process is an essential part of applications that involve large-scale data processing and analysis, such as for business intelligence, machine learning, and IoT [25, 32].

The structure of an ETL process is shown in Fig. 1a. Data from multiple sources are transformed using a well-defined set of syntactic rules that are applied in a step-by-step fashion, and loaded into a sink which is a data store. Typically, the computation steps of the transformation process are represented as a Directed Acyclic Graph (DAG)—each node in the DAG is a transformation step, and edges connect one transformation step into another. There are no cycles in the transformation steps as they do not lead to any desired outcomes in the data transformation.

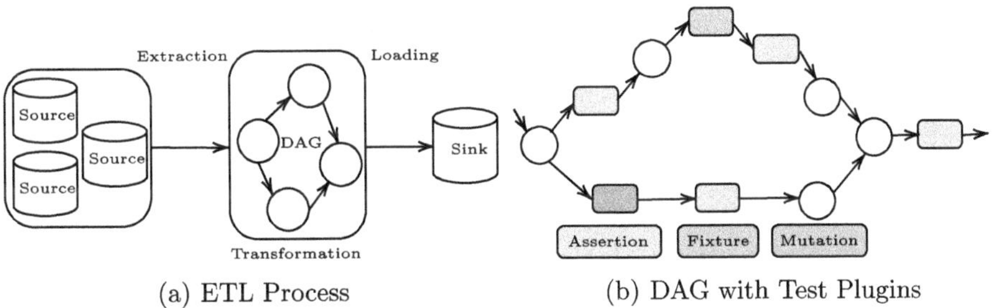

(a) ETL Process   (b) DAG with Test Plugins

Fig. 1. ETL Process and Transformation Stage.

Commercial and Open Source ETL Platforms. There are several commercial platforms available for end-to-end ETL workflow development, including Informatica [19,25], IBM DataStage [18], Oracle Data Integrator [4], Google Cloud Dataflow [17], Azure Data Factory [3], and AWS Glue [12]. Open-source ETL platforms include Apache Spark [15], Apache Beam [13], Apache Airflow [2], Apache NiFi [14], and Cask Data Application Platform (CDAP) [16]. All these platforms provide features for extracting data, computation steps for transforming data, and subsequently loading data into a data warehouse for deriving business intelligence. They are used by applications across several business domains, they cater to ETL processes being designed and used by non-technical users too.

Supported Features for Low-Code Pipeline Development. A subset of the ETL platforms discussed above provide features for developing end-to-end ETL data pipelines with minimal programming, using visual interfaces with drag-and-drop elements. These platforms make the otherwise complex ETL workflow development easier and usable by professionals from different business domains.

Most of the platforms provide support for validating the ETL workflow that has been defined using syntactic checks and some elementary testing techniques. Features to test the functionality of a workflow across the transformation steps are lacking but are critical to ensure the correctness of the ETL processes. Our work fills this gap partially, by developing a test automation framework for testing low-code ETL workflows through the use of assertions, fixtures and the well-known mutation testing.

3   Related Work

Chen et al. [22], Golfarelli and Rizzi [24], and Kim et al. [26] have provided insights on the evolution of ETL and related aspects, such as business intelligence, data warehousing and aspects of data science such as skills, tool usage, challenges, and best practices. Data cleaning and the associated challenges have been explained by Rahm and Do in [30]. Vassiliadis et al. [35] and Vassiliadis [34] have presented detailed insights into ETL processes and associated tools, particularly regarding early work in these fields. The recent study by Patel and Patel [29] provides a literature survey on the progressive growth of ETL tools and their features.

For real-world data and the associated pipelines, Homayouni et al. [25] provide another comprehensive overview highlighting the challenges to be addressed for functional testing of ETL. Our work is motivated by the problems highlighted in this survey, and we address several testing challenges related to data cleaning as mentioned therein.

Recently, various studies [23,25,32] have highlighted the challenges associated with ETL testing, particularly on functional testing. Based on these, we have identified the following key considerations for the proposed plugins:

Need for Domain-Specific Tests. Most test assertions that are useful for data quality validation are based on rules that are specific to the business domain of an ETL data pipeline. Therefore, general-purpose testing tools allow users to write custom validation rules [25].

Effectiveness of Test Assertions. Homayouni et al. [25] have noted, "the effectiveness of test assertions in detecting faults is not evaluated by current approaches". To this end, the proposed framework facilitates the generation of mutants to enable mutation testing. A number of data samples could be mutated to represent invalid values. Subsequently, assertions in the pipeline can be used to detect these and a mutation score could be computed to measure their effectiveness.

Generalization Across Platforms. Use of common constructs such as assertions and fixtures ensures that the design remains adaptable for implementing plugins for any ETL development platform, as long as the platform supports custom extensions/plugins, and offers appropriate APIs for development. Also, we used generic expressions (predicates) for specifying assertions, ensuring that they can extend across platforms with minimal effort.

4 Low-Code ETL Testing Framework

Our proposed framework focuses on the transformation stage of ETL processes. The transformation stage in an ETL process is modeled as a Directed Acyclic Graph (DAG), as shown in Fig. 1b. Each node in this DAG represents one transform that is applied to the records in this process. Typical transformation steps involve changing (a part of) data in some way or the other, starting from simple re-labeling, re-naming, dropping parts of the data, to slightly more complex steps that execute scripts which transform data.

The proposed framework consists of three plugins for the transformation stage, namely, *Assertion*, *Fixture*, and *Mutation*. These plugins can be added as transform steps, thus facilitating data validation and quality tests within the existing low-code ETL data pipeline. This implementation ensures that our proposed framework remains platform-agnostic. By adding them as a part of the transformation stage, we retain the basic feature of the platform continuing to be low-code while providing useful testing features to the user. The given ETL workflow is not altered in the process of inserting the test plug-ins, ensuring the same underlying ETL pipeline functionality. The framework has been prototyped on the CDAP open-source platform. The source code for our prototype is available at https://github.com/easytest-etl/prototype.

4.1 Assertions

Algorithm 1 describes the general steps of an assertion transform. The assertion transform evaluates the records that pass through it, i.e. those that reach this node in the DAG, based on a rule expression. Records that satisfy the rule expression are passed to the next step without any change; otherwise, the assertion fails, and an exception is thrown. The rule for the assertion can be specified using propositional logic predicates. Furthermore, these expressions can also access contextual information set by fixtures that precede the assertion. For the prototype on CDAP, we have used the Java Expression Language (JEXL) library from Apache Commons [20] to implement the rule expressions.

The *initialize()* function is called when the pipeline starts. During initialization, the name of the assertion is obtained from plugin properties, and we query the shared map for this assertion to check whether this assertion is enabled, or has been disabled by a preceding fixture, and the *assertionEnabled* flag is set accordingly. Then, the rule expression is validated for syntactic correctness. During execution of the pipeline, the *transform(record)* function is called with each record that passes through this step, and if the assertion is enabled, the rule expression is evaluated. If the record does not satisfy the rule, an exception is thrown; otherwise, the record is passed to the next step. Finally, when the pipeline execution finishes, the *dispose()* function is called and the entries for this assertion in the shared map are cleared.

4.2   Fixtures

Fixture transforms provide functionality for sharing key-value pairs and other data objects between multiple assertions in the pipeline. Furthermore, assertions can be controlled dynamically at runtime through fixtures. Both assertions and fixtures access a shared map to facilitate this. The shared map can either be implemented within the plugin itself using a `Map` data structure or use an external in-memory storage backend. Moreover, this mechanism could be utilized to orchestrate the testing of different pipelines across the ETL workflow. The working of the fixture transform is described in Algorithm 2.

The *initialize()* function is called when the pipeline starts. During initialization, a unique identifier for this fixture is obtained using the fixture name that is set in plugin properties. The *setup()* function is called only once, i.e. when the first record reaches this step in the DAG. This ensures that the fixture values are set up only if this step is reachable. Finally, when the pipeline finishes execution, the entries set by this fixture in the shared map are cleared.

4.3   Mutation

This plugin is used for mutation testing (fuzz testing) on the ETL pipeline. Several standard mutation operators have been adapted to cater to changes in data in an ETL pipeline, and implemented in the prototype to test the effectiveness of the assertions. A working of the mutation plugin is shown in Algorithm 3 using uniform field weights. These operators mutate records in various ways depending on the data type of the fields. For example, string values could be truncated or set to null. Similarly, integer values could be mutated to zero or negative, or very small or large values (outliers). Floating-point numbers could be mutated to high precision decimals to check for rounding errors, for instance.

In the prototype, the plugin facilitates fuzz testing based on a number of random samples. The sample selection follows a Bernoulli distribution and the

```
assertionEnabled ← False;
assertionName ← getAssertionName();
function initialize is
    assertionMapId ← getSharedMapId(assertionName);
    assertionEnabled ← SharedMap.getBoolean(assertionMapId, "enabled");
    expression ← getRuleExpression();
    expression.validate();
end_function
function transform(record) is
    if assertionEnabled is False then
        return record;
    end
    if evaluate(expression, record, assertionMapId) is False then
        FailPipeline with AssertionException;
    end
    return record;
end_function
function dispose is
    SharedMap.clear(assertionMapId);
end_function
```
Algorithm 1: Assertion

probability of selection can be configured from the plugin properties. Furthermore, weight values could be assigned to the fields selected for mutation. In samples selected for mutation, fields with higher mutation weight are mutated more often than those with lower mutation weight, thereby enabling frequent mutations on fields that are more likely to be erroneous in the original data. The *initialize()* function is called when the pipeline starts. During initialization, we obtain the selection probability p and mutation weights for the fields that are set in plugin properties. Subsequently, we initialize a Bernoulli sampler instance with probability set to p. The *transform(record)* is called for each record that passes through this step. If the Bernoulli trial results in an affirmation, then the *mutate(record, weights)* function is called, which applies mutation operators to the fields according to their data type and mutation weight. Otherwise, the record is passed to the next step without any change.

4.4   Examples

We have implemented example ETL pipelines on CDAP using publicly available datasets for Railways, Weather, and Air Quality Index from within India and Fuel Consumption [11] and Monkeypox disease [1] from the global Our World in Data (OWID) datasets [7]. Table 1 gives a summary of all four datasets, including parameters like total size, record count, data frequency, etc. The datasets qualify several properties of big data, making ETL pipelines on them non-trivial, and motivating the need for functional testing of the pipelines. Integration of the proposed test plugins facilitate data validation and quality test as part of the

```
fixtureName ← getFixtureName();
function initialize is
    mapId ←
      getSharedMapId(fixtureName);
end_function
function setup is
    map ←
      SharedMap.getMap(mapId);
    foreach (key, value) in
      getFixtureValues() do
      | map.put(key, value);
    end
end_function
function dispose is
  | SharedMap.clear(mapId);
end_function
```
Algorithm 2: Fixture

```
function initialize is
    p ←
      getSelectionProbability();
    weights ←
      getDefaultWeights();
    dist ← Bernoulli(p);
end_function
function transform(record) is
    if dist.sample() is 1 then
      | return mutate(record,
      |   weights);
    end
  | return record;
end_function
```
Algorithm 3: Mutation

Table 1. Summary of Dataset Parameters

Dataset	Size	Records	Columns	Frequency	Geographical Spread
Railways	16 MB	1,86,125	12	N/A	8151 Indian Railway Stations (Pan-India)
Weather-AQI	9.4 MB	2,37,956	6	Daily	8 major Indian cities
Fuel-OWID	977 KB	21,812	131	Yearly	50+ countries across various continents
Monkeypox-OWID	6.6 MB	1,17,125	15	Daily	140 countries

low-code ETL data pipeline. These examples are carefully chosen to illustrate the usefulness of the functional testing features proposed in this paper. We describe two of the applications below, along with the test features as applicable.

Railways Data Pipeline Example. Railway datasets, such as schedule [8], operating statistics [9], and traffic density statistics [10] were collected for this pipeline. Our transformations are basic applications, such as distance-based/time-based fare calculations, number of days along the route for the train runs, etc. Corresponding to these transforms, we have implemented the following assertions for data validation and quality testing of the pipeline: (1) *Ensure Columns:* Perform some basic data validation checks, such as missing or null values on fields such as train name and station. (2) *Ensure Validity:* Validate numeric data on station sequence and distance fields. (3) *Check target spec:* Validate computed fields, such as the train fare. A rule expression in JEXL corresponding to an assertion for this dataset is: `input['Distance'] != null && input['Distance'] >= 0`. Such an assertion will hold if the value is not null and contains non-negative value for the `Distance` field. This assertion validates a simple but important property of the distance being non-null and non-negative, important for several queries related to railway routes.

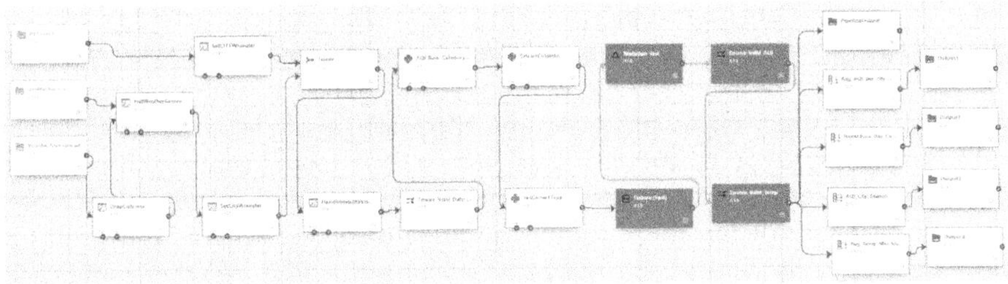

Fig. 2. ETL Workflow for Weather-AQI Dataset with Test Plugins.

Table 2. Summary of Pipelines and Plugins

Dataset	Assertion	Fixture	Mutation
Railways	Check for missing or null values, validity of computed values, station sequence, and distance	Add a key-value pair to indicate reachability of transform step	Not applicable
Weather-AQI	Check for missing or null values, validity of AQI range, dates, city, and temperature	Enable or disable test plugins individually based on the input data	Mutate AQI samples to negative values
Fuel-OWID	Nullity checks for ISO code, energy per capita, year, and logical checks for fuel consumption	Enable or disable test plugins individually based on the input data.	Remove some required fields, insert invalid ISO codes, mutate consumption share to negative values
Monkeypox-OWID	Nullity checks for ISO code, verify name-code mapping for countries, and validate date for seasonal grouping	Set correct name-code mapping for countries	Mutate values to zero, drop values from some fields.

Weather and AQI Data Pipeline Example. Weather (rainfall and temperature) data [6] and Air Quality Index (AQI) data [5] for major Indian cities were collected for this pipeline. Our transformations integrate AQI and weather data to produce various aggregates, such as seasonal average AQI, average yearly AQI across cities, count of annual hazardous AQI days across cities, average monthly rainfall and temperature over the past eight years, etc. Our test plugins have been used for this example as follows: (1) Assertion plugins have been used to identify missing/invalid data. Primarily, nullity checks were performed on AQI, dates, and city columns, such as `input['DATE'] !="" && input['CITY']!= ""` and `input['AQI'] != null && (input['AQI']>=0 && input['AQI']<=500)` and logical checks on temperature data: `input['Tmp_Min'] < input['Tmp_Max']`. (2) Mutation plugin was used to test the effectiveness of the assertions. Specifically, a number of AQI samples were mutated to have negative values, consequently, the subsequent assertion for AQI validity check failed. (3) Fixtures were used to control (enable/disable) the individual test plugins based on the input data flowing through them. Figure 2 shows the AQI-Weather pipeline in CDAP Studio. Table 2 gives a summary of all four example pipelines and the tests performed using the created plugins for each of them.

5   Conclusion

In this work, we propose a general framework for plugins to facilitate functional testing of low-code ETL workflows, and demonstrate a prototype implementation based on the CDAP open-source framework. There are various ways this framework could be extended to support additional features. Some of the future research directions include: (1) Implementation of these plugins on popular open source platforms such as Apache NiFi and others, (2) A comprehensive study on effective mutation operators for data anomalies in datasets across different business domains and use cases, (3) Additional plugins that are useful for real-time data pipelines, and (4) Exploring possible ways to implement mutations directly in the underlying DAG specification, thus facilitating the validation of the control flow. Furthermore, validation studies for the proposed framework based on publicly available ETL use cases are planned.

Some limitations of our prototype implementation using CDAP include lack of assertions for aggregate records, lack of support for dynamic values within fixtures, etc. These limitations are primarily due to the underlying architecture of CDAP and APIs thereof, and could be addressed if these plugins are implemented for platforms that offer more flexibility in their plugin architecture.

Acknowledgments. The authors thank Mr. Naveen Kumawat and Mr. Sayan Hrik, research interns at IIT Madras, for setting up examples to validate the CDAP plugins.

References

1. 2022 Monkeypox Outbreak Data (2025). https://github.com/owid/monkeypox
2. Apache Airflow. https://airflow.apache.org/use-cases/etl_analytics/
3. Azure Data Factory. https://azure.microsoft.com/products/data-factory
4. Data Integrator. https://oracle.com/middleware/technologies/data-integrator.html
5. Indian AQI Data: CPCB (2025). https://airquality.cpcb.gov.in/ccr/#/caaqm-dashboard-all/caaqm-landing/aqi-repository
6. Nikhil, V.J.: Weather Data (2025). https://data.opencity.in/dataset/daily-temperature-70-years-data-for-major-indian-cities

7. Our World in Data (2025). https://ourworldindata.org/
8. Train Schedule Data. https://data.gov.in/catalog/indian-railways-train-time-table
9. Train Statistics. https://data.gov.in/catalog/operating-statistics-trains
10. Train Traffic Density. https://data.gov.in/catalog/train-traffic-density-statistics
11. OWID Energy Data (2023). https://ourworldindata.org/energy
12. Amazon Web Services Glue (2024). https://aws.amazon.com/glue/
13. Apache Beam® (2024). https://beam.apache.org/
14. Apache NiFi (2024). https://nifi.apache.org/
15. Apache Spark™ (2024). https://spark.apache.org/
16. Cask Data Application Platform (2024). https://github.com/cdapio/cdap
17. Google Cloud Dataflow (2024). https://cloud.google.com/products/dataflow
18. IBM DataStage (2024). https://www.ibm.com/products/datastage
19. Informatica (2024). https://www.informatica.com/products/cloud-data-integration
20. JEXL Overview (2024). https://commons.apache.org/proper/commons-jexl/
21. QuerySurge (2024). https://www.querysurge.com/
22. Chen, H., Chiang, R., Storey, V.: Business intelligence and analytics: from big data to big impact. MIS Q. **36**, 1165–1188 (2012)
23. Gao, J., Xie, C., Tao, C.: Big data validation and quality assurance–issuses, challenges, and needs. In: IEEE SOSE 2016, pp. 433–441. IEEE (2016)
24. Golfarelli, M., Rizzi, S.: From star schemas to big data: 20+ years of data warehouse research. In: Flesca, S., Greco, S., Masciari, E., Saccà, D. (eds.) A Comprehensive Guide Through the Italian Database Research Over the Last 25 Years. SBD, vol. 31, pp. 93–107. Springer, Cham (2018). https://doi.org/10.1007/978-3-319-61893-7_6
25. Homayouni, H., Pourebadi, M.M., Nguyen, S.T., Hashemi, M., Shirazi, H.: Comprehensive functional ETL testing methodologies for real-world data. In: 2024 IEEE QRS-C, pp. 11–20. IEEE (2024)
26. Kim, M., Zimmermann, T., DeLine, R., Begel, A.: Data scientists in software teams: State of the art and challenges. In: IEEE/ACM 40th International Conference on Software Engineering (ICSE), pp. 585–585 (2018). https://doi.org/10.1145/3180155.3182515
27. Kromer, M.: Assert data transformation in mapping data flow (2024). https://learn.microsoft.com/en-us/azure/data-factory/data-flow-assert
28. Li, L., Wu, Z.: How can no/low code platforms help end-users develop ML applications? - a systematic review. In: HCI International 2022 - Late Breaking Papers: Interacting with eXtended Reality and Artificial Intelligence, pp. 338–356. Springer, Cham (2022)
29. Patel, M., Patel, D.B.: Progressive growth of ETL tools: a literature review of past to equip future. In: Rising Threats in Expert Applications and Solutions, pp. 389–398. Springer, Singapore (2021)
30. Rahm, E., Do, H.: Data cleaning: problems and current approaches. IEEE Data Eng. Bull. **23**, 3–13 (2000)
31. Sambasivan, N., Kapania, S., Highfill, H., Akrong, D., Paritosh, P., Aroyo, L.M.: "Everyone wants to do the model work, not the data work": data cascades in high-stakes AI. In: Proceedings of the CHI Conference on Human Factors in Computing Systems. ACM, New York (2021)
32. Simitsis, A., Skiadopoulos, S., Vassiliadis, P.: The history, present, and future of ETL technology. In: DOLAP, pp. 3–12 (2023)

33. Tobin, D.: Streamline your data pipeline with no-code ETL tools (2023)
34. Vassiliadis, P.: A survey of extract-transform-load technology. Int. J. Data Warehouse. Min. **5**, 1–27 (2009)
35. Vassiliadis, P., Vagena, Z., Skiadopoulos, S., Karayannidis, N., Sellis, T.: Arktos: a tool for data cleaning and transformation in data warehouse environments. IEEE Data Eng. Bull. **23** (2001)

A Spark-Based Pipeline for Real-Time Multi-Faceted Analysis of Public Sentiment During the Indian Recession

Deshana Vikas Shah[1] and S. Saravanan[2]($\boxtimes$)

[1] Department of Computer and Communication Engineering, Amrita School of Engineering, Chennai, Amrita Vishwa Vidyapeetham, Chennai, India
`ch.en.u4cce22039@ch.students.amrita.edu`
[2] Department of Computer Science and Engineering, Amrita School of Computing, Chennai, Amrita Vishwa Vidyapeetham, Chennai, India
`s_saravanan@ch.amrita.edu`

Abstract. Public sentiment on social media offers valuable insights into economic trends; however, real-time analysis of recession-related tweets using Spark Structured Streaming remains relatively unexplored. Governments require such analytics to assess public distress, predict economic behavior, and implement timely interventions. This study aims to bridge that gap by analyzing a synthetic dataset that mimics Indian Twitter reactions to the 2023 recession. Employing Apache Spark, we developed a scalable pipeline to classify tweets based on sentiment, sarcasm, depressive tone, sector relevance, and suggestions. The dataset is divided into an 80:20 ratio for training and evaluation, and machine learning models are trained using Spark MLlib. The Random Forest model achieved an impressive 98.88% accuracy in sentiment classification, 99.99% in suggesting recommendations, and 90.00% in depression detection model; XGBoost recorded 100% in sarcasm detection; and Naive Bayes reached 84.49% accuracy in sector classification, surpassing the 80.00% performance threshold. A real-time Flask dashboard visualizes these predictions through dynamic charts, facilitating immediate policy responses. This work illustrates how Spark Structured Streaming can effectively process live social media data on a scale, while providing governments with a proactive tool to monitor sentiment during recession and formulate data-driven strategies, hence resulting in better decisions that are more considerate of people's needs.

Keywords: Apache Spark · Sentiment Analysis · Flask · Machine Learning · Stream Processing · Recession

1 Introduction

In today's interconnected world, social media platforms like X (formerly Twitter), Facebook, LinkedIn and Instagram provide insights into public sentiment [1, 2]. These platforms apprehend trending topics, emotional responses, economic concerns, and public

© The Author(s), under exclusive license to Springer Nature Switzerland AG 2026
R. Gupta et al. (Eds.): BDA 2025, LNCS 16041, pp. 359–369, 2026.
https://doi.org/10.1007/978-3-032-15134-6_27

discussions, especially during periods of financial uncertainty, such as recessions. This data facilitates governments and policymakers to monitor sentiment in real-time, identify early warning signals, and respond effectively to economic challenges [3, 4].

Currently, enough research has not been done on sentiment analysis of recession-related tweets using Apache Spark Structured Streaming in conjunction with a real-time interactive dashboard. Traditional data pipelines often depend on static datasets and batch processing and they may not adequately address the rapid and high-volume nature of social media content. Hence, a real-time, scalable solution is required to leverage public discourse for timely insights during economic downturns [5–7]. Governments can leverage such analysis to observe public sentiment, identify requests for policy changes, detect mental health trends, and assess which economic sectors are facing heightened concern [8–10]. However, the informal tone and unstructured nature of social media posts present challenges for traditional analytics.

This research outlines a Spark-based streaming architecture for real-time sentiment analysis during the 2023 recession in India. A synthetic dataset comprising 449,870 tweets was generated using AI models based on content from an existing Kaggle dataset [11], with original columns removed to ensure privacy. The dataset was partitioned into an 80:20 train-test ratio and annotated across five analytical dimensions: sentiment (positive, neutral, negative), sarcasm, depressive tone, referenced economic sector, and presence of constructive suggestions.

Here, sentiment analysis classifies whether the emotional tone expressed of the given text is positive or negative or neutral. This is required to find the overall public feeling regarding the recession. The aim of sarcasm analysis is to find instances where the expressed sentiment is contrary to the literal meaning of the words, often conveyed through irony or mockery. Hence, identifying sarcasm is important for accurate sentiment understanding. Depressive tone analysis aims to find language patterns and keywords that indicate sadness, hopelessness, or other signs associated with depression in the text. This can support monitoring potential mental health trends during economic hardship. Sector classification categorizes tweets based on the specific economic sector they reference, such as finance, manufacturing, agriculture, or technology. This is required to know which industry is facing recession. Suggestion analysis is to identify tweets that propose constructive solutions, policy recommendations, or ways to mitigate the negative impacts of the recession. This is required to provide valuable insights for policymakers.

Three machine learning models that are built by XGBoost, Random Forest, and Naive Bayes algorithms are evaluated for each classification task. The model demonstrating the highest accuracy in each dimension (all exceeding 80%) was selected. These models were integrated into a real-time Spark Structured Streaming pipeline to facilitate live classification and visualization of incoming tweets. The results are presented on a Flask-based dashboard with dynamically updating charts, enabling policymakers and analysts to monitor public sentiment and discourse as it evolves.

2 Related Works

2.1 Social Media for Economic Forecasting

Recent advancements indicate the significant potential of social media for economic monitoring. Malhi, A. et al., [12] developed a real-time UK economic forecasting framework using Twitter sentiment analysis and a GNN-GRU model on Azure. Their research achieved 94.19% accuracy in identifying recession indicators, outperforming existing models and highlighting deep learning's potential for economic monitoring. Similarly, Ningsih, M. R. et al., [13] conducted sentiment analysis on Indonesian public sentiment about the 2023 global recession to aid Sustainable Development Goals assessments. They developed an ensemble machine learning model using Logistic Regression, Decision Tree, Random Forest, and SVM, trained on VADER-labeled data with features like CountVectorizer, BM25, and Word Embedding. The model achieved 95.02% accuracy, surpassing previous methods, though language limitations were noted for future improvement. Eka Putra, F. P.et al., [14] studied public sentiment on the 2023 recession using Twitter data. They analyzed tweets tagged with "Recession" and found that 94% were neutral, 4% positive, and 2% negative, indicating overall public indifference toward the recession. Mandlik, N. et al., [15] examined recession trends by analyzing data from Twitter, Reddit, and The New York Times through a three-stage data science pipeline: collection, filtering, and analysis. They found that social media exhibited greater keyword variability, while news articles provided better geographic and political context. Sentiment scores varied by platform, suggesting the data's potential as recession indicators. Nathanael, G. K. [16] analyzed Twitter perceptions of the 2023 global recession using Python and NVIVO12. The author of [16] found 46% of users expressed neutral sentiments and 44% negative. Negative themes included economic projections, politics, and employment, while cryptocurrency was noted for its positive outlook, reflecting a generally pessimistic view of the recession. Together, these studies emphasize the importance of region-specific and multi-faceted analyses, a need that our study on the Indian recession aims to fulfill.

2.2 Mental Health Correlates in Economic Discourse

Bhargava, C. et al. [17] aimed to improve early depression detection using Twitter data, emphasizing overlooked linguistic features. They utilized a hybrid CNN-LSTM model along with a rule-based Vader sentiment analyzer. Results showed that the CNN-LSTM model surpassed the LSTM model in accuracy and precision for identifying depressive traits in tweets. Gudiato, C., & Horhoruw, L. F. M. [18] performed a sentiment analysis on Twitter data to detect suicidal ideation using a Naive Bayes classifier, achieving 72% accuracy. It had a precision of 0.91 for negative and neutral categories, while recall for positive tweets was high at 0.97. Improvements are necessary for negative and neutral sentiments to enhance early detection of mental health issues. Ullah, W. et al., [19] studied depression detection on Twitter using 1.6 million tweets and NLP techniques. They applied the Grey Relational Grade method and machine learning models like Logistic Regression, SVM, XGBoost, and Random Forest, achieving up to 97% accuracy in real-time mental health monitoring. Nonetheless, these studies are predominantly focused on the detection of depression, neglecting important factors·such as sarcasm and suggestion mining, which are vital components in public crisis response. Our model, in contrast, integrates these elements through a multi-task classification approach.

2.3 Technical Foundations for Real-Time Analysis

A scalable infrastructure is crucial for modern sentiment analysis. Ismail et al. [20] developed a Kafka-Spark pipeline for processing data in real-time. However, it was limited to basic sentiment classification. Our work expands on these technical foundations by incorporating multi-dimensional analysis capabilities. Specifically, this work presents a multi-dimensional analysis across five distinct aspects: sentiment, sarcasm, depression, suggestion presence, and sector classification, enhancing the depth of recession-related tweet analysis. A synthetic dataset was generated using AI tools by carefully modeling after a real Kaggle dataset to maintain realistic proportions and characteristics. A real-time, interactive dashboard was developed using Flask, enabling dynamic visualization of analyzed tweets through continuously updating pie charts. The study focuses on the context of the Indian recession 2023, which has not been systematically explored in previous research. Additionally, model comparisons were conducted across tasks, with the optimal algorithm selected from Random Forest, Naive Bayes, and XGBoost for each classification objective to ensure robust and task-specific model performance like the methodology presented in [19].

3 System Architecture

The architecture of the proposed system is designed to enable real-time sentiment analysis of social media content related to the 2023 recession, utilizing a scalable and efficient big data pipeline as shown in Fig. 1. Input data can be collected from three major social media platforms: Facebook (posts), X, formerly Twitter (tweets), and LinkedIn (posts). These platforms serve as significant sources of public opinion during economic downturns, where individuals and communities express concerns, experiences, and suggestions online. For government agencies and policymakers, analyzing this data can provide insights into public sentiment, identify sector-specific distress signals, and support evidence-based decision-making. To manage high-velocity, multi-source data, Apache Kafka is employed for real-time data ingestion. Kafka ensures reliable and fault-tolerant streaming of data into the pipeline. Once ingested, the data flows into the Apache Spark Structured Streaming environment, where it undergoes preprocessing steps such as text normalization, tokenization, and feature extraction. Spark's MLlib library [21] is utilized to apply trained machine learning models for five key classification tasks: sentiment detection, sarcasm detection, depressive tone identification, suggestion analysis, and economic sector classification. Processed and raw data are stored in MongoDB [22], a flexible NoSQL database optimized for real-time applications. This dual-storage approach guarantees real-time insights and historical data availability for further analysis or auditing. The continuous nature of the data stream facilitates ongoing updates and retraining opportunities as new data is collected. Analytical results are presented through a real-time, web-based dashboard, which provides a visual representation of public discourse. The dashboard features five dynamic pie charts, each representing one of the key analytical dimensions—sentiment, sarcasm, suggestions, depression, and sector relevance. These charts are updated in real-time as new data is streamed, processed, and classified, offering an overview of public responses to the recession across digital platforms.

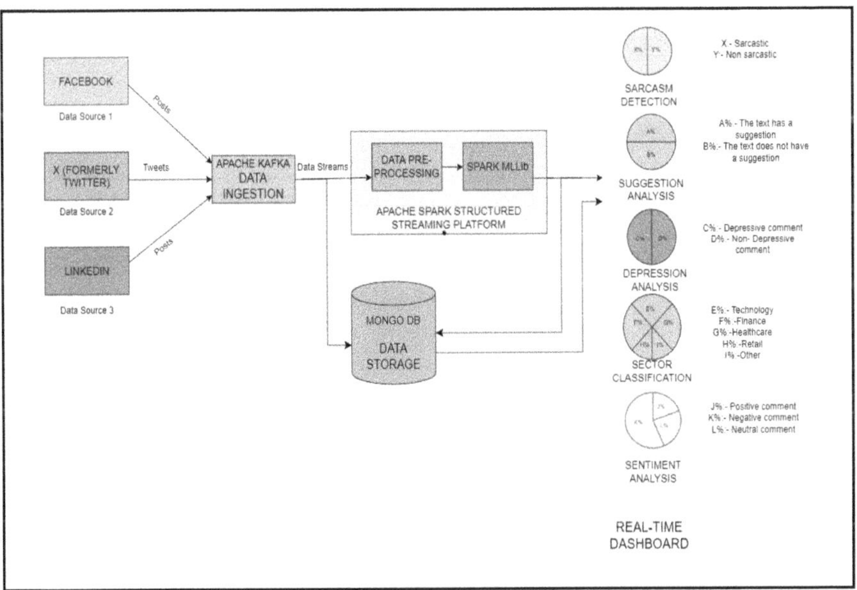

Fig. 1. Proposed System Architecture for Real-time Sentiment Analysis of Social Media Data Related to the 2023 Recession

4 Implementation

This section delineates the comprehensive steps undertaken to construct the real-time tweet classification system. It encompasses dataset preparation, preprocessing methodologies, machine learning model training, and the visualization of results in real-time. A variety of tools and frameworks, including Apache Spark, Pandas, scikit-learn, and Flask, were employed throughout the various phases of the project to ensure efficiency, scalability, and user interactivity. The implementation is structured into distinct phases to facilitate clarity and a systematic understanding of the process.

4.1 Dataset Preparation

1. The Kaggle dataset [11] was utilized to create 449,870 synthetic tweets using advanced AI language models. The generated synthetic data can mimic the statistical distributions and thematic diversity of real Twitter discourse, aiming to train models robust enough for noisy, real-world data. We guided the AI models to incorporate linguistic nuances, diverse emotional ranges, and Indian contextual elements.

Further, the dataset was created using advanced AI models to generate entirely novel content, completely bypassing the use of identifiable personal information from real users.

2. Next, we initiate a Spark Session with the application name "ContentAnalysisPipeline." The analysis functions assess the text content using various methods, including sentiment analysis via TextBlob polarity thresholds (a Python library for performing sentiment analysis by calculating the polarity), detection of depressive keywords, sarcasm identification using regex patterns, suggestion extraction, and classification in specific domains; each function analyzes the raw text and generates either categorical or boolean outputs.

3. Spark UDFs encapsulate these functions with defined return types, enabling distributed execution and integrating custom logic within Spark's framework while ensuring scalability for large-scale text processing.

4. Transformations are applied to the DataFrame by introducing new columns that reflect each analysis metric, enhancing the original "content" column with additional attributes such as sentiment labels, boolean indicators for depressive or sarcastic content, suggestions, and sector classifications.

5. The produced categories include: Sentiment (Positive, Negative, Neutral), Depressive (True or False), Sarcasm detection (True or False), Suggestion detection (True or False), Sector Classification (Technology, Finance, Healthcare, Retail, or other (if no relevant keywords are detected)).

4.2 Real-Time Stream Tweets Classification

1. After enriching the dataset, it was converted to a Spark DataFrame to facilitate parallel and distributed processing. This improved dataset is separated into training (80%) and testing (20%) portions, while maintaining consistency through a predetermined random seed. The training dataset consists of 359,896 entries, whereas the testing dataset comprises 89,794 entries. The training dataset has a size of 41.0MB, and the testing dataset is 10.3MB in size.

2. Next, we train five models using machine learning algorithms in Spark MLlib, one for each classification type, using three different machine learning algorithms: Random Forest, Naïve Bayes, and XGBoost. All the models developed attain a test accuracy exceeding the 80% benchmark. We determine the most effective models for each classification.

4.3 Real-Time Dashboard for Displaying the Classification Results

1. The system features a Flask-based web interface that automatically refreshes every two seconds, simulating real-time processing. The frontend displays tweet content alongside five analytical dimensions: sentiment, depressive tone, sarcasm, sector classification, and suggestions, all organized in a clear format. A responsive design ensures optimal viewing across a variety of devices, with the results of tweet analysis presented in distinctly defined containers.

2. Each tweet is processed through parallel machine learning inference utilizing five pretrained models. This pipeline executes text classification for sentiment polarity (three

classes), detection of depressive content (binary), identification of sarcasm (binary), economic sector categorization (multi-class), and the identification of suggestions (binary). All predictions are generated synchronously within the 2-s refresh interval to guarantee real-time performance.
3. The system also dynamically generates five pie charts that visualize the distribution of predictions within the current 10-tweet window. Matplotlib renders each chart as a base64-encoded PNG, featuring distinct color schemes and percentage labels for immediate visual interpretation. The visualization module automatically normalizes counts to percentages, ensuring consistent scaling across the various analysis dimensions.

5 Results

The generated dataset includes 359,896 and 89,974 records for the training set and the test set respectively, with sizes of 40 MB and 10.3 MB, respectively. A snippet of the dataset is provided in Fig. 2. The Random Forest model for sentiment analysis, as shown in Table 1, achieved a training accuracy of 98.83% and a testing accuracy of 98.88% thus making it the best model for sentiment analysis with a recall and F1-Score of 97.00% and precision of 98.00%. The XGBoost model for sarcasm detection, as shown in Table 2, achieved 100% accuracy on both the training and testing datasets, along with a precision, recall, and F1-score of 100%. This performance indicates its effectiveness as a model for detecting sarcasm. The Random Forest model, as shown in Table 3, is identified as the most effective model for Suggestion Detection, achieving a train and test accuracy of 99.99%. The model demonstrates a precision and F1-score of 100%, while the recall is measured at 99.00%. Naïve Bayes model is ideal for Sector classification as it offers a solid train and test accuracy of 84.69% and 84.49% respectively, while offering a precision of 86.23%, recall of 84.49%, and F1-score of 82.89%, as shown in Table 4. The Random Forest model demonstrates high performance in Depression Detection, achieving a train accuracy of 96.25% and test accuracy of 90.00% with a precision of 91.00%, a recall of 90.00%, and an F1-score of 89.00%, as shown in Table 5.

	content	sentiment	is_depressive	is_sarcastic	has_suggestion	sector
1	with negative outlook, on expectations of decline in steel demand in India.	negative	False	False	False	technology
2	India's growth to take a knock	neutral	False	False	False	other
3	& so much more...	positive	False	False	False	other
4	Recruitment business was never so good	positive	False	False	False	other
5	#economy #mpc #ratehikes #rbi #inflation #recession"	neutral	False	False	False	finance
6	her #excuse "the possible #recession" by @Chairmaniba @IDFS_India &	neutral	False	False	False	technology
7	talks about the reasons for the impending prediction of recession in India	neutral	False	False	False	other
8	g forward to the #panel discussion on "Rapid #Digitalization – Hoped For	neutral	False	False	False	other
9	e worried? #globalslowdown #slowdown #recession #economy #investing	neutral	False	False	True	finance
10	#Business #Economy #Finance #GreatLockDown #Recession #Covid19	neutral	False	False	False	finance
11	Read More: https://t.co/7BGKRumU21	positive	False	False	False	other
12	zationDisaster strike in #ModiNoteGate #recession https://t.co/TZqALiwS46	neutral	False	False	False	other
13	for such Deteriorating condition, 130+ Poor Countries deleting fm World...	neutral	False	False	False	technology
14	Full respect to you Sridhar Vembu 🙏	positive	False	False	False	other
15	https://t.co/1MjWIahd1O"	neutral	False	False	False	other
16	e we having some dry powder ready for the fireworks in the stock market?	neutral	False	False	False	finance
17	2. India did not print trillions to deal with covid	neutral	False	False	False	other
18	mand or supply? How does this affect #government actions and policies?	neutral	False	False	False	other
19	n to limit #climatechange	@RBI_India @RBIGov http://t.co/17jPFWPH0R	neutral	False	False	True	technology
20	https://t.co/11WPg88y"	neutral	False	False	False	other

Fig. 2. Sample of the Synthetically Generated Tweet Dataset from Kaggle dataset [11]

Table 1. Performance Comparison of Sentiment Analysis Models on Recession Tweets

Algorithm	Train Acc	Test Acc	Precision	Recall	F1-Score
Xgboost	96.94%	96.90%	97.00%	97.00%	97.00%
Random Forest	98.83%	98.88%	98.00%	97.00%	97.00%
Naïve Bayes	92.95%	92.89%	93.28%	92.88%	91.82%

Table 2. Performance Comparison of Sarcasm Detection Models on Recession Tweets

Algorithm	Train Acc	Test Acc	Precision	Recall	F1-Score
Xgboost	100%	100%	100%	100%	100%
Random Forest	100%	100%	100%	99.99%	100%
Naïve Bayes	99.97%	99.96%	99.95%	99.95%	99.95%

Table 3. Performance Comparison of Suggestion Detection Models on Recession Tweets

Algorithm	Train Acc	Test Acc	Precision	Recall	F1-Score
Xgboost	99.98%	99.98%	100%	99.00%	100%
Random Forest	99.99%	99.99%	100%	99.00%	100%
Naïve Bayes	99.46%	99.44%	99.42%	99.43%	99.32%

Table 4. Performance Comparison of Sector Classification Models on Recession Tweets

Algorithm	Train Acc	Test Acc	Precision	Recall	F1-Score
Xgboost	84.33%	84.18%	87.00%	84.00%	82.00%
Random Forest	80.42%	80.56%	82.00%	81.00%	78.00%
Naïve Bayes	84.69%	84.49%	86.23%	84.49%	82.89%

Table 5. Performance Comparison of Depression Detection Models on Recession Tweets

Algorithm	Train Acc	Test Acc	Precision	Recall	F1-Score
Xgboost	80.63%	82.50%	86.00%	82.00%	79.00%
Random Forest	96.25%	90.00%	91.00%	90.00%	89.00%
Naïve Bayes	81.87%	82.50%	86.00%	82.00%	79.00%

The final implementation involves not only streaming and analysis of the tweets but also visualisation of the results through an interactive dashboard generated by using

Flask and Matplotlib. Figure 3 illustrates the streaming of tweets in real-time. The dashboard presented in Fig. 4 provides a dynamic, user-friendly view of evolving trends and refreshes itself every 2 s as it streams and analyses new tweets. For enhanced clarity and immediate interpretability for policymakers, the dashboard prominently features a series of interactive pie charts. Each pie chart visually represents the distribution of classifications across key categories: A Sentiment Pie Chart provides a clear breakdown of positive, negative, and neutral public sentiment, allowing for quick assessment of overall mood. Similarly, dedicated pie charts illustrate the proportion of sarcastic versus non-sarcastic tweets, the prevalence of depressive tones, the distribution of discussions across various economic sectors, and the frequency of policy suggestions.

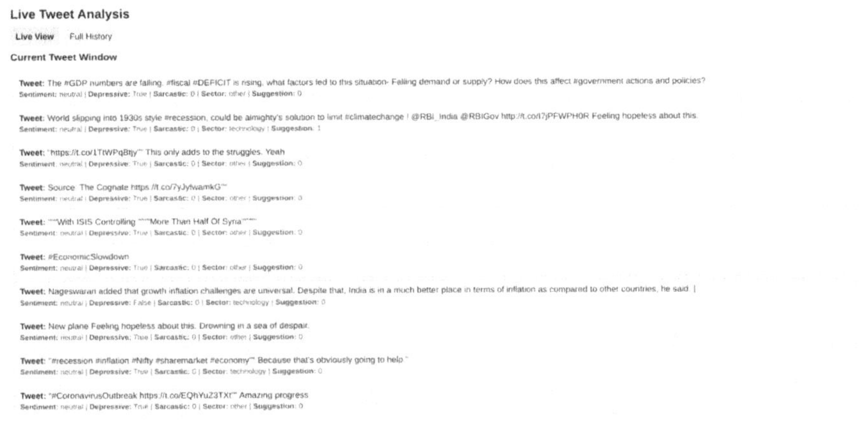

Fig. 3. Live streaming of recession related tweets

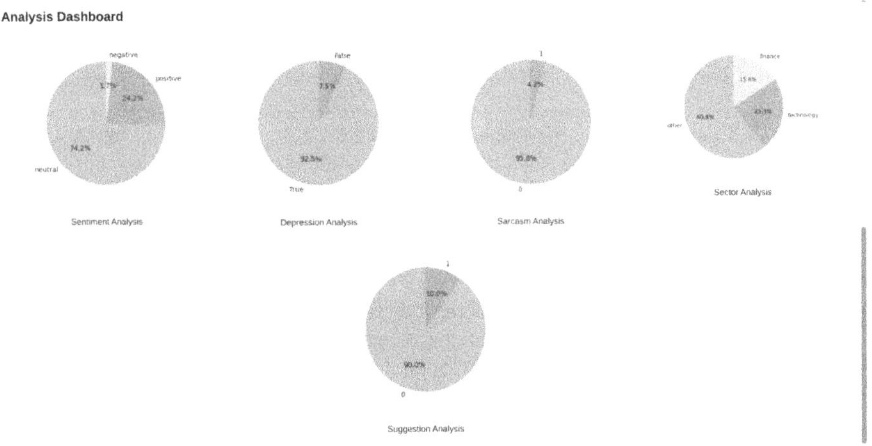

Fig. 4. Real-time Pie Chart Visualizing the Distribution of Sarcasm, Sentiment, Depression, Sector Classification, and Suggestions in Streaming Tweets Related to Recession.

6 Conclusion

This research presents a real-time sentiment analysis pipeline for recession-related tweets, built on Apache Spark Structured Streaming and powered by machine learning. Using a synthetic dataset inspired by real-world data, tweets were categorized into five analytical dimensions—sentiment, sarcasm, depressive tone, sector relevance, and suggestions—capturing a broad spectrum of socio-economic signals. The models demonstrated strong performance: Random Forest achieved 98.88% accuracy for sentiment, 99.99% for suggestions and 90.00% for depression; XGBoost reached 100% for sarcasm; and Naive Bayes delivered 84.49% accuracy for sector classification. These results confirm the system's ability to detect complex linguistic patterns in high-velocity data streams. A Flask-based real-time dashboard was developed to visualize these insights as they happen, enabling continuous monitoring of public discourse. This study is extremely useful for governments, as such a system acts more as a strategic tool than a technical solution. In economic downturns, public sentiment shifts rapidly. Traditional data sources lag, but social media offers immediate, actionable feedback. By adopting real-time sentiment analysis, governments can better understand citizen concerns, mental health trends, and which sectors are under the most stress, allowing for timely, targeted, and data-informed interventions in times of crisis.

References

1. Drus, Z., Khalid, H.: Sentiment analysis in social media and its application: systematic literature review. Procedia Comput. Sci. **161**, 707–714 (2019)
2. Rangarajan, P.K., et al.: The social media sentiment analysis framework: deep learning for sentiment analysis on social media. Int. J. Electr. Comput. Eng. (IJECE) **14**(3), 1-x1 (2024)
3. Dionisius, Putra, H.D., Ardhytama, N., Purnama, Y.: Analysis of public sentiment about economic recession threat in Indonesia on Twitter (X). In: 2024 International Conference on Informatics, Multimedia, Cyber and Information System (ICIMCIS), Jakarta, Indonesia, pp. 133–138 (2024). https://doi.org/10.1109/ICIMCIS63449.2024.10957495
4. Xu, Q.A., Chang, V., Jayne, C.: A systematic review of social media-based sentiment analysis: emerging trends and challenges. Decis. Anal. J. **3**, 100073 (2022)
5. Vishal, K.S., Srinidhi, S.K., Shashank, U.S., Samhitha, S.S., Venugopalan, M.: A scalable model for text mining using sentiment analysis on PySpark. In: 2023 4th International Conference on Intelligent Technologies (CONIT), pp. 1–8. IEEE (2024)
6. Reddy, P.N., Alapati, R., Radha, D.: Real-time tweets analysis using machine learning and big-data. In: 2024 IEEE North Karnataka Subsection Flagship International Conference (NKCon), pp. 1–6. IEEE (2024)
7. Gokulan, S., Reddy, M., Varshith, M., Venugopalan, M.: TweetFeel: analyzing emotions in the twittersphere. In: 2024 International Conference on Knowledge Engineering and Communication Systems (ICKECS), vol. 1, pp. 1–6. IEEE (2024)
8. Verma, S.: Sentiment analysis of public services for smart society: literature review and future research directions. Gov. Inf. Quart. **39**(3), 101708 (2022). https://doi.org/10.1016/j.giq.2022.101708
9. Chauhan, P., Sharma, N., Sikka, G.: The emergence of social media data and sentiment analysis in election prediction. J. Ambient Intell. Hum. Comput. **12**, 2601–2627 (2021). https://doi.org/10.1007/s12652-020-02423-y

10. Aswathy, A., Prabha, R., Gopal, L.S., Pullarkatt, D., Ramesh, M.V.: An efficient twitter data collection and analytics framework for effective disaster management. In 2022 IEEE Delhi Section Conference (DELCON), pp. 1–6. IEEE (2022)
11. Bris, V.: Tweets about Recession in India 2023. Kaggle.com (2023). https://www.kaggle.com/datasets/aiotsir/tweets-about-recession-in-india-2023. Accessed 25 Apr 2025
12. Malhi, A., Naiseh, M., Jangra, K.: Real-time Twitter data sentiment analysis to predict the recession in the UK using graph neural networks. In: 2024 International Wireless Communications and Mobile Computing (IWCMC), pp. 1595–1600. IEEE (2024)
13. Ningsih, M.R., Wibowo, K.A.H., Dullah, A.U., Jumanto, J.: Global recession sentiment analysis utilizing VADER and ensemble learning method with word embedding. J. Soft Comput. Explor. **4**(3), 142–151 (2023)
14. Eka Putra, F.P., Maulana, F.I., Akbar, N.M., Febriantoro, W.: Twitter sentiment analysis about economic recession in Indonesia. Bull. Soc. Inform. Theory Appl. **7**(1), 1–7 (2023). https://doi.org/10.31763/businta.v7i1.592
15. Mandlik, N., Barhanpurkar, K., Bhandwaldar, H., Goyal, S.B., Rajawat, A.S., Rane, S.: Exploring recession indicators: analyzing social network platforms and newspapers textual datasets. In Applied Data Science and Smart Systems, pp. 180–186. CRC Press (2025)
16. Nathanael, G.K.: Understanding recession response by Twitter users: a text analysis approach. Heliyon **10**(1) (2024)
17. Bhargava, C., Al, E.: Depression detection using sentiment analysis of tweets. Turk. J. Comput. Math. Educ. (TURCOMAT) **12**(11), 5411–5418 (2021)
18. Gudiato, C., Horhoruw, LF. M.: Tweet sentiment analysis to detect suicide intentions through a machine learning approach to big data using the naive bayes method. G-Tech: J. Teknol. Terapan **9**(1), 381–389 (2025)
19. Ullah, W., Oliveira-Silva, P., Nawaz, M., Zulqarnain, R.M., Siddique, I., Sallah, M.: Identification of depressing tweets using natural language processing and machine learning: application of grey relational grades. J. Radiat. Res. Appl. Sci. **18**(1), 101299 (2025)
20. Ismail, A., Sazali, F.H., Jawaddi, S.N.A., Mutalib, S.: Stream ETL framework for twitter-based sentiment analysis: leveraging big data technologies. Expert Syst. Appl. **261**, 125523 (2025)
21. Meng, X., et al.: Mllib: machine learning in apache spark. J. Mach. Learn. Res. **17**(1), 1235–1241 (2016)
22. Kristina, C., Michael, D.: MongoDB: the Definitive Guide. O'Reilly Media Inc, California (2010)

Author Index